Annals of Mathematics Studies

Number 42

ANNALS OF MATHEMATICS STUDIES

Edited by Robert C. Gunning, John C. Moore, and Marston Morse

1. Algebraic Theory of Numbers, *by* HERMANN WEYL
3. Consistency of the Continuum Hypothesis, *by* KURT GÖDEL
6. The Calculi of Lambda-Conversion, *by* ALONZO CHURCH
10. Topics in Topology, *by* SOLOMON LEFSCHETZ
11. Introduction to Nonlinear Mechanics, *by* N. KRYLOFF *and* N. BOGOLIUBOFF
15. Topological Methods in the Theory of Functions of a Complex Variable, *by* MARSTON MORSE
16. Transcendental Numbers, *by* CARL LUDWIG SIEGEL
17. Problème Général de la Stabilité du Mouvement, *by* M. A. LIAPOUNOFF
19. Fourier Transforms, *by* S. BOCHNER *and* K. CHANDRASEKHARAN
20. Contributions to the Theory of Nonlinear Oscillations, Vol. I, *edited by* S. LEFSCHETZ
21. Functional Operators, Vol. I, *by* JOHN VON NEUMANN
22. Functional Operators, Vol. II, *by* JOHN VON NEUMANN
24. Contributions to the Theory of Games, Vol. I, *edited by* H. W. KUHN *and* A. W. TUCKER
25. Contributions to Fourier Analysis, *edited by* A. ZYGMUND, W. TRANSUE, M. MORSE, A. P. CALDERON, *and* S. BOCHNER
26. A Theory of Cross-Spaces, *by* ROBERT SCHATTEN
27. Isoperimetric Inequalities in Mathematical Physics, *by* G. POLYA *and* G. SZEGO
28. Contributions to the Theory of Games, Vol. II, *edited by* H. W. KUHN *and* A. W. TUCKER
29. Contributions to the Theory of Nonlinear Oscillations, Vol. II, *edited by* S. LEFSCHETZ
30. Contributions to the Theory of Riemann Surfaces, *edited by* L. AHLFORS *et al.*
31. Order-Preserving Maps and Integration Processes, *by* EDWARD J. MCSHANE
32. Curvature and Betti Numbers, *by* K. YANO *and* S. BOCHNER
33. Contributions to the Theory of Partial Differential Equations, *edited by* L. BERS, S. BOCHNER, *and* F. JOHN
34. Automata Studies, *edited by* C. E. SHANNON *and* J. MCCARTHY
35. Surface Area, *by* LAMBERTO CESARI
36. Contributions to the Theory of Nonlinear Oscillations, Vol. III, *edited by* S. LEFSCHETZ
37. Lectures on the Theory of Games, *by* HAROLD W. KUHN. In press
38. Linear Inequalities and Related Systems, *edited by* H. W. KUHN *and* A. W. TUCKER
39. Contributions to the Theory of Games, Vol. III, *edited by* M. DRESHER, A. W. TUCKER *and* P. WOLFE
40. Contributions to the Theory of Games, Vol. IV, *edited by* R. DUNCAN LUCE *and* A. W. TUCKER
41. Contributions to the Theory of Nonlinear Oscillations, Vol. IV, *edited by* S. LEFSCHETZ
42. Lectures on Fourier Integrals, *by* S. BOCHNER
43. Ramification Theoretic Methods in Algebraic Geometry, *by* S. ABHYANKAR

LECTURES ON FOURIER INTEGRALS

BY

Salomon Bochner

WITH AN AUTHOR'S SUPPLEMENT ON

Monotonic Functions, Stieltjes Integrals, and Harmonic Analysis

---

TRANSLATED FROM THE ORIGINAL BY

Morris Tenenbaum
and
Harry Pollard

PRINCETON, NEW JERSEY
PRINCETON UNIVERSITY PRESS
1959

L. C. Card 59-5589

Printed in the United States of America
by Westview Press, Boulder, Colorado

Princeton University Press On Demand Edition, 1985

TRANSLATORS' PREFACE

In undertaking this translation of Bochner's classical book and its supplement (Monotone Funktionen, Stieltjessche Integrale und harmonische Analyse, Mathematische Annalen, Volume 108 (1933), pp. 378-410), our main purpose was to make generally available to the present generation of group-theorists and practitioners in distributions the historical and concrete problems which gave rise to these disciplines. Here can be found the theory of positive definite functions, of the generalized Fourier integral, and even forms of the important theorems concerning the reciprocal of Fourier transforms.

The translators are grateful to Professor Bochner for his encouragement in this work and for his many valuable suggestions.

Morris Tenenbaum
Harry Pollard

Cornell University

CONTENTS

Page

CHAPTER I: BASIC PROPERTIES OF TRIGONOMETRIC INTEGRALS............ 1

§1. Trigonometric Integrals Over Finite Intervals.......... 1
§2. Trigonometric Integrals Over Infinite Intervals........ 5
§3. Order of Magnitude of Trigonometric Integrals.......... 10
§4. Uniform Convergence of Trigonometric Integrals......... 13
§5. The Cauchy Principal Value of Integrals................ 18

CHAPTER II: REPRESENTATION — AND SUM FORMULAS................... 23

§6. A General Representation Formula....................... 23
§7. The Dirichlet Integral and Related Integrals........... 27
§8. The Fourier Integral Formula........................... 31
§9. The Wiener Formula..................................... 35
§10. The Poisson Summation Formula......................... 39

CHPATER III: THE FOURIER INTEGRAL THEOREM........................ 46

§11. The Fourier Integral Theorem and the Inversion Formulas 46
§12. Trigonometric Integrals with e^{-x}..................... 51
§13. The Absolutely Integrable Functions. Their Faltung and Their Summation.............................. 54
§14. Trigonometric Integrals with Rational Functions........ 63
§15. Trigonometric Integrals with e^{-x^2}.................... 67
§16. Bessel Functions...................................... 70
§17. Evaluation of Certain Repeated Integrals.............. 74

CHAPTER IV: STIELTJES INTEGRALS.................................. 78

§18. The Function Class $\mathfrak{P}$ 78
§19. Sequences of Functions of $\mathfrak{P}$ 85
§20. Positive-Definite Functions........................... 92
§21. Spectral Decomposition of Positive-Definite Functions. An Application to Almost Periodic Functions......... 97

CHAPTER V: OPERATIONS WITH FUNCTIONS OF THE CLASS $\mathfrak{F}_0$............ 104

§22. The Question.. 104
§23. Multipliers... 108
§24. Differentiation and Integration....................... 114
§25. The Difference-Differential Equation.................. 120
§26. The Integral Equation................................. 130
§27. Systems of Equations.................................. 134

CHAPTER VI: GENERALIZED TRIGONOMETRIC INTEGRALS.................. 138

§28. Definition of the Generalized Trigonometric Integrals.. 138
§29. Further Particulars About the Functions of $\mathfrak{F}_k$......... 145
§30. Further Particulars About the Functions of $\mathfrak{T}_k$......... 153
§31. Convergence Theorems.................................. 160
§32. Multipliers... 166
§33. Operator Equations.................................... 173
§34. Functional Equations.................................. 178

CHAPTER VII: ANALYTIC AND HARMONIC FUNCTIONS..................... 182

§35. Laplace Integrals..................................... 182
§36. Union of Laplace Integrals............................ 189
§37. Representation of Given Functions by Laplace Integrals. 194
§38. Continuation. Harmonic Functions...................... 202

CONTENTS

Page

§39. Boundary Value Problems for Harmonic Functions.......... 208

CHAPTER VIII: QUADRATIC INTEGRABILITY.......................... 214

§40. The Parseval Equation..................................... 214
§41. The Theorem of Plancherel................................. 219
§42. Hankel Transform.. 224

CHAPTER IX: FUNCTIONS OF SEVERAL VARIABLES..................... 231

§43. Trigonometric Integrals in Several Variables.............. 231
§44. The Fourier Integral Theorem.............................. 239
§45. The Dirichlet Integral.................................... 249
§46. The Poisson Summation Formula............................. 255

APPENDIX.. 264

Concerning Functions of Real Variables........................ 264
Measurability... 264
Summability... 266
Differentiability.. 270
Approximation in the Mean..................................... 271
Complex Valued Functions...................................... 276
Extension of Functions.. 277
Summation of Repeated Integrals............................... 279

REMARKS - QUOTATIONS... 281

MONOTONIC FUNCTIONS, STIELTJES INTEGRALS AND HARMONIC ANALYSIS...... 292

Introduction.. 292

I: MONOTONIC FUNCTIONS... 295
§1. Definition of the Monotonic Functions...................... 295
§2. Continuity Intervals....................................... 299
§3. Sequences of Monotonic Functions.......................... 303

II: STIELTJES INTEGRALS.. 307

§4. Definition and Important Properties........................ 307
§5. Uniqueness and Limit Theorems............................. 312

III: HARMONIC ANALYSIS... 316

§6. Fourier-Stieltjes Integrals................................ 316
§7. Uniqueness and Limit Theorems............................. 320
§8. Positive-Definite Functions................................ 325
§9. Spectral Decomposition of Square Integrable Functions..... 328

SYMBOLS - INDEX... 332

LECTURES ON FOURIER INTEGRALS

CHAPTER I

BASIC PROPERTIES OF TRIGONOMETRIC INTEGRALS

§1. Trigonometric Integrals Over Finite Intervals

1. We denote as trigonometric integrals expressions of the form

$$\Phi(\alpha) = \int_a^b f(x) \cos \alpha x \, dx \tag{1}$$

or

$$\Psi(\alpha) = \int_a^b f(x) \sin \alpha x \, dx \quad . \tag{2}$$

It is frequently more convenient to use the exponential factor $e^{i\alpha x}$ in place of the trigonometric factors $\cos \alpha x$ and $\sin \alpha x$. The trigonometric integral will then read

$$J(\alpha) = \int_a^b f(x) \, e^{i\alpha x} \, dx \text{ [1]} \quad .$$

For typographical simplification we shall always denote the function $e^{i\xi}$ by $e(\xi)$. Hence we shall write $J(\alpha)$ as

$$J(\alpha) = \int_a^b f(x) e(\alpha x) \, dx \; . \tag{3}$$

It is also customary to denote trigonometric integrals as Fourier integrals [1] because J. J. Fourier provided the first incentive to the study of these integrals [2].

---

[1] We shall also frequently use the symbols Γ, H, J to denote respectively the gamma function, the Hankel function and the Bessel function. These special uses at times will be evident from the context.

Whenever the contrary is not evident from the context, a "number" will be a complex number, and a function a complex function of a real variable. A function $f(x)$ will therefore be an expression of the form $f_1(x) + if_2(x)$ where $f_1(x)$ and $f_2(x)$ are real valued functions as usually defined. For dealing with such functions, cf. Appendix 12. We shall assume once and for all that each function which occurs under an integral sign, will first of all be integrable on each finite interval, and we shall take as a basis the integral concept of Lebesque. Thus we assume that, automatically, any given function is measurable (Lebesgue) in its entire extent and "summable" (Lebesgue) in every finite interval.

2. If the limits of integration a and b are the same for the integrals (1) to (3), then

$$J(\alpha) = \Phi(\alpha) + i\Psi(\alpha) ,$$

and

$$2\Phi(\alpha) = J(\alpha) + J(-\alpha); \qquad 2i\Psi(\alpha) = J(\alpha) - J(-\alpha) .$$

Because of the similarity in construction of $\Phi(\alpha)$ and $\Psi(\alpha)$, we shall frequently prove a statement for only one of the three integrals, and when the transfer is an obvious one, assume its correctness for the other two. Since, in addition,

$$\Phi(-\alpha) = \Phi(\alpha); \qquad \Psi(-\alpha) = -\Psi(\alpha)$$

and

$$J(-\alpha) = \overline{J_1(\alpha)}\,[1] ,$$

where

$$J_1(\alpha) = \int_a^b \overline{f(x)}e(\alpha x)\, dx$$

with the function $f_1(x) = \overline{f(x)}$, it will suffice for the study of the functions $\Phi(\alpha)$, $\Psi(\alpha)$ and $J(\alpha)$ to limit ourselves to one of the half lines $\alpha \geqq 0$ or $\alpha \leqq 0$. As a rule, we shall favor the right half line.

3. At least one of the two limits of the definite integral with which we shall be concerned, will in general, be infinite. To simplify writing, we shall omit the upper integration limit when its value is $+\infty$, and the lower limit when its value is $-\infty$. The integral

$$\int_0 f(x) \cos \alpha x\, dx$$

[1] The crossbar means "conjugate-complex".

will therefore extend, for example, over the interval $[0, \infty]$[1], and the integral

$$\int f(x)e(\alpha x)\,dx \tag{4}$$

over the interval $[-\infty, \infty]$.

4. A basic property of the trigonometric integrals with which we shall immediately concern ourselves is that they become, "in general", arbitrarily small for large values of α. In this section we shall restrict ourselves to the case where the limits of integration are finite, and only to a discussion of the integral (3).

If the function $f(x)$ has no qualifications attached to it, then the integral (3) is merely a special case of (4). The integral (3) becomes (4) if, outside of (a, b), the function $f(x)$ is extended by means of zero values; i.e., if $f(x)$ is extended, by means of the stipulation "$f(x) = 0$ if $x \neq (a, b)$", to a function defined in $[-\infty, \infty]$. This assertion is not true, however, if $f(x)$ has restrictions assigned to it. For example, if it is required that $f(x)$ be differentiable, then (3) is a special case of (4) only if $f(x)$ vanishes for $x = a$ and $x = b$; since only then is the new function which arises by assigning zero values outside (a, b) differentiable in $[-\infty, \infty]$ (cf. Appendix 8). And if $f(x)$, after its extension in $[-\infty, \infty]$ is intended to be continuously differentiable, then not only must it be continuously differentiable in (a, b) but the function and its derivative must also both vanish for $x = a$ and $x = b$.

Our assertion states that[2]

$$J(\alpha) \longrightarrow 0 \quad \text{as} \quad \alpha \longrightarrow \pm\infty \quad . \tag{5}$$

If $f(x)$ is differentiable in (a, b) and if we denote by M, a bound of $f(x)$ and also of

$$\int_a^b |f'(x)|dx \quad ,$$

then it follows from

---

[1] (λ, μ) will mean the interval $\lambda \leq x \leq \mu$; $[\lambda, \mu]$ the interval $\lambda < x < \mu$. Mixed brackets will also be employed so that $(\lambda, \mu]$ will mean $\lambda \leq x < \mu$.

[2] We shall write for the limit, with no difference in meaning, either $\lim f(\xi) = h$ or $f(\xi) \longrightarrow h$.

$$J(\alpha) = \frac{1}{i\alpha}\,[f(b)e(\alpha b) - f(a)e(\alpha a)] - \frac{1}{i\alpha}\int_a^b f'(x)e(\alpha x)\,dx$$

that

$$|J(\alpha)| \leqq \frac{3M}{|\alpha|} \tag{6}$$

and from this (5) follows. If we write

$$J(\alpha) = \int_a^c + \int_c^b = J_1(\alpha) + J_2(\alpha)$$

and if (5) is valid for $J_1(\alpha)$ and $J_2(\alpha)$, then it is evidently also valid for $J(\alpha)$. A similar reasoning would apply if more intervals were involved. Hence (5) is valid for a piecewise differentiable function, in particular for a piecewise constant function ("step function").

By a limiting process it is now possible to prove that (5) is valid for any (integrable) function. Let $f(x)$ and $f_1(x)$ be two functions such that

$$\int_a^b |f(x) - f_1(x)|dx \leqq \epsilon \quad . \tag{7}$$

Then for the corresponding integrals $J(\alpha)$ and $J_1(\alpha)$, one has

$$|J(\alpha) - J_1(\alpha)| = \left| \int_a^b (f(x) - f_1(x))e(\alpha x)\,dx \right|$$

$$\leqq \int_a^b |f(x) - f_1(x)|dx \leqq \epsilon \quad .$$

Let (5) be satisfied for $f_1(x)$. Hence there exists an $\alpha(\epsilon)$ such that for $|\alpha| > \alpha(\epsilon)$

$$|J_1(\alpha)| \leqq \epsilon \quad .$$

Therefore for $|\alpha| \geqq \alpha(\epsilon)$,

$$|J(\alpha)| \leqq |J_1(\alpha)| + |J(\alpha) - J_1(\alpha)| \leqq 2\epsilon \quad .$$

But to each (integrable) function $f(x)$ and to each ϵ, one can specify a step-function $f_1(x)$ which satisfies (7), (Appendix 10). From this it follows that:

<u>For each function</u>[1] $f(x)$

$$\int_a^b f(x)e(\alpha x)\,dx \longrightarrow 0 \quad \text{as} \quad \alpha \longrightarrow \pm\infty .$$

<u>An analogous relation also holds for the functions</u> $\Phi(\alpha)$ and $\Psi(\alpha)$ [3].

5. We observe that $J(\alpha)$ is a continuous function, and this fact can be proved as follows:

$$|J(\alpha + \rho) - J(\alpha)| \leq \int_a^b |f(x)||e(\rho x) - 1|dx \leq M(\rho) \int_a^b |f(x)|dx \quad ,$$

where $M(\rho)$ is the maximum of $|e(\rho x) - 1|$ in the interval (a, b). But if $\rho \longrightarrow 0$, then $M(\rho) \longrightarrow 0$. — $\Phi(\alpha)$ and $\Psi(\alpha)$ are also continuous functions, cf. 2[2].

§2. <u>Trigonometric Integrals Over Infinite Intervals</u>.

1. We say that the function $g(x)$ is integrable in $[a, \infty]$, if the integral

$$\int_a^A g(x)\,dx$$

approaches a finite limit as $A \longrightarrow \infty$. We denote this limit by

$$\int_a g(x)\,dx \quad . \tag{1}$$

We shall also say that the integral (1) "exists" or that it "converges".

---

[1] Since the function $f(x)$ occurs under the integral sign, it will be tacitly assumed, as agreed to in our previous statement, that it is integrable.

[2] Paragraph 2 of the present section is meant. Each section is divided into several paragraphs. A simple number denotes a paragraph, and a round bracketed number a formula. Therefore (5) denotes the formula (5). If the paragraph or formula is quoted from other sections, then the number of the section is stated in advance. Thus §51, 3 denotes paragraph 3 of §51, and §51, (9), the formula (9) of §51.

Whenever a function $g(x)$ has a certain property in a subinterval $[A, \infty]$ or $[-\infty, B]$ of its interval of definition, then we shall also say that it has this property as $x \longrightarrow \infty$ or as $x \longrightarrow -\infty$.

Since for each $A > a$, the integral (1) along with

$$\int_A^{\infty} g(x)\, dx \tag{2}$$

either converges or does not converge, it follows that the function $g(x)$ is integrable in $[a, \infty]$ if it is integrable as $x \longrightarrow \infty$.

It is a basic property of the Lebesgue integral, that in a finite interval, each integrable function is also absolutely integrable. Hence each of the functions considered heretofore is, in each finite interval, absolutely integrable. The same assertion, however, cannot be made if the interval of integration is infinite. If $g(x)$ is integrable as $x \longrightarrow \infty$ in the sense of our definition, then $|g(x)|$ need not be also integrable as $x \longrightarrow \infty$, although the converse does hold. Next, if $f(x)$ is absolutely integrable in $[a, \infty]$ then because $|f(x) \sin \alpha x| \leqq |f(x)|$, the integral

$$\Psi(\alpha) = \int_a^{\infty} f(x) \sin \alpha x\, dx \tag{3}$$

converges for all values of α. Again $\Psi(\alpha) \longrightarrow 0$ as $\alpha \longrightarrow \pm\infty$. This is deducible from

$$|\Psi(\alpha)| \leqq \left| \int_a^A f(x) \sin \alpha x\, dx \right| + \int_A^{\infty} |f(x)|\, dx .$$

Since the second integral on the right is independent of α, it can be made, by a suitable choice of A, smaller than ϵ. With A fixed, the first integral will become smaller than ϵ for $|\alpha| \geqq \alpha(\epsilon)$. Hence for $|\alpha| \geqq \alpha(\epsilon)$

$$|\Psi(\alpha)| \leqq 2\epsilon .$$

Corresponding assertions are valid for $\Phi(\alpha)$ and $J(\alpha)$.

For example, let $f(x) = e^{-kx}$, $k > 0$ and $a = 0$, and let us calculate $J(\alpha)$, the simplest of the three. From

$$\int_0^A e^{-(k-i\alpha)x}\, dx = \frac{1}{k - i\alpha}\left(1 - e^{-kA} e^{(\alpha A)}\right) ,$$

one obtains, by letting $A \longrightarrow \infty$, and by separating the real and imaginary parts

$$\int_0 e^{-kx} \cos \alpha x \, dx = \frac{k}{k^2 + \alpha^2}; \qquad \int_0 e^{-kx} \sin \alpha x \, dx = \frac{\alpha}{k^2 + \alpha^2} \tag{4}$$

Both expressions actually approach zero as $\alpha \longrightarrow \pm \infty$ [4].

As regards behavior at infinity, an important class of functions which need not be absolutely integrable are monotonic functions. Let the (real valued) function $f(x)$ converge monotonically to zero as $x \longrightarrow \infty$, i.e., let it be monotonic in a certain interval $[A, \infty]$, and convergent to zero as $x \longrightarrow \infty$. Since we already have at our command integrals over finite intervals, we can assume, therefore, that the point $x = A$ coincides with the initial point $x = a$. A function monotonic in $[a, \infty]$ which converges to zero as $x \longrightarrow \infty$ is, in its entire range, either positive and decreasing, or negative and increasing. Since an increasing function becomes, by a change of sign, a decreasing one, we need consider only the decreasing one.

2. We shall need the following theorem of analysis; the so-called second mean value theorem of integration. If, in the interval (a, b), the function $\varphi(x)$ is continuous, and the function $p(x)$ is positive and monotonically decreasing, then in the interval (a, b) there is a value c between a and b for which

$$\int_a^b p(x)\varphi(x) \, dx = p(a) \int_a^c \varphi(x) \, dx .$$

In particular, let $\varphi(x) = \sin \alpha x$, $\alpha > 0$. From

$$\left| \int_a^c \sin \alpha x \, dx \right| \leqq \frac{2}{\alpha}$$

it follows that

$$\left| \int_a^b p(x) \sin \alpha x \, dx \right| \leqq \frac{2p(a)}{\alpha} . \tag{5}$$

Now in $(a, \infty]$, let the function $p(x)$ decrease monotonically to zero. From

$$\left| \int_A^{A'} p(x) \sin \alpha x \, dx \right| \leqq \frac{2}{\alpha} p(A) \qquad \alpha > 0, \tag{6}$$

in conjunction with the fact that $p(A) \longrightarrow 0$ as $A \longrightarrow \infty$, it follows that the integral

$$\Psi(\alpha) = \int_a p(x) \sin \alpha x \, dx$$

is convergent for $\alpha > 0$. We can now allow A' in (6) to become infinite, and we have

$$\left| \int_A p(x) \sin \alpha x \, dx \right| \leqq \frac{2}{\alpha} p(A) \quad .$$

Hence it follows that $\Psi(\alpha) \longrightarrow 0$ as $\alpha \longrightarrow \infty$. Summarizing, we formulate the following theorem.

THEOREM 1. If in $[a, \infty]$, the function $f(x)$ under consideration, as $x \longrightarrow \infty$; either

1. is absolutely integrable, or
2. converges monotonically to zero,

then the integrals $\Phi(\alpha)$, $\Psi(\alpha)$, $J(\alpha)$ exist for

1. all α or
2. all $\alpha \neq 0$,

and converge to zero as $\alpha \longrightarrow \pm \infty$ [5].

The restriction $\alpha \neq 0$, made under 2 applies only to $\Phi(\alpha)$ and $J(\alpha)$. For a function decreasing monotonically to zero, it is not necessary that the integral

$$\int_a f(x) \, dx \, ,$$

which should represent the value $\Phi(0)$ or $J(0)$, converge (for example $f(x) = 1/x$).

Now let $f(x)$ be representable in the form

$$f(x) = g(x) \sin px \quad ,$$

where p is a constant, and $g(x)$ approaches zero monotonically. By means of the relation

$$2\Psi(\alpha) = \int_a g(x) \cos (\alpha - p)x \, dx - \int_a g(x) \cos (\alpha + p)x \, dx \quad ,$$

one recognizes again that, with the possible exceptions of $\alpha = p$ and

$\alpha = -p$, the integral $\Psi(\alpha)$ exists and converges to zero as $\alpha \longrightarrow \pm\infty$. The same assertion holds for

$$f(x) = g(x) \cos px \quad ,$$

and more generally for

$$f(x) = g(x) \sin (px + q) \quad ,$$

where p and q are constants.

THEOREM 1a. The assumption 2 in Theorem 1 can be generalized by setting

$$f(x) = g(x) \sin (px + q) \quad ,$$

where p and q are constants, and $g(x)$ approaches zero monotonically as $x \longrightarrow \infty$. However, the integrals need not converge for the values $\alpha = \pm p$ [5].

THEOREM 1b. A further generalization of the theorem results if the factor $\cos \alpha x$ or $\sin \alpha x$ in $\Phi(\alpha)$ and $\Psi(\alpha)$, is replaced by $\cos \alpha(x - t)$ or $\sin \alpha(x - t)$, where t is an additional constant [5].

This generalization can be justified by the transformation $y = x - t$.

3. Analogous statements are valid for a left half line $[-\infty, b]$, and for the entire interval $[-\infty, \infty]$. We call the integral

$$\int g(x)\, dx$$

convergent if both integrals

$$\int_0 g(x)\, dx \quad \text{and} \quad \int^0 g(x)\, dx$$

converge. In this sense, we shall later on attach to the function $f(x)$, the special integral

$$E(\alpha) = \frac{1}{2\pi} \int f(x) e(-\alpha x)\, dx$$

and denote it as the (<u>Fourier</u>) <u>transform</u> [6] of $f(x)$. The integral $E(\alpha)$

is therefore normalized somewhat differently from the integral $J(\alpha)$, namely

$$J(\alpha) = 2\pi E(-\alpha) \quad .$$

From the above, we see that $E(\alpha)$ exists for all $\alpha \neq 0$, and converges to zero as $\alpha \longrightarrow \pm \infty$, provided $f(x)$ is either absolutely convergent or approaches zero monotonically not only as $x \longrightarrow \infty$ but also as $x \longrightarrow -\infty$.

4. If $a > 0$, then the integral

$$\int_a \frac{\sin \alpha x}{x}\, dx$$

falls under Theorem 1, because in the interval $(a, \infty]$, the function $f(x) = 1/x$ decreases monotonically to zero. On the other hand $f(x)$ is not integrable in the interval $[0, a]$, and therefore not in $[0, \infty]$. Although the whole integrand $x^{-1}\sin \alpha x$ is regular there, and hence the integral

$$\Psi(\alpha) = \int_0 \frac{\sin \alpha x}{x}\, dx$$

exists for all α, yet $\Psi(\alpha)$ is not convergent to zero as $\alpha \longrightarrow \infty$. The transformation $\alpha x = \xi$, $\alpha > 0$ yields for example

$$\int_0 \frac{\sin \alpha x}{x}\, dx = \int_0 \frac{\sin \xi}{\xi}\, d\xi \quad .$$

Hence $\Psi(\alpha)$ is constant for $\alpha > 0$, and this constant, as we shall see later in §4, 3 is different from zero.

§3. Order of Magnitude of Trigonometric Integrals

1. The question arises whether an assertion can be made with regard to the rapidity with which $\Phi(\alpha)$ and $\Psi(\alpha)$ decrease to zero as $\alpha \longrightarrow \infty$. According to Lebesgue, if the function $f(x)$ is only known to be (absolutely) integrable, no statement of this kind can be made even if the interval happens to be finite. Rather, it can be shown that these integrals can decrease to zero arbitrarily slowly [7]. The situation changes however, if more precise information about the function $f(x)$ is available. If $f(x)$ is monotonically decreasing in (a, b) or monotonically decreasing to zero in $(a, \infty]$, then by §2, (5), there exists a constant A, such that for $\alpha > 0$

$$|\Psi(\alpha)| \leq A\,\alpha^{-1}$$

which can be written with the familiar Landau symbol

$$\Psi(\alpha) = O(\alpha^{-1}) \quad . \tag{1}$$

We recall the meaning of this symbol. Let $\varphi(\xi) > 0$ as $\xi \longrightarrow \infty$. Then $f(\xi) = O[\varphi(\xi)]$ states that the quotient

$$f(\xi)/\varphi(\xi)$$

is bounded as $\xi \longrightarrow \infty$; and $f(\xi) = o[\varphi(\xi)]$ states that it approaches zero. If $f(\xi) = O[\varphi(\xi)]$, and $f_1(\xi) = O[\varphi_1(\xi)]$, where $\varphi(\xi) \leq \varphi_1(\xi)$, and if $h(\xi)$ and $h_1(\xi)$ are bounded as $\xi \longrightarrow \infty$, then $fh + f_1h_1 = O(\varphi_1)$. Analogous statements are valid for the o-relation.

If $f(x)$ is differentiable, then (1) is valid for an interval (a, b), cf. §1, (6); if $f(x)$ has an absolutely integrable derivative and approaches zero as $\xi \longrightarrow \infty$, then (1) is valid for an interval $[a, \infty]$. The last statement can be verified by means of the usual partial integration formula (Appendix 8):

$$\int_a f(x)\sin\alpha x dx = \frac{1}{\alpha}\; f(a)\cos\alpha a + \frac{1}{\alpha}\int_a f'(x)\cos\alpha x dx \quad .$$

2. That (1) holds on the one hand for monotonic and on the other hand for differentiable functions is no accident. There is in fact the following connection between them. If one knows that (1) holds for monotonically decreasing functions, then it follows immediately that it also holds for monotonically increasing functions, and that it holds generally for functions which can be represented as linear combinations (with complex coefficients) of monotonic functions. We shall denote, as usual, these last functions as <u>functions of bounded variation</u>. For our purposes, we shall not need the "true" concept of bounded variation. It will be sufficient for us to show directly, that each function which has an absolutely integrable derivative is of bounded variation in the sense stated above. Since if $f_1'(x) + i\, f_2'(x)$ is absolutely integrable, $f_1'(x)$ and $f_2'(x)$ are also, we need to prove our assertion only for real valued functions. Let $f(x)$ have an integrable derivative in (a, b). Then we can set

$$\begin{aligned} f(x) &= f(b) + \int_x^b \frac{|f'(\xi)| - f'(\xi)}{2}\,d\xi - \int_x^b \frac{|f'(\xi)| + f'(\xi)}{2}\,d\xi \\ &= f(b) + h_1(x) - h_2(x). \end{aligned} \tag{2}$$

Both functions $h_1(x)$ and $h_2(x)$, which still depend on the parameter b, are monotonically decreasing since the integrands are non-negative. If one is dealing with the interval $[a, \infty]$, and if $f'(x)$ itself is absolutely integrable, then one applies (2) for some fixed b in $[a, \infty]$. As $x \longrightarrow \infty$, the limit on the right side of (2) exists, and hence also the limit of $f(x)$. We denote this limit by $f(\infty)$. If, therefore, with x fixed in (2), we now let $b \longrightarrow \infty$, we obtain

$$f(x) = f(\infty) + h_1(x) - h_2(x) \quad ,$$

with $h_1(x)$ and $h_2(x)$ monotonically decreasing functions, and with the parameter b in these functions having now the value $+\infty$, Q.E.D. Also, $\lim h_1(x) = \lim h_2(x) = 0$ as $x \longrightarrow \infty$. If therefore $f(x)$ decreases to zero as $x \longrightarrow \infty$, one can then represent $f(x)$ as the difference of two monotonic functions which also decrease to zero [8].

3. The inquiry can be extended to include the case in which $f(x)$ has infinite discontinuities at isolated points. In an interval (a, b) and for any c, let

$$f(x) = \frac{g(x)}{|x-c|^{\mu}}$$

where $g(x)$ is of bounded variation and μ is a positive number < 1. One can then easily show that

$$\int_a^b f(x) \begin{matrix}\cos\\ \sin\end{matrix} \alpha x dx = O(|\alpha|^{-\mu}) \quad .$$

We shall not, however, prove this [9].

4. Let the function $f(x)$ be differentiable in $[-\infty, \infty]$, and together with its derivatives be absolutely integrable. Since $f'(x)$ is absolutely integrable, the integral

$$g(x) = \int^x f'(\xi)\, d\xi$$

exists, and since $g'(x) = f'(x)$, we have $f(x) = g(x) + c$. As $x \longrightarrow -\infty$, $g(x) \longrightarrow 0$. If now c were $\neq 0$, then $f(x)$ could not be absolutely integrable as $x \longrightarrow -\infty$. Hence $f(x) = g(x)$, and in particular $f(x) \longrightarrow 0$ as $x \longrightarrow -\infty$. If we now put

$$\int f'(\xi)\, d\xi = c_1 \quad ,$$

then

$$f(x) = c_1 - \int_x f'(\xi)\,d\xi \; .$$

Again we obtain $c_1 = 0$, and hence $f(x) \longrightarrow 0$ also as $x \longrightarrow -\infty$. Let us now consider the transform

$$E(\alpha) = \frac{1}{2\pi}\int f(x)e(-\alpha x)\,dx \quad .$$

By partial integration, we have

$$i\alpha E(\alpha) = \frac{1}{2\pi}\int f'(x)e(-\alpha x)\,dx \quad .$$

Applying Theorem 1 (cf. also §2, 3) to the integrand $f'(x)$ instead of to $f(x)$, we see that $i\alpha E(\alpha)$ converges to zero as $\alpha \longrightarrow \pm\infty$, i.e.,

$$E(\alpha) = o(|\alpha|^{-1}) \quad .$$

If the function $f'(x)$ has in its turn an absolutely integrable derivative, then by the same reasoning, one obtains

$$i\alpha E(\alpha) = o(|\alpha|^{-1})$$

and therefore

$$E(\alpha) = o(|\alpha|^{-2}) \quad .$$

Continuing in this manner, we obtain the following general theorem:

> If the function $f(x)$ is k-times differentiable in $[-\infty, \infty]$, $k = 0, 1, 2, \ldots,$[1] and together with its first k derivatives, is absolutely integrable, then for its transform we have
>
> $$E(\alpha) = o(|\alpha|^{-k}) \quad .$$

§4. Uniform Convergence of Trigonometric Integrals

1. Consider the convergent integral

$$\int_a g(x, \lambda)\,dx$$

which depends on a parameter λ. The integral is called uniformly

---

[1] By the o-th derivative of a function, we mean the function itself.

convergent, if to each ϵ, one can find an $A(\epsilon)$, such that for $A > A(\epsilon)$ and for all considered values λ

$$\left| \int_A g(x, \lambda)\, dx \right| \leq \epsilon .$$

An analogous definition is valid for an integral extending over $[-\infty, \infty]$, cf. §2, 3.

Uniform convergence will occur if there exists an absolutely integrable function $\gamma(x)$ for which $|g(x,\lambda)| < |\gamma(x)|$. Hence

THEOREM 2. For an absolutely integrable function $f(x)$, the integral §2, (3), is uniformly convergent for all α. If $f(x) \longrightarrow 0$ monotonically as $x \longrightarrow \infty$, then the uniform convergence for $|\alpha| \geq \alpha_0 (> 0)$ follows from §2, 2. A similar statement is valid for $\Phi(\alpha)$. More generally, if

$$f(x) = g(x)\sin(px + q)$$

and $g(x) \longrightarrow 0$ monotonically, then $\Phi(\alpha)$ and $\Psi(\alpha)$ are uniformly convergent in each interval (α_1, α_2) which does not contain the points $\alpha = +p$ and $\alpha = -p$. The same assertion, moreover, holds for the general integral

$$(1) \qquad \int_a f(x)\cos \alpha(x - t)\, dx; \quad \int_a f(x)\sin \alpha(x - t)\, dx .$$

Furthermore, each of these integrals, in each interval in which it converges uniformly, is a continuous function of α.

The last statement is easily verified. Consider an α-interval in which the function

$$\Psi_n(\alpha) = \int_a^{a+n} f(x)\sin \alpha x\, dx$$

converges uniformly to $\Psi(\alpha)$ as $n \longrightarrow \infty$. Since $\Psi_n(\alpha)$ is continuous everywhere in α, cf. §1, 5, the limiting function $\Psi(\alpha)$, by a well known theorem, is also continuous in this interval. A similar statement is valid for the other integrals.

2. Uniformly convergent integrals can be differentiated and integrated under the integral sign with respect to the parameter, in accordance

with the following rules [10]:

a) If a function $g(x, \lambda)$ is continuous for $a \leq x < \infty$ and $\lambda_0 \leq \lambda \leq \lambda_1$, and if the integral

$$G(\lambda) = \int_a g(x, \lambda)\, dx$$

converges uniformly, then the function $G(\lambda)$ is continuous in (λ_0, λ_1).

b) Also

$$\int_{\lambda_0}^{\lambda_1} G(\lambda)\, d\lambda = \int_a \left[\int_{\lambda_0}^{\lambda_1} g(x, \lambda)\, d\lambda\right] dx \quad .$$

c) Moreover, if the function $g(x, \lambda)$ is differentiable at every point with respect to λ, and if the function

$$g_\lambda(x, \lambda) = \frac{\partial g(x, \lambda)}{\partial \lambda}$$

is itself continuous and has a uniformly convergent integral, then the function $G(\lambda)$ is differentiable, and

$$G'(\lambda) = \int_a g_\lambda(x, \lambda)\, dx \quad .$$

3. As a first application, we shall integrate the first equation of §2, (4) with respect to α between the limits 0 and α. This gives

$$(2) \qquad \int_0 e^{-kx} \frac{\sin \alpha x}{x}\, dx = \text{arc tan } \frac{\alpha}{k} \qquad k > 0 \ .$$

For fixed $\alpha > 0$ the integral on the left is uniformly convergent in $0 \leq k < \infty$ as seen from the relation

$$\left| \int_A \frac{e^{-kx}}{x} \sin \alpha x dx \right| \leq \frac{2}{\alpha} \frac{e^{-kA}}{A} \leq \frac{2}{\alpha} \frac{1}{A} \quad .$$

Hence by (a)

$$\int_0 \frac{\sin x}{x}\, dx = \lim_{k \to 0} \int_0 e^{-kx} \frac{\sin x}{x}\, dx = \lim_{k \to 0} \text{arc tan } \frac{1}{k} \quad .$$

But this last limit, since $k > 0$, has the value $\pi/2$. Therefore [11]

$$\int_0 \frac{\sin x}{x}\,dx = \frac{\pi}{2}\,, \tag{3}$$

and hence

$$\int \frac{\sin x}{x}\,dx = \pi\,. \tag{4}$$

We shall frequently use the above rules tacitly. On occasion, we shall employ them under more general conditions than those formulated above. But in this event, only the evaluation of some definite integrals will be involved, of which no essential use will be made subsequently.

4. By (4) and §2, 4, we have [12]

$$\frac{1}{\pi}\int \frac{\sin \alpha x}{x}\,dx = \begin{cases} 1, & \alpha > 0 \\ 0, & \alpha = 0 \\ -1, & \alpha < 0\,. \end{cases} \tag{5}$$

In this case, one speaks of a discontinuous integral. The integral

$$D(\lambda) = \frac{1}{\pi}\int \frac{\sin x}{x}\cos \lambda x\,dx \tag{6}$$

can be reduced to the above, if one sets

$$2\sin x \cos \lambda x = \sin(1-\lambda)x + \sin(1+\lambda)x\,. \tag{7}$$

A simple separation of cases yields the values

$$D(\lambda) = \begin{cases} 1\,, & |\lambda| > 1 \\ 1/2, & |\lambda| = 1 \\ 1\,, & |\lambda| < 1\,. \end{cases} \tag{8}$$

The integral (6) is called the <u>Dirichlet discontinuous factor</u>. If $\cos \lambda x$ is replaced by $\sin \lambda x$, then the integrand becomes an odd function, and the integral insofar as it converges, vanishes. Summarizing we have for $|\lambda| \neq 1$

$$\frac{1}{\pi}\int \frac{\sin x}{x}\,e(\lambda x)\,dx = D(\lambda)\,. \tag{9}$$

For $\rho > 0$ and real σ, the transformation $x = \rho\xi$ gives

$$\frac{1}{\pi}\int \frac{\sin \rho x}{x}\,e(\sigma x)\,dx = D\left(\frac{\sigma}{\rho}\right)\,, \tag{10}$$

and the more general transformation $x = \rho(\xi - t)$ gives

$$\frac{1}{\pi}\int \frac{\sin \rho(x-t)}{x - t}\, e(\sigma x)\, dx = D(\tfrac{\sigma}{\rho})e(\sigma t) \quad . \tag{11}$$

Multiplying (9) by $e(-\lambda a)$, and taking the real part, one obtains for the integral

$$h(\lambda) = \frac{1}{\pi}\int \frac{\sin x}{x} \cos \lambda(x - a)\, dx ,$$

the value

$$h(\lambda) = D(\lambda) \cos \lambda a \quad . \tag{12}$$

We are now in a position to calculate the value of the integral

$$H(\lambda) = \frac{1}{\pi}\int \frac{\sin x}{x}\, \frac{\sin \lambda(x-a)}{x - a}\, dx \quad .$$

Since $h(\lambda)$ results from $H(\lambda)$ by formal differentiation, we shall make use of the rules under 2. The integral $H(\lambda)$, is, by Theorem 2, uniformly convergent for all λ, since the factor $\sin \lambda(x - a)$ is absolutely integrable as $x \longrightarrow \pm\infty$. The integral $h(\lambda)$, by Theorem 2, converges uniformly in each closed interval which does not contain the points $+1$ and -1. Therefore, with the possible exception of the points $+1$ and -1, the function $H(\lambda)$ is differentiable and

$$H^1(\lambda) = h(\lambda) \quad .$$

Since $H(\lambda)$ is continuous everywhere, it follows that

$$H(\lambda) = H(0) + \int_0^\lambda h(\lambda)\, d\lambda = \int_0^\lambda h(\lambda)\, d\lambda \quad ,$$

and hence by (8) and (12)

$$H(\lambda) = \frac{\sin a}{a}, \quad \frac{\sin \lambda a}{a}, \quad -\frac{\sin a}{a}$$

according as

$$\lambda \geqq 1, \quad |\lambda| \leqq 1, \quad \lambda \leqq -1 \quad .$$

For the special case $a = 0$, the integral

$$\frac{1}{\pi}\int \frac{\sin x}{x}\, \frac{\sin \lambda x}{x}\, dx \quad ,$$

in the same intervals, has the values $1, \lambda, -1$. In particular

$$\frac{1}{\pi}\int\left(\frac{\sin x}{x}\right)^2 dx = 1 \quad . \tag{13}$$

Making use of (7), we obtain [13]

$$\frac{1}{\pi}\int\left(\frac{\sin x}{x}\right)^2 \cos \lambda x \, dx = \begin{cases} 1 - \frac{|\lambda|}{2}, & |\lambda| \leqq 2 \\ 0 \quad , & |\lambda| \geqq 2 \quad . \end{cases} \tag{14}$$

Integration with respect to λ from 0 to λ, gives for $\lambda > 0$

$$\frac{1}{\pi}\int\left(\frac{\sin x}{x}\right)^2 \frac{\sin \lambda x}{x} dx = \begin{cases} \lambda - \frac{\lambda^2}{4}, & 0 \leqq \lambda \leqq 2 \\ 1 \quad , & 2 \leqq \lambda \quad . \end{cases} \tag{15}$$

In particular

$$\frac{1}{\pi}\int\left(\frac{\sin x}{x}\right)^3 dx = \frac{3}{4} \quad . \tag{16}$$

Continuing this process, one would find, in general, for integral $p \geqq 2$, that the function

$$C_p(\lambda) = \int\left(\frac{\sin x}{x}\right)^p \cos \lambda x \, dx \tag{17}$$

vanishes outside of a sufficiently large interval (namely for $|\lambda| \geqq p$). We note finally that

$$\frac{1}{\pi}\int\left(\frac{\sin x}{x}\right)^4 dx = \frac{2}{3} \quad . \tag{18}$$

§5. The Cauchy Principal Value of Integrals

1. Let c be a given interior point in an interval (a, b), and $f(x)$ a function with the following property. For (sufficiently small) $\epsilon > 0$, let $f(x)$ be integrable in the intervals $(a, c - \epsilon)$ and $(c + \epsilon, b)$, and let the sum

$$\int_a^{c-\epsilon} f(x)\,dx + \int_{c+\epsilon}^{b} f(x)\,dx$$

approach a limit as $\epsilon \longrightarrow 0$. We denote this limiting value as the Cauchy principal value of the integral

$$\int_a^b f(x)\,dx \quad .$$

If $f(x)$ is integrable in the neighborhood of the point c, and hence integrable in the entire interval (a, b), then the Cauchy principal value exists and equals the usual value of the integral. A similar definition is valid if more than one singular point exists, namely points $c_1, \ldots, c_k$, and also if the integration limit a or b is not finite. A special case of the Cauchy principal value occurs when the point c by chance coincides with the end point a; that is $f(x)$ is integrable for $\epsilon > 0$ in the interval $(a + \epsilon, b)$, and

$$\lim_{\epsilon \to 0} \int_{a+\epsilon}^{b} f(x)\,dx$$

exists. — If $f(x)$ is integrable in each finite interval, and

$$\lim_{N \to \infty} \int_{-N}^{N} f(x)\,dx$$

exists, then it is also customary to speak of the Cauchy principal value of the integral

$$\int f(x)\,dx \quad .$$

In this case, the individual integrals $\int_0$ and $\int^0$ do not need to converge.

2. The integrals

$$\varphi(\alpha) = \int_a^b \frac{f(x)}{x - t} \cos \alpha x\,dx, \qquad \psi(\alpha) = \int_a^b \frac{f(x)}{x - t} \sin \alpha x\,dx \quad ,$$

are of especial interest. Here t is a point of the interval (a, b) and $f(x)$ is, first of all, integrable. First let $t = 0$, $a = -1$, $b = 1$. The integral $\psi(\alpha)$ then exists in the usual sense. The existence of $\varphi(\alpha)$ is tied up with the question of the limiting value of

$$\lim_{\epsilon \to 0} \int_{\epsilon}^{1} \frac{f(x) - f(-x)}{x} \cos \alpha x\,dx \quad .$$

It exists if the function

$$g(x) = \frac{f(x) - f(-x)}{x}$$

is integrable in $[0, 1]$. In this event, it is also true by Theorem 1 that

$$\varphi(\alpha) \longrightarrow 0 \quad \text{as} \quad |\alpha| \longrightarrow \infty \ .$$

We observe that $g(x)$ is integrable in $[0, 1]$, if $f(x)$ is by chance continuously differentiable, since by the definition of the derivative

$$\lim_{x \to 0} \frac{f(x) - f(-x)}{2x} = f'(0) \ .$$

Hence $g(x)$ is even continuous at the critical point $x = 0$.

For $t \neq 0$, one obtains, more generally, that the integrals $\varphi(\alpha)$ and $\psi(\alpha)$ exist, whenever the function

$$\frac{f(t+x) - f(t-x)}{x}$$

is integrable in some interval $0 < x \leqq x_0$. And this function will be integrable, if again $f(x)$ is by chance continuously differentiable.

3. As Hardy has shown in detailed investigations [14], one can operate with Cauchy principal values largely as with ordinary integrals. Consider, for example, the principal value of

$$\psi(\alpha) = \int_{1-a}^{1+a} \frac{f(x)}{x - 1} \sin \alpha x \, dx \tag{1}$$

where $f(x)$ is a continuously differentiable function. Writing it as

$$\sin \alpha \int_0^a \frac{f(1+x) - f(1-x)}{x} \cos \alpha x \, dx + \cos \alpha \int_0^a [f(1+x) + f(1-x)] \frac{\sin \alpha x}{x} \, dx,$$

one recognizes that $\psi(\alpha)$ is differentiable. Carrying out the differentiation, a simple transformation leads to

$$\psi'(\alpha) = \int_{1-a}^{1+a} \frac{x \, f(x)}{x - 1} \cos \alpha x \, dx \ .$$

In other words, the value of $\psi'(\alpha)$ results from differentiating (1) under the integral sign. The same assertion also holds for

$$\varphi(\alpha) = \int_{1-a}^{1+a} \frac{f(x)}{x - 1} \cos \alpha x \, dx \ .$$

We will apply this in order to calculate the value of

$$\Psi(\alpha) = \int_0^\infty \frac{\sin \alpha x}{(1-x^2)x}\, dx \quad .$$

We divide the integral into the three sums

$$\int_0^{1-a} + \int_{1-a}^{1+a} + \int_{1+a}^{\infty} = \Psi_1 + \Psi_2 + \Psi_3 \quad ,$$

with $0 < a < 1$. As we have just seen, Ψ_2 is differentiable arbitrarily often under the integral sign. The same holds for Ψ_1. Formally differentiating Ψ_3 once or twice, we obtain the integrals

$$\int_{1+a}^{\infty} \frac{\cos \alpha x}{1 - x^2}\, dx \quad \text{or} \quad - \int_{1+a}^{\infty} \frac{x \sin \alpha x}{1 - x^2}\, dx \quad .$$

By Theorem 2, therefore, $\Psi'(\alpha)$ for all α, and $\Psi''(\alpha)$ for $\alpha \neq 0$ exist and are continuous, and

$$\Psi'(\alpha) = \int_0^\infty \frac{\cos \alpha x}{1 - x^2}\, dx, \quad \Psi''(\alpha) = - \int_0^\infty \frac{x \sin \alpha x}{1 - x^2}\, dx \quad .$$

Hence for $\alpha > 0$,

$$\psi(\alpha) + \psi''(\alpha) = \int_0^\infty \frac{\sin \alpha x}{x}\, dx = \frac{\pi}{2} \quad .$$

The general solution of this differential equation is

$$\Psi(\alpha) = \frac{\pi}{2} + A \sin \alpha + B \cos \alpha \tag{2}$$

where A and B are constants. Since $\Psi(\alpha)$ is continuous everywhere, by letting $\alpha \longrightarrow 0$, (2) has the value $B = -\pi/2$. Similarly by letting $\alpha \longrightarrow 0$ in $\psi'(\alpha) = A \cos \alpha - B \sin \alpha$, we obtain

$$A = \int_0^\infty \frac{dx}{1 - x^2}$$

Now writing for the last integral

$$\frac{1}{2}\int_0^2 \frac{dx}{1 - x} + \frac{1}{2}\int_0^2 \frac{dx}{1 + x} + \int_2^\infty \frac{dx}{1 - x^2} \quad ,$$

there results

$$A = 0 + \frac{1}{2}\log 3 - \frac{1}{2}\log\frac{2+1}{2-1} = 0 \quad .$$

A simple transformation yields finally [15] for $a > 0$, $\alpha > 0$:

$$(3)\qquad \int_0 \frac{\sin\alpha x}{x(a^2-x^2)}\,dx = \frac{\pi}{2a^2}(1-\cos a\alpha), \qquad \int_0 \frac{\cos\alpha x}{a^2-x^2}\,dx = \frac{\pi}{2a}\sin a\alpha \;,$$

$$\int_0 \frac{x\sin\alpha x}{a^2-x^2}\,dx = -\frac{\pi}{2}\cos a\alpha \quad .$$

4. These integrals could have been calculated more quickly by complex integration. In a similar manner, the following formula [16], can be obtained, also for $a > 0$ and $0 < R(\lambda) < 2$:

$$(4)\qquad \int_0 \frac{y^{\lambda-1}e(\alpha y)}{a^2-y^2}\,dy = -i\frac{\pi}{2}a^{\lambda-2}e(\alpha a) + e(\lambda\,\pi/2)\int_0 \frac{x^{\lambda-1}e^{-\alpha x}}{a^2+x^2}\,dx \quad .$$

$$(5)\qquad \int_0 \frac{y^{\lambda-1}e(\alpha y)}{a^2+y^2}\,dy = -i\frac{\pi}{2}a^{\lambda-2}e^{-a\alpha}e(\lambda\,\pi/2) + e(\lambda\,\pi/2)\int_0 \frac{x^{\lambda-1}e^{-\alpha x}}{a^2-x^2}\,dx \quad .$$

5. For later needs, we shall now prove that the real part of

$$(6)\qquad \int_{-\alpha}^{\alpha} \frac{e^{-x}}{x+i\epsilon}\,dx, \qquad \alpha > 0$$

converges, as $\epsilon \longrightarrow 0$, to the principal value of

$$(7)\qquad \int_{-\alpha}^{\alpha} \frac{e^{-x}}{x}\,dx \quad .$$

The real part of the difference of (7) and (6) amounts to

$$\epsilon^2\int_0^{\alpha} \frac{e^{-x}-e^{x}}{x}\cdot\frac{1}{x^2+\epsilon^2}\,dx \quad .$$

And since $x^{-1}(e^{-x}-e^{x})$ is bounded in $(0, \alpha)$, its absolute value is smaller than a constant times

$$\epsilon^2\int_0 \frac{1}{x^2+\epsilon^2}\,dx = \epsilon\,\frac{\pi}{2} \quad .$$

CHAPTER II

REPRESENTATION — AND SUM FORMULAS

§6. A General Representation Formula

1. Let $K(\xi)$ and $f(\xi)$ be given functions in the interval $[-\infty, \infty]$. Under suitable conditions, the integral

$$(1) \qquad f_n(x) = \int f\left(x + \frac{\xi}{n}\right) K(\xi)\, d\xi$$

exists for $n > n_0$. Letting $n \longrightarrow \infty$ under the integral sign, we may expect to obtain

$$(2) \qquad f(x) \int K(\xi)\, d\xi = \lim_{n \to \infty} f_n(x) \quad .$$

That is

$$(2') \qquad f(x) \int K(\xi)\, d\xi = \lim_{n \to \infty} \int f\left(x + \frac{\xi}{n}\right) K(\xi)\, d\xi \quad .$$

We call this relation a <u>representation of the function</u> $f(x)$ <u>by means of</u> the kernel $K(\xi)$. We can also write for (1)

$$(3) \qquad f_n(x) = n \int f(x + \xi) K(n\xi)\, d\xi \quad ,$$

or

$$(4) \qquad f_n(x) = n \int f(\xi) K[n(\xi - x)]\, d\xi \quad .$$

The relation (2) is to be expected only if $f(\xi)$ is continuous at $x = \xi$. But a representation is still possible in the general case where $f(\xi)$ has the right and left limits $f(x + 0)$ and $f(x - 0)$. In this case, the expression (2) is to be replaced by

$$(5) \qquad f(x + 0) \int_0 K(\xi)\, d\xi + f(x - 0) \int^0 K(\xi)\, d\xi = \lim_{n \to \infty} f_n(x) \quad .$$

If $K(\xi)$ is even, i.e., $K(-\xi) = K(\xi)$, and if in addition

$$\int K(\xi)\, d\xi = 1 \quad , \tag{6}$$

then it reads[1]

$$\frac{1}{2}\,[f(x+0) + f(x-0)] = \lim_{n\to\infty} \int f\left(x + \frac{\xi}{n}\right) K(\xi)\, d\xi \quad . \tag{7}$$

THEOREM 3 [17]. For the validity of (5) at points x for which $f(x+0)$ and $f(x-0)$ exist, one of the two following assumptions is sufficient.

(a) $f(\xi)$ is bounded, $|f(\xi)| \leq G$, and $K(\xi)$ is absolutely integrable.

(b) $f(\xi)$ is absolutely integrable, $K(\xi)$ is absolutely integrable and bounded, and $K(\xi) = o(|\xi|^{-1})$ as $|\xi| \longrightarrow \infty$.

PROOF. Setting $f_1(\xi) = f(\xi)$ for $\xi \geq x$, and $= 0$ for $\xi < x$, it is evident that $f_1(x)$ also satisfies assumption (a) or (b). Hence (5) reads

$$f(x+0) \int_0 K(\xi)\, d\xi = \lim_{n\to\infty} \int_0 f\left(x + \frac{\xi}{n}\right) K(\xi)\, d\xi \quad . \tag{8}$$

Similarly one obtains

$$f(x-0) \int^0 K(\xi)\, d\xi = \lim_{n\to\infty} \int^0 f\left(x + \frac{\xi}{n}\right) K(\xi)\, d\xi \quad . \tag{9}$$

Conversely, (5) is a consequence of (8) and (9). It is, therefore, sufficient to prove formulas (8) and (9), and since they are symmetrically constructed, we need concentrate on only one of them. We choose (8).

PROOF OF (a). The integral $f_n(x)$ exists for all n. Let

$$\varphi(n) = \int_0 \left[f\left(x + \frac{\xi}{n}\right) - f(x+0) \right] K(\xi)\, d\xi$$

Then, for $A > 0$

$$|\varphi(n)| \leq \int_0^A \left| f\left(x + \frac{\xi}{n}\right) - f(x+0) \right| |K(\xi)|\, d\xi + 2G \int_A |K(\xi)|\, d\xi \quad .$$

---

[1] If $\int K(\xi) d\xi \neq 0$, then it can be normalized by multiplying $K(\xi)$ by a constant.

Since $K(\xi)$ is absolutely integrable, the second term on the right can be made, by a suitable choice of A, smaller than an arbitrary number $\epsilon > 0$. The first term is

$$\leq \delta(n) \int_0^A |K(\xi)|d\xi \quad ,$$

where $\delta(n)$ denotes the upper limit of $|f(x + t) - f(x + 0)|$ in the interval $0 < t \leq An^{-1}$. But with A fixed, the length of this interval becomes arbitrarily small with n^{-1}, and by the definition of the limit $f(x + 0)$, $\delta(n)$ also becomes arbitrarily small. Therefore for fixed A, the first term is smaller than ϵ for $n \geq n(\epsilon)$. Hence $|\varphi(n)| \leq \epsilon + \epsilon = 2\epsilon$ for $n \geq n(\epsilon)$. Expressed in another way

$$\varphi(n) \longrightarrow 0 \quad \text{as} \quad n \longrightarrow \infty \quad .$$

PROOF OF (b). For $c > 0$, we write

$$n \int_0 f(x + \xi)K(n\xi)\, d\xi = n \int_0^c + n \int_c = g_n(x) + h_n(x) \quad .$$

Since the limit $f(x + 0)$ exists, $f(x + \xi)$ is bounded in a certain interval $0 < \xi \leq c$. If for this c, one sets the function $f(x + \xi)$ equal to zero outside of this interval, and momentarily denotes the resulting function by $f(x + \xi)$, then $g_n(x)$ agrees with the corresponding expression (3). Now using the proof of (a), we obtain as $n \longrightarrow \infty$

$$g_n(x) \longrightarrow f(x + 0) \int_0 K(\xi)\, d\xi \quad .$$

We must still show that $h_n(x)$ exists for $n > n_0$, and that $h_n(x) \longrightarrow 0$ as $n \longrightarrow \infty$. For $c \leq \xi < \infty$, we have $n\xi \geq nc$. Since $nc \longrightarrow \infty$, and

$$K(\xi) = o(|\xi|^{-1}) \quad ,$$

there exists, for $n > n_0$, an ϵ_n with $\epsilon_n \longrightarrow 0$, such that

$$|K(n\xi)| < \epsilon_n (n\xi)^{-1}, \qquad c \leq \xi < \infty \quad .$$

Therefore $|n\int_c| \leq \epsilon \int_c \xi^{-1}|f(x + \xi)|d\xi$, and hence our assertion in regard to $h_n(x)$ follows.

2. The Fêjer kernel [18]

$$K(\xi) = \frac{1}{\pi}\left(\frac{\sin \xi}{\xi}\right)^2 \tag{10}$$

has special significance. The corresponding representation becomes, provided $f(x)$ is bounded or absolutely integrable,

$$(11) \qquad \frac{f(x+0) + f(x-0)}{2} = \lim_{n \to \infty} \frac{1}{\pi n} \int f(\xi) \left[\frac{\sin n(\xi - x)}{\xi - x} \right]^2 d\xi \quad .$$

More generally, one has for $p > 1$

$$(12) \qquad \frac{f(x+0) + f(x-0)}{2} = \lim_{n \to \infty} \frac{n^{1-p}}{c_p} \int f(\xi) \left[\frac{\sin n(\xi - x)}{\xi - x} \right]^p d\xi \quad ,$$

where

$$c_p = \int \left(\frac{\sin \xi}{\xi} \right)^p d\xi \quad .$$

If $p = 1$, i.e., if

$$(13) \qquad K(\xi) = \frac{1}{\pi} \frac{\sin \xi}{\xi} \quad ,$$

then (12) is the so-called Fourier integral formula, which must be treated separately since it does not come under Theorem 3, cf. §8.

3. We shall now prove an important criterion for the case of the Féjer kernel. Let $f(\xi)$ be absolutely integrable, and the value of $f(x)$ finite. If we set

$$\int_0 [f(x + \xi) - f(x)] \frac{(\sin n\xi)^2}{\pi n\xi^2} d\xi = \int_0^\delta + \int_\delta = g_n(x) + h_n(x) \quad ,$$

then

$$\pi |h_n(x)| \leq \int_\delta \frac{|f(x+\xi)|}{n\xi^2} d\xi + \int_\delta \frac{|f(x)|}{n\xi^2} d\xi \leq \frac{1}{n\delta^2} \int_\delta |f(x + \xi)| \, d\xi + \frac{|f(x)|}{n\delta} \quad .$$

Hence

$$(14) \qquad \lim_{n \to \infty} h_n(x) = 0 \qquad \text{for fixed } \delta .$$

To evaluate $g_n(x)$, we introduce the function

$$F(\xi) = \int_0^\xi |f(x + \xi) - f(x)| \, d\xi \quad ,$$

and assume for the point x under consideration, that

(15) $$\frac{1}{\xi} F(\xi) \longrightarrow 0 \quad \text{as} \quad \xi \longrightarrow 0 .$$

Making use of the inequality

$$\left| \frac{\sin x}{x} \right| < \frac{2}{1+x} \qquad x > 0 ,$$

(which can be easily proved by considering the two cases $0 < x \leq 1$ and $1 < x$), one obtains

(16)
$$\frac{\pi}{4} |g_n(x)| \leq \int_0^\delta \frac{nF'(\xi)d\xi}{(1+n\xi)^2} = \frac{n\delta}{(1+n\delta)^2} \frac{1}{\delta} F(\delta) + 2 \int_0^\delta \frac{1}{\xi} F(\xi) \frac{n^2\xi}{(1+n\xi)^3} d\xi .$$

If, in addition, we take (15) into consideration and make use of the relation

$$\int_0^\delta \frac{n^2 \xi d\xi}{(1+n\xi)^3} \leq \int_0 \frac{xdx}{(1+x)^3} = \frac{1}{2} ,$$

we find from (16) that for each $\epsilon > 0$, there is a δ, such that

(17) $$\overline{\lim_{n \to \infty}} |g_n(x)| \leq \epsilon .$$

Since ϵ can be made arbitrarily small, it follows from (14) and (17) that

$$g_n(x) + h_n(x) \longrightarrow 0 .$$

Hence at each point x for which (15) is valid, we have

(18) $$f(x) = \lim_{n \to \infty} \frac{1}{\pi n} \int f(\xi) \left[\frac{\sin n(\xi - x)}{\xi - x} \right]^2 d\xi .$$

But Lebesgue has proved that for any arbitrary (integrable) function $f(x)$, (15) is valid for almost all points x [19]. We have, therefore,

> THEOREM 4. For an absolutely integrable function $f(x)$, the representation (18) exists for almost all points x of the entire interval [19].

§7. The Dirichlet Integral and Related Integrals

1. Let $f(x)$ be a given function in $(0, \infty]$, which is positive and monotonically decreasing. We denote its limit at the end point $x = 0$, as is usual, by $f(+0)$. For any p, $0 < p \leq 1$, the kernel

(1) $$K(x) = \frac{\sin x}{x^p}$$

does not fall under Theorem 3. However, by Theorem 1, the integrals

(2) $$F(n) = \int_0 f(\tfrac{x}{n}) \frac{\sin x}{x^p}\, dx ,$$

exist for all $n > 0$. We shall now show that the relation

(3) $$\lim F(n) = f(+0) \int_0 \frac{\sin x}{x^p}\, dx , \qquad n \longrightarrow \infty ,$$

is also true. Let

$$\psi(n) = F(n) - f(+0) \int_0 \frac{\sin x}{x^p}\, dx .$$

Then

$$\psi(n) = \int_0^A \left[f(\tfrac{x}{n}) - f(+0)\right] \frac{\sin x}{x^p}\, dx + \left[\int_A \frac{1}{x^p} f(\tfrac{x}{n}) \sin x\, dx - \int_A \frac{1}{x^p} f(+0) \sin x\, dx\right]$$

$$= \psi_1(n) \qquad + \psi_2(n) .$$

By §2, (5)

$$|\psi_2(n)| \leqq \frac{2}{A^p}\left[f(\tfrac{A}{n}) + f(+0)\right] \leqq \frac{4f(+0)}{A^p} .$$

Hence for suitable A, $|\psi_2(n)| \leqq \epsilon$. But with A fixed, $|\psi_1(n)| \leqq \epsilon$ for $n \geqq n_0$, a result which can be arrived at in a manner similar to that used in evaluating the first term in the proof of Theorem 3, assumption (a). From this (3) follows.

2. In particular, we obtain for $p = 1$

(4) $$f(+0) = \lim_{n \to \infty} \frac{2}{\pi} \int_0 f(x) \frac{\sin nx}{x}\, dx .$$

If the function $f(x)$ is positive and monotonically decreasing in the finite interval $(0, a)$, then it can be extended by zero values in $[a, \infty]$. Hence

(5) $$f(+0) = \lim_{n \to \infty} \frac{2}{\pi} \int_0^a f(x) \frac{\sin nx}{n}\, dx, \qquad a > 0 .$$

This formula is valid in particular for $f(x) = 1$, and therefore also for $f(x) = c$, where c is any arbitrary constant. Further, if it holds for $f_1(x)$ and $f_2(x)$, then it also holds for $f_1(x) + f_2(x)$. The following theorem now follows easily.

THEOREM 5. The relation (5) is valid for each function of bounded **variation**.

COROLLARY. For $0 < a < 2\pi$

$$(6) \qquad 0 = \lim_{n \to \infty} \frac{1}{\pi} \int_0^a f(x) \left(\frac{1}{\sin \frac{x}{2}} - \frac{2}{x} \right) \sin nx \, dx \quad .$$

This result follows from Theorem 1, since the factor of $\sin nx$ is integrable. With the addition of (5) and (6), we have [20]

$$(7) \qquad f(+0) = \lim_{n \to \infty} \frac{1}{\pi} \int_0^a f(x) \frac{\sin nx}{\sin \frac{x}{2}} \, dx, \qquad 0 < a < 2\pi \quad .$$

3. We return to the function $f(x)$ specified in (1). For the kernel

$$(8) \qquad K(x) = \frac{\cos x}{x^p} \quad , \qquad 0 < p < 1,$$

one obtains

$$(9) \qquad \lim_{n \to \infty} \int_0 f(\tfrac{x}{n}) K(x) \, dx = f(+0) \int_0 K(x) \, dx \quad .$$

We have, for the present, removed the value $p = 1$, because for this value, $K(x)$ is not integrable in the neighborhood of the point $x = 0$. But it is indeed possible to remove this point by introducing, for **fixed** $a > 0$, the kernel: $K(x) = x^{-1} \cos x$ for $x \geq a$ and $= 0$ for $x < a$. Then (9) again holds. To prove this, we proceed just as we did in 1. However for the component $\psi_1(n)$, a slight change is necessary. It will now read

$$\psi_1(n) = \int_a^A \left[f(\tfrac{x}{n}) - f(+0) \right] K(x) \, dx \quad ,$$

and here again, one finds that it becomes arbitrarily small with n^{-1}. If (9) is valid for the kernels $K_1(x)$ and $K_2(x)$, then it will also be valid for $K = K_1 + K_2$. Bringing in Theorem 3, assumption (a), one finds that (9) is valid for each kernel, which, as $x \longrightarrow \infty$, can be written as

(10) $$K(x) = a\frac{\cos x}{x^p} + b\frac{\sin x}{x^q} + H(x) \quad ,$$

where $p > 0$, $q > 0$ and $H(x)$ is absolutely integrable as $x \longrightarrow \infty$, and instead of requiring that $f(x)$ be positive and monotonically decreasing, it is possible to allow $f(x)$ to be a function of bounded variation in $(0, \infty]$. For example, one may allow $f(x)$ to have an absolutely integrable derivative in $[0, \infty]$.

In particular, each function $K_0(x)$ has the property (10), which, as $x \longrightarrow \infty$, admits an asymptotic expansion

(11) $$\frac{\cos x}{x^p}\left(a_0 + \frac{a_1}{x} + \frac{a_2}{x^2} + \cdots\right) + \frac{\sin x}{x^q}\left(b_0 + \frac{b_1}{x} + \frac{b_2}{x^2} + \cdots\right)$$

with $p > 0$ and $q > 0$ (as for example the Besel function $J_\nu(x)$ for $\Re(\nu) > -1$). By this is meant that for each $m > 0$, the difference

$$K_0(x) - \frac{\cos x}{x^p}\sum_{\mu=0}^{m}\frac{a_\mu}{x^\mu} - \frac{\sin x}{x^q}\sum_{\mu=0}^{m}\frac{b_\mu}{x^\mu}$$

has the order of magnitude $O(x^{-m-1})$ as $x \longrightarrow \infty$.

Hence $K_0(x) \longrightarrow 0$ as $x \longrightarrow \infty$, and the integral

(12) $$K_1(x) = \int_x K_0(\xi)\, d\xi$$

exists, and is a function of the same character as $K_0(x)$. This fact is recognized without difficulty by the formula $(r > 0, \alpha \neq 0)$

$$\int_x \frac{e(\alpha\xi)}{\xi^r}\, d\xi = -\frac{1}{i\alpha}\frac{e(\alpha x)}{x^r} + \frac{r}{i\alpha}\int_x \frac{e(\alpha\xi)}{\xi^{r+1}}\, d\xi \quad .$$

One can thus repeat the process (12), resulting in functions

(13) $$K_{\mu+1}(x) = \int_x K_\mu(\xi)\, d\xi, \qquad \mu = 0, 1, 2, \ldots \quad ,$$

all of which are similar in character to $K_0(x)$.

THEOREM 6. For $\lambda = 1, 2, 3, \ldots$, we consider the integral

$$F(n) = \int_0 f(\tfrac{x}{n})x^\lambda K_0(x)\, dx \quad .$$

In order that $F(n)$ exist for $n > 0$ and approach a finite limit as $n \longrightarrow \infty$, it is sufficient that

1. $f(x)$ be λ-times differentiable in $(0, \infty]$, and $f^{(\lambda)}(x)$ be continuous in $(0, \infty]$, and that

2. each of the functions $g(x) = x^\lambda f(x)$, $g'(x), g''(x), \ldots, g^{(\lambda)}(x)$, have an absolutely integrable derivative in $[0, \infty]$. This limit has the value

$$\lambda! K_{\lambda+1}(0) f(+0) \quad .$$

PROOF. By assumption 1, it easily follows that

$$g(0) = g'(0) = \ldots = g^{(\lambda-1)}(0) = 0, \qquad g^{(\lambda)}(0) = \lambda! f(+0) \quad .$$

By assumption 2, it follows that the functions $g(x), g'(x), \ldots, g^{(\lambda)}(x)$ are bounded as $x \longrightarrow \infty$. Writing

$$F(n) = n^\lambda \int_0 g(\tfrac{x}{n}) K_0(x)\, dx \ ,$$

it follows that the integral $F(n)$ exists. Hence one can integrate it partially λ-times one after another, and obtain

$$F(n) = \int_0 g^{(\lambda)}(\tfrac{x}{n}) K_\lambda(x)\, dx \quad .$$

By (9), we have

$$\lim_{n \to \infty} F_n = g^{(\lambda)}(0) \int_0 K_\lambda(x)\, dx \ , \qquad \text{Q.E.D.}$$

§8. The Fourier Integral Formula

1. This formula reads

$$(1) \qquad \frac{1}{2}\,[f(x+0) + f(x-0)] = \lim_{n \to \infty} \frac{1}{\pi} \int f(x+\xi) \frac{\sin n\xi}{\xi}\, d\xi \quad .$$

It will suffice to discuss the part formula

$$f(x+0) = \lim_{n \to \infty} \frac{2}{\pi} \int_0 f(x+\xi) \frac{\sin n\xi}{\xi}\, d\xi \quad .$$

Let us consider, for fixed x and variable n, the integral

$$\Phi(n) = \frac{2}{\pi}\int_0^{\infty} f(x + \xi)\,\frac{\sin n\xi}{\xi}\,d\xi \quad .$$

For arbitrary $a > 0$, decompose it into the two sums

$$\Phi_1(n) = \frac{2}{\pi}\int_0^a f(x + \xi)\,\frac{\sin n\xi}{\xi}\,d\xi, \qquad \Phi_2(n) = \frac{2}{\pi}\int_a^{\infty} \frac{f(x+\xi)}{\xi}\,\sin n\xi\,d\xi \quad .$$

We can make use of §2 to evaluate $\Phi_2(n)$. By Theorem 1, $\Phi_2(n)$ exists for $n > 0$ and converges to zero as $n \longrightarrow \infty$, provided the function

$$\frac{f(x+\xi)}{\xi} \tag{3}$$

is either absolutely integrable or converges monotonically to zero as $\xi \longrightarrow \infty$. By the valuation $(A > |x| + 1)$

$$\int_A^{\infty} \left|\frac{f(x+\xi)}{\xi}\right| d\xi = \int_{A+x}^{\infty} \left|\frac{f(\xi)}{\xi}\right| \left|\frac{\xi}{\xi - x}\right| d\xi \leqq \frac{A}{A - |x|}\int_{+1}^{\infty} \left|\frac{f(\xi)}{\xi}\right| d\xi \quad ,$$

one observes that the first condition mentioned above can be modified to one requiring that the function

$$\frac{f(\xi)}{\xi} \tag{4}$$

be absolutely integrable as $\xi \longrightarrow \infty$. In this form, the condition is independent of the considered point x, and refers only to the infinite behavior of (4). It is also possible to modify the above second condition by requiring the function (4) to approach zero monotonically as $\xi \longrightarrow \infty$. In this case, the function (3) itself need not be monotonic, but one recognizes by the decomposition

$$\frac{f(x+\xi)}{\xi} = \frac{f(x+\xi)}{x + \xi} + x\,\frac{f(x+\xi)}{x + \xi}\cdot\frac{1}{\xi}$$

that it is the sum of two functions, each of which approaches zero monotonically as $\xi \longrightarrow \infty$. By Theorem 1 (a), it is even possible, as is easily verified, to generalize the second condition by requiring that the function (4) should be representable in the form

$$g(\xi)\,\sin\,(p\xi + q), \qquad\qquad p > 0 \;,$$

where $g(\xi)$ approaches zero monotonically as $\xi \longrightarrow \infty$. But in this case the integral $\Phi_2(n)$ need not exist for $n = + p$.

We must still state conditions under which the relation

$$\lim_{n \to \infty} \Phi_1(n) = f(x + 0)$$

holds. In §7, we became acquainted with the most important of these conditions, namely that $f(x + \xi)$ be of bounded variation in the interval $0 < \xi < a$. We shall not, however, enter into a discussion of still other conditions which are deduced in the theory of Fourier series. Summarizing, we have the following.

THEOREM 7 [21]. If the function $f(\xi)$ is of bounded variation in the neighborhood of the point x, then the formula

$$\frac{1}{2} [f(x + 0) + f(x - 0)] = \lim_{n \to \infty} \frac{1}{\pi} \int f(\xi) \frac{\sin n(\xi - x)}{\xi - x} d\xi \tag{5}$$

is valid, provided the function $f(\xi)$ not only as $\xi \longrightarrow \infty$, but also as $\xi \longrightarrow -\infty$, fulfills one of the following conditions:

α) it is absolutely integrable

β) it is monotonically convergent to zero, or more generally, is representable in the form $g(\xi) \sin(p\xi + q)$, where $g(\xi)$ converges monotonically to zero.

REMARK. It is obviously possible to generalize condition β) by requiring that the function (4) or $g(\xi)$ be representable as a linear combination of functions converging monotonically to zero. This representation is possible, for example, if the function in question converges to zero as $\xi \longrightarrow \infty$, and has an absolutely integrable derivative. A kind of special case of condition β) is therefore the following:

γ) it converges to zero and has an absolutely integrable derivative, or the function $g(\xi)$ has these properties.

Hardy [22] has shown that the requirement that (4) have an absolutely integrable derivative as $\xi \longrightarrow \infty$ is equivalent to the requirement that $\xi^{-1} f'(\xi)$ be absolutely integrable as $\xi \longrightarrow \infty$.

EXAMPLES. 1. $f(x) = x^{-1} \sin x$. Then by §4, 4, for $n \geq 1$

$$\frac{1}{\pi}\int f(\xi)\,\frac{\sin n(\xi - x)}{\xi - x}\,d\xi = \frac{\sin x}{x} = f(x) .$$

2. $f(x) = \cos x$. Then by §4, 4, for $n > 1$

$$\frac{1}{\pi}\int f(\xi)\,\frac{\sin n\xi}{\xi}\,d\xi = 1 = f(0) .$$

3. $f(x) = e^{-kx}$ for $x > 0$ and $= 0$ for $x < 0$. Then by §4, 3

$$\frac{1}{\pi}\int f(\xi)\,\frac{\sin n\xi}{\xi}\,d\xi = \frac{1}{\pi}\arc\tan\frac{n}{k}, \qquad k > 0 .$$

As $n \longrightarrow \infty$, the right side actually has the limit $\frac{1}{2}[f(+0) + f(-0)]$.

4. By our theorem, for example, the

$$\lim_{n\to\infty}\frac{1}{\pi}\int_0 \frac{1}{\xi^p}\,\frac{\sin n(\xi - x)}{\xi - x}\,d\xi, \qquad 0 < p < 1 ,$$

exists for $x \neq 0$ and $= x^{-p}$ for $x > 0$, and $= 0$ for $x < 0$.

2. THEOREM 8. If $f(x)$ has the same infinite behavior as in Theorem 7, and if, for almost all x in the interval (a, b), the expression

$$\frac{1}{\pi}\int f(\xi)\,\frac{\sin n(\xi - x)}{\xi - x}\,d\xi \tag{6}$$

is convergent as $n \longrightarrow \infty$, then the limit function is, for almost all x in (a, b), identical with $f(x)$.

PROOF. We decompose (6) into the terms

$$\int_a^b + \int^a + \int_b = B_1 + B_2 + B_3 .$$

If in (6) $f(x)$ is replaced by zero in (a, b), B_2 and B_3 results. Hence by Theorem 7, $B_2 + B_3$ converges to zero for all x in $[a, b]$. Therefore, by the hypothesis, the expression

$$\varphi(n, x) = \frac{1}{\pi}\int_a^b f(\xi)\,\frac{\sin n(\xi - x)}{\xi - x}\,d\xi$$

is convergent for almost all x in (a, b) as $n \longrightarrow \infty$. It remains to show that it converges to $f(x)$. We form

$$\psi(n, x) = \frac{1}{2n}\int_0^{2n} \varphi(\nu, x)\, d\nu \quad .$$

Since we can interchange the order of integration with respect to ν and ξ (Appendix 7, 10), we obtain

$$\psi(n, x) = \frac{1}{\pi n}\int_a^b f(\xi)\left[\frac{\sin n(\xi - x)}{\xi - x}\right]^2 d\xi \quad .$$

If $\varphi(n)$ is bounded for $0 < n < \infty$, and attains a limit as $n \longrightarrow \infty$, then

$$\psi(n) = \frac{1}{n}\int_0^n \varphi(\nu)\, d\nu$$

approaches the same limit; as per a general theorem (Appendix 17). For this reason

$$\lim \varphi(n, x) = \lim \psi(n, x), \qquad n \longrightarrow \infty \quad ,$$

for almost all x in (a, b). But by Theorem 4, $\lim \psi(n, x) = f(x)$ for almost all x in (a, b); hence also $\lim \varphi(n, x) = f(x)$ for almost all x in (a, b).

§9. The Wiener Formula [23]

1. Recently, Norbert Wiener has compared the formula

$$\lim_{n \to \infty} \int_0 f(\tfrac{x}{n})K(x)\, dx = f(+0)\int_0 K(x)\, dx$$

with the case where $n \longrightarrow 0$. Assuming that the "mean value" of the function $f(x)$

$$\mathfrak{M}\{f\} = \lim_{x \to \infty} \frac{1}{x}\int_0^x f(\xi)\, d\xi \tag{1}$$

exists (and is finite), the Wiener formula then reads

$$\lim_{n \to 0} \int_0 f(\tfrac{x}{n})K(x)\, dx = \mathfrak{M}\{f\}\int_0 K(x)\, dx \quad . \tag{2}$$

THEOREM 9. For the validity of (2), the following assumptions are sufficient:

1) that $K(x)$ be differentiable in $(0, \infty]$, and that there be a constant H such that

$$|x^2K(x)| \leq H \quad \text{for} \quad 1 \leq x < \infty \quad . \tag{3}$$

2) that there be a constant G such that

$$\frac{1}{x}\int_0^x |f(\xi)|\,d\xi \leq G \quad \text{for} \quad 0 < x < \infty \quad . \tag{4}$$

PROOF. Subtracting from the given function $f(x)$ its mean value, there results a new function which again satisfies 2, and whose mean value vanishes, i.e.,

$$\lim_{x \to \infty} \frac{1}{x}\int_0^x f(\xi)\,d\xi = 0 \quad . \tag{5}$$

Since formula (2) holds for each constant function $f(x)$, and is "additive", it is sufficient to prove our theorem for functions satisfying (5). However, two preliminary remarks are required for the actual proof.

α) Introducing the function

$$\Phi(x) = \int_0^x |f(\xi)|\,d\xi \quad , \tag{6}$$

and taking (4) into consideration, we obtain for $0 < A < B$

$$\int_A^B \frac{|f(x)|}{x^2}\,dx = \int_A^B \frac{d\Phi}{x^2} = \frac{\Phi(B)}{B^2} - \frac{\Phi(A)}{A^2} + 2\int_A^B \frac{\Phi(x)}{x^3}\,dx \leq \frac{G}{B} + 2G\int_A \frac{dx}{x^2} \leq \frac{3G}{A} \quad .$$

Hence also

$$\int_A \frac{|f(x)|}{x^2}\,dx \leq \frac{3G}{A} \quad . \tag{7}$$

By (3), we find for $A \geq 1$, with $x = n\xi$

$$\left|\int_A f\left(\frac{x}{n}\right)K(x)\,dx\right| \leq \frac{3GH}{A} \tag{8}$$

β) Introducing the function

$$\Phi_n(t) = \int_0^t f(\tfrac{x}{n})\,dx = t\left\{\frac{n}{t}\int_0^{t/n} f(x)\,dx\right\} ,$$

and making use of (4) and (5), we obtain

$$(9) \qquad |\Phi_n(t)| \leqq Gt \quad \text{and} \quad \lim_{n \to 0} \frac{\Phi n(t)}{t} = 0 .$$

Also, (5) implies as follows. To each $a > 0$ and $\eta > 0$, it is possible to determine an $n_o > 0$ such that for $0 < n \leqq n_o$, and for $t > a$

$$|\Phi_n(t)| \leqq \eta t .$$

We are now ready for the proof itself. Because of (5) we need show that

$$(10) \qquad \int_0 f(\tfrac{x}{n})K(x)\,dx$$

converges to zero as $n \longrightarrow 0$. Let an $\epsilon > 0$ be given. Decomposing the integral (10) into the terms

$$\int_0^A \quad \text{and} \quad \int_A$$

we determine, by reason of (8), a fixed $A \geqq 1$ such that

$$\left|\int_A\right| \leqq \epsilon .$$

We now set

$$(11) \qquad \int_0^A = \int_0^A K(x)\,d\Phi_n(x) = \Phi_n(A)K(A) - \int_0^A \Phi_n(x)K'(x)\,dx .$$

By (9), there is an n_1 such that for $0 < n \leqq n_1$

$$|\Phi_n(A)K(A)| \leqq \epsilon . .$$

There still remains the right side integral in (11). We decompose it into

$$\int_0^a + \int_a^A .$$

On the one hand, by (9)

$$\left|\int_0^a\right| \leqq Ga \int_0^a |K'(x)|\, dx \quad ,$$

which can be made smaller than ϵ for a suitable fixed a. On the other hand, for this a and for

$$\eta = \epsilon\left[\int_a^A x|K'(x)|\, dx\right]^{-1} \quad ,$$

we determine an n_0 in accordance with observation β. Then

$$\left|\int_a^A\right| \leqq \eta \int_a^A x|K'(x)|\, dx \leqq \epsilon \quad .$$

Hence for $0 < n \leqq$ minimum (n_0, n_1)

$$\left|\int_0 f(\tfrac{x}{n})K(x)\, dx\right| \leqq \epsilon + \epsilon + \epsilon + \epsilon = 4\epsilon \quad ,$$

thus completing the proof.

For example, if $f(x)$ is a bounded function whose mean value exists, then

$$\lim_{n \to 0} \frac{2}{\pi} \int_0 f(x) \frac{\sin^2 nx}{nx^2}\, dx = \mathfrak{M}\{f\} \quad .$$

From this, one easily finds the following:

2. <u>If $f(x)$ is a given function in $[-\infty, \infty]$, then by its "mean value", we shall understand the limit</u>

$$\mathfrak{M}\{f\} = \lim_{X \to \infty} \frac{1}{2X} \int_{-X}^{X} f(\xi)\, d\xi \tag{12}$$

insofar as this limit exists. If $f(x)$ has such a mean value, and is bounded, say, then

$$\lim_{n \to 0} \frac{1}{\pi} \int f(x) \frac{\sin^2 nx}{nx^2}\, dx = \mathfrak{M}\{f\} \quad . \tag{13}$$

§10. The Poisson Summation Formula [24]

1. Let the function $f(x)$ be defined for all x. We form the transformation[1]

$$\varphi(\alpha) = \int f(x)e(2\pi\alpha x)\,dx \quad . \tag{1}$$

Then the Poisson formula reads (here α and k are integers)

$$\sum_{\alpha=-\infty}^{+\infty} \varphi(\alpha) = \sum_{k=-\infty}^{+\infty} f(k) \quad . \tag{2}$$

Before proving (2) under suitable assumptions for $f(x)$, we shall formally deduce from it certain other formulas. Until further notice, λ and μ will be any two positive numbers for which

$$\lambda\mu = 1.$$

Replacing $f(x)$ by $f(\frac{x}{\lambda})$, there results from (2)

$$\sqrt{\lambda}\sum_{-\infty}^{+\infty} \varphi(\alpha\lambda) = \sqrt{\mu}\sum_{-\infty}^{+\infty} f(k\mu) \quad . \tag{3}$$

Let t be any number in the interval $0 \leq t < 1$. We consider the function $F(x) = f(t\mu + x)$, and denote its transform by $\Phi(\alpha)$. Hence

$$\Phi(\alpha\lambda) = \int f(t\mu + x)e(2\pi\alpha\lambda x)\,dx = e(-\,2\pi\alpha t)\varphi(\alpha\lambda) \;,$$

and (3) becomes therefore

$$\sqrt{\lambda}\sum_{-\infty}^{+\infty} e(-\,2\pi\alpha t)\varphi(\alpha\lambda) = \sqrt{\mu}\sum_{-\infty}^{+\infty} f(t\mu + k\mu) \quad . \tag{4}$$

Now let $f(x)$ be a function defined in the interval $(0, \infty]$. We introduce the integrals

$$\psi(\alpha) = 2\int_0 f(x)\cos(2\pi\alpha x)\,dx \quad , \tag{5}$$

$$\chi(\alpha) = 2\int_0 f(x)\sin(2\pi\alpha x)\,dx \quad . \tag{6}$$

---

[1] This transformation is normalized somewhat differently from the proper transformation $E(\alpha)$, cf. §2, 3.

If the function $f(x)$ is extended in the interval $[-\infty, 0]$ so as to become an even function, then $\varphi(\alpha) = \psi(\alpha)$ and one obtains

$$(7)\quad \begin{aligned} &\sqrt{\lambda}\left\{\psi(0) + 2\sum_{\alpha=1}^{\infty} \cos(2\pi\alpha t)\cdot\psi(\alpha\lambda)\right\} \\ &= \sqrt{\mu}\left\{f(t\mu) + \sum_{k=1}^{\infty}\left[f(k\mu + t\mu) + f(k\mu - t\mu)\right]\right\}. \end{aligned}$$

For example, this gives for $t = 0$ and $t = \frac{1}{2}$, the special cases

$$(8)\quad \sqrt{\lambda}\left\{\psi(0) + 2\sum_{\alpha=1}^{\infty}\psi(\alpha\lambda)\right\} = \sqrt{\mu}\left\{f(0) + 2\sum_{k=1}^{\infty} f(k\mu)\right\},$$

$$(9)\quad \sqrt{\lambda}\left\{\psi(0) + 2\sum_{\alpha=1}^{\infty}(-1)^{\alpha}\psi(\alpha\lambda)\right\} = 2\sqrt{\mu}\sum_{k=0}^{\infty} f\left((k + \tfrac{1}{2})\mu\right).$$

But if the function $f(x)$ is extended by an odd function, then $\varphi(\alpha) = i\chi(\alpha)$, and one obtains by (4)

$$(10)\quad 2\sqrt{\lambda}\sum_{\alpha=1}^{\infty}\sin 2\pi\alpha t\cdot\chi(\alpha\lambda) = \sqrt{\mu}\left\{f(t\mu) + \sum_{k=1}^{\infty}\left[f(k\mu + t\mu) - f(k\mu - t\mu)\right]\right\}.$$

For $t = 0$, there seems to be a contradiction because the left side has the value 0 and the right side the value $\sqrt{\mu}\, f(0\mu)$. This discrepancy, however, disappears when one realizes that after completion of $f(x)$, the entire function has two limits of opposite sign at the point $t = 0$. For this reason the value 0 is to be understood for $f(0)$. —Replacing λ and μ by $\lambda/2$ and 2μ, and modifying somewhat the running indices, one obtains for $t = 1/4$

$$(11)\quad \sqrt{\lambda}\sum_{\alpha=0}^{\infty}(-1)^{\alpha}\chi\left[\left(\alpha + \tfrac{1}{2}\right)\lambda\right] = \sqrt{\mu}\sum_{k=0}^{\infty}(-1)^{k} f\left((k + \tfrac{1}{2})\mu\right).$$

2. Now for the proof itself. We shall assume in this paragraph that each function $f(x)$ has the limits $f(x + 0)$ and $f(x - 0)$ at each point x, and that at each such point, the value of the function is $\frac{1}{2}[f(x + 0) + f(x - 0)]$. If, therefore, for example, a function $g(x)$ is given in an interval $(0, p)$ where p is an integer, and if for the

purpose of applying (2), it is extended by zero values to a function $f(x)$ defined everywhere, then the expression

$$\frac{1}{2}[g(0) + g(p)] + \sum_{1}^{p-1} g(k)$$

is to replace the right sum in (2).

Further, in this paragraph we call a series

$$\sum_{-\infty}^{+\infty} A_\nu \tag{12}$$

convergent, if the partial sums

$$s_n = \sum_{-n}^{n} A_\nu \tag{13}$$

approaches a limit as $n \longrightarrow \infty$. Analogously, we call an integral extending over the whole interval $[-\infty, \infty]$, convergent if its Cauchy principal value exists, cf. §5, 1.

Hence for integers α and integers $p > 0$,

$$\int_{-p-1/2}^{p+1/2} f(x)e(2\pi\alpha x)\,dx = \int_{-1/2}^{1/2} \left(\sum_{-p}^{p} f(x+k) \right) e(2\pi\alpha x)\,dx \quad . \tag{14}$$

If the series

$$\sum_{-\infty}^{+\infty} f(x+k) \tag{15}$$

converges uniformly in the interval $-\frac{1}{2} \leq x < \frac{1}{2}$, call $g(x)$ the limit function, then the expression on the right of (14) is convergent as $p \longrightarrow \infty$. Hence the expression on the left is also, i.e., the integral $\varphi(\alpha)$ exists and

$$\varphi(\alpha) = \int_{-1/2}^{1/2} g(x)e(2\pi\alpha x)\,dx \quad .$$

From this it follows that

$$\sum_{-n}^{n} \varphi(\alpha) = \int_{-1/2}^{1/2} g(x)\,\frac{\sin(2n+1)\pi x}{\sin \pi x}\,dx \quad .$$

If furthermore $g(x)$ is of bounded variation, then the right side converges to $g(0)$ as $n \longrightarrow \infty$, cf. §7, 2. Hence the left side also converges, and if one now inserts the value of $g(0)$, (2) results. Applying a formal transformation stated in 1, we obtain

THEOREM 10. For the validity of formula (4), it is sufficient that the series

$$\sum_{k=-\infty}^{+\infty} f(x + t\mu + k\mu) \tag{16}$$

converge uniformly in

$$-\frac{1}{2} \leqq x < \frac{1}{2} \quad , \tag{17}$$

and that its sum be of bounded variation in this interval. Both series in (4) are, under this hypothesis, themselves convergent.

3. We shall now have two special cases reducible to this criterion.

a) Let the function $f(x)$ be positive, monotonically decreasing to zero, and integrable in the interval $\ell - 2 \leqq x < \infty$. For $k \geqq \ell$ and $x \geqq -1$, we then have

$$f(x + k) \leqq \int_{k-1}^{k} f(x + \xi)\, d\xi \leqq \int_{k-2}^{k-1} f(\xi)\, d\xi$$

and hence for $p \geqq \ell$ and arbitrary $r > 0$

$$\sum_{p}^{p+r} f(x + k) \leqq \int_{k-2} f(\xi)\, d\xi \quad .$$

Therefore the series

$$\sum_{k=\ell}^{\infty} f(x + k) \tag{18}$$

converges uniformly in $-1 \leqq x < \infty$. Moreover, since all terms of the sum are monotonically decreasing, the function representing the sum is also monotonically decreasing.

b) Let the function $f(x)$ be differentiable in the interval

$\ell - 1 \leqq x < \infty$, and let the integrals

$$\int_{\ell-1} f(\xi)\,d\xi \quad \text{and} \quad \int_{\ell-1} |f'(\xi)|\,d\xi \tag{19}$$

be finite. Hence

$$\left|\sum_{p}^{m} f(x+k) - \int_{p}^{m+1} f(x+\xi)\,d\xi\right| \leqq \sum_{p}^{m}\left|\int_{0}^{1} [f(x+k) - f(x+k+\xi)]\,d\xi\right|$$

If we now set, for $0 \leqq \xi < 1$,

$$|f(x+k) - f(x+k+\xi)| = \left|\int_{0}^{\xi} f'(x+k+\eta)\,d\eta\right| \leqq \int_{k}^{k+1} |f'(x+\eta)|\,d\eta \quad ,$$

we obtain

$$\left|\sum_{p}^{m} f(x+k) - \int_{p}^{m+1} f(x+\xi)\,d\xi\right| \leqq \sum_{p}^{m}\int_{k}^{k+1} |f'(x+\eta)|\,d\eta = \int_{p}^{m+1} |f'(x+\eta)|\,d\eta \quad ,$$

and from this there finally results

$$\left|\sum_{p}^{m} f(x+k)\right| \leqq \left|\int_{p}^{m+1} f(x+\xi)\,d\xi\right| + \int_{p}^{m+1} |f'(x+\eta)|\,d\eta \quad .$$

Because of the convergence of the integrals (19), this implies that the series (18) converges uniformly in the interval (17). We denote its sum by $f_1(x)$. If $f_1(x)$ is real, then $f_1(x) - f_1(0)$ can be represented in (17) as the difference of the monotonic functions

$$\sum_{1}^{\infty}\int_{0}^{x} \frac{|f'(\xi+k)| \pm f'(\xi+k)}{2}\,d\xi \quad .$$

If $f_1(x)$ is not real, then its real and imaginary parts separately will have such a representation each.

From this observation, we reach the following conclusion.

THEOREM 10a. For the validity of formula (4), it is sufficient that $f(x)$ be of bounded variation in a finite interval and satisfy one of the following conditions both as $x \longrightarrow \infty$ and as $x \longrightarrow -\infty$:

1) it is monotonic and absolutely integrable

2) it is integrable and has an absolutely integrable derivative.

Both series in (4) are automatically convergent [25].

REMARK IN REGARD TO 2. The function $f(x)$ itself need not be absolutely integrable as $x \longrightarrow \infty$. Thus, for example, the function $x^{-1} \sin \sqrt{x}$ comes under 2.

EXAMPLES. 1. The best known application of (3) is the formula [26]

$$\frac{1}{\sqrt{x}} \sum_{-\infty}^{+\infty} e^{-k^2 \pi / x} = \sum_{-\infty}^{+\infty} e^{-k^2 x \pi} \quad .$$

It results from the relation

$$\int e^{-\xi^2} e(\alpha\xi)\, d\xi = \sqrt{\pi}\, e^{-\alpha^2/4}$$

which we shall prove later, §15, (9).

2. If $f(x) = e^{-px}$, $p > 0$, then by §2, (4)

$$\psi(\alpha) = \frac{2p}{p^2 + (2\pi\alpha)^2}, \qquad \chi(\alpha) = \frac{2(2\pi\alpha)}{p^2 + (2\pi\alpha)^2}$$

By (7) and (10), we obtain the formulas [27]

$$(20) \quad \frac{1}{2\lambda} \cosh p\mu \left(\tfrac{1}{2} + t\right) \cdot \sinh^{-1} \frac{p\mu}{2} = \frac{1}{p} + 2 \sum_{\alpha=1}^{\infty} \cos 2\pi\alpha t \frac{p}{p^2 + (2\pi\alpha\lambda)^2} \quad ,$$

$$(21) \quad \frac{1}{2\lambda} \sinh p\mu \left(\tfrac{1}{2} - t\right) \cdot \sinh^{-1} \frac{p\mu}{2} = 2 \sum_{\alpha=1}^{\infty} \sin 2\pi\alpha t \frac{2\pi\alpha\lambda}{p^2 + (2\pi\alpha\lambda)^2}$$

One notes that the second can be obtained from the first by differentiation with respect to t. They are valid for $p > 0$, $\lambda > 0$, $\mu > 0$, $\lambda\mu = 1$, $0 < t < 1$. For special values of the parameter, numerous well known formulas can be obtained. For example, if $t = 1/4$, we obtain from (21)

$$\frac{1}{2\lambda} \left(e^{\frac{p\mu}{4}} + e^{-\frac{p\mu}{4}} \right)^{-1} = 2 \sum_{\beta=0}^{\infty} (-1)^{\beta} \frac{2\pi(2\beta+1)\lambda}{p^2 + [2\pi(2\beta+1)\lambda]^2} \quad .$$

From this one can derive the formula [28]

$$(22)\qquad \sum_{k=-\infty}^{+\infty}\left(e^{2k\pi q}+e^{-2k\pi q}\right)^{-1}=\frac{1}{2q}\sum_{k=-\infty}^{+\infty}\left(e^{\frac{k\pi}{2q}}+e^{-\frac{k\pi}{2q}}\right)^{-1}.$$

3. With the aid of (20) and (21), the following trigonometric integrals can be evaluated [29].

$$\int_0 \frac{e^{\omega x}-e^{-\omega x}}{e^{\pi x}-e^{-\pi x}}\cos\alpha x\,dx=\frac{\sin\omega}{e^{\alpha}+2\cos\omega+e^{-\alpha}},\qquad -\pi<\omega<\pi,$$

$$\int_0 \frac{e^{\omega x}+e^{-\omega x}}{e^{\pi x}-e^{-\pi x}}\sin\alpha x\,dx=\frac{1}{2}\,\frac{e^{\alpha}+e^{-\alpha}}{e^{\alpha}+2\cos\omega+e^{-\alpha}},\qquad -\pi<\omega<\pi,$$

$$\int_0 \frac{\cos\omega x}{e^{\pi x}-e^{-\pi x}}\sin\alpha x\,dx=\frac{1}{4}\,\frac{e^{\alpha}-e^{-\alpha}}{e^{\omega}+e^{-\omega}+e^{\alpha}+e^{-\alpha}},$$

$$\int_0 \frac{\sin\alpha x}{e^{2\pi x}-1}\,dx=\frac{1}{4}\coth\frac{\alpha}{2}-\frac{1}{2\alpha}.$$

4. As a last application of the Poisson formula, we consider the function $f(x) = x^{-s}$ for $0 < x < \infty$, $0 < s < +1$. Hence, cf. §12, 1

$$\chi(\alpha)=2\int_0 \frac{\sin 2\pi\alpha x}{x^{s}}\,dx=\frac{2\Gamma(1-s)\cos 1/2\, s\pi}{(2\pi\alpha)^{1-s}}$$

We have therefore by (11)

$$(23)\qquad \sum_0^{\infty}\frac{(-1)^n}{(2n+1)^s}=\left(\frac{2}{\pi}\right)^{1-s}\cos\frac{1}{2}s\pi\Gamma(1-s)\sum_0^{\infty}\frac{(-1)^n}{(2n+1)^{1-s}},$$

which is a special case of the functional equation for the Dirichlet L-function [30]. Although the function just considered does not come under Theorem 10a, nevertheless the application of the summation formula is justifiable. We remark without proof the following. As far as the behavior of $f(x)$ outside a finite interval is concerned, it is sufficient for the validity of (10), that $f(x)$ approach zero monotonically as $x \longrightarrow \infty$. As far as its behavior in a finite interval is concerned it is sufficient for the validity of (4) that $f(x)$ be of bounded variation in the neighborhood of the "lattice points" $t_\mu + k\mu$, $k = 0, \pm 1, \pm 2, \ldots$, and otherwise be integrable only.

CHAPTER III

THE FOURIER INTEGRAL THEOREM

§11. The Fourier Integral Theorem and the Inversion Formulas

1. We denote the formula

$$\frac{1}{2}\,[f(x+0)+f(x-0)] = \frac{1}{\pi}\int_0 d\alpha \int f(\xi)\cos\alpha(\xi - x)\,d\xi \tag{1}$$

as a Fourier integral theorem.

THEOREM 11. It is sufficient for the validity of (1), that $f(\xi)$ be of bounded variation in the neighborhood of x, and that one of the following conditions be fulfilled not only as $\xi \longrightarrow +\infty$, but also as $\xi \longrightarrow -\infty$.[1]

1) $f(\xi)$ be absolutely integrable,

2) $\frac{f(\xi)}{\xi}$ be absolutely integrable and $f(\xi)$ be convergent monotonically to zero, or more generally be representable in the form $g(\xi)\sin(p\xi + q)$, where $g(\xi)$ approaches zero monotonically [31].

In case 2), the integral with respect to α, is a Cauchy principal-value with at most two singular points.

PROOF. Writing the integral on the right of (1) in the form

$$\lim_{n\to\infty}\int_0^n d\alpha \int f(\xi)\cos\alpha(\xi - x)\,d\xi \tag{2}$$

[1] It is, therefore, sufficient for example, that $f(x)$ satisfy condition 1. as $\xi \longrightarrow \infty$ and satisfy condition 2. as $\xi \longrightarrow -\infty$.

and interchanging the order of integration with respect to α and ξ, we obtain

$$(3) \qquad \lim_{n \to \infty} \int f(\xi) \frac{\sin n(\xi - x)}{\xi - x} d\xi \quad .$$

Because of Theorem 7, it is necessary only to show that this interchange of the order of integration is admissible. Formula (1) is additive; i.e., if it holds for $f_1(\xi)$ and $f_2(\xi)$, then it also holds for $f_1(\xi) + f_2(\xi)$. For its proof, we can therefore assume that $f(\xi)$ vanishes outside of a half line $[a, \infty]$.

PROOF of 1). By hypothesis, $f(\xi)$ is absolutely integrable. Hence because

$$|f(\xi) \cos \alpha(\xi - x)| \leq f(\xi)$$

the equality

$$\int_\sigma^n d\alpha \int_a f(\xi) \cos \alpha(\xi - x)\, d\xi = \int_a d\xi \int_\sigma^n f(\xi) \cos \alpha(\xi - x)\, d\alpha$$

is valid for each finite interval (σ, n) by a basic general theorem (Appendix 7, 10), Q.E.D.

PROOF of 2). For finite numbers σ, n, a, b,

$$\int_\sigma^n d\alpha \int_a^b f(\xi) \cos \alpha(\xi - x)\, d\xi = \int_a^b d\xi \cdot \int_\sigma^n f(\xi) \cos \alpha(\xi - x)\, d\alpha$$

follows from 1). For fixed x, it is here admissible to allow $b \longrightarrow \infty$ provided the integral

$$\varphi(\alpha) = \int_a f(\xi) \cos \alpha(\xi - x)\, d\xi$$

is uniformly convergent in $\sigma \leq \alpha \leq n$. In fact, if we set

$$\varphi_b(\alpha) = \int_a^b f(\xi) \cos \alpha(\xi - x)\, d\xi \quad ,$$

then

$$\int_\sigma^n \varphi_b(\alpha)\, d\alpha \longrightarrow \int_\sigma^n \varphi(\alpha)\, d\alpha \quad .$$

We now apply Theorem 2. If $f(\xi)$ approaches zero monotonically, then each interval (ϵ, n) for which $0 < \epsilon < n$, is suitable as an α-interval. We, therefore, then have

$$\int_\epsilon^n d\alpha \int_a f(\xi)\cos \alpha(\xi - x)\, d\xi = \int_a f(\xi) \frac{\sin n(\xi - x)}{\xi - x}\, d\xi - \int_a f(\xi) \frac{\sin \epsilon(\xi - x)}{\xi - x}\, d\xi$$

Since $(\xi - x)^{-1} f(\xi)$ is, together with $\xi^{-1} f(\xi)$, also absolutely integrable as $\xi \longrightarrow \infty$, it follows by Theorem 2 again, that the second integral on the right is continuous in ϵ, and hence converges to zero as $\epsilon \longrightarrow 0$. Therefore the integral on the left has a limit as $\epsilon \longrightarrow 0$. This limit is

$$(4) \qquad \lim_{\epsilon \to 0} \int_\epsilon^n d\alpha \int_a f(\xi) \cos \alpha(\xi - x)\, d\xi = \int_a f(\xi) \frac{\sin n(\xi - x)}{\xi - x}\, d\xi$$

Q.E.D.

In the more general case where $f(\xi) = g(\xi) \sin (p\xi + q)$, $p > 0$, it is necessary to interchange the order of integration in the repeated integral

$$\left(\int_0^{p-\epsilon} + \int_{p+\epsilon}^n \right) d\alpha \int_a f(\xi)\cos \alpha(\xi - x)\, d\xi$$

It then appears that a limit exists as $\epsilon \longrightarrow 0$, and this limit equals the expression on the right of (4), Q.E.D.

2. For the sake of brevity, we shall write $f(x)$ in place of $\frac{1}{2}[f(x + 0) + f(x - 0)]$. If the question of convergence is disregarded, then (1) may be interpreted as follows. Each "random" function defined in $[-\infty, \infty]$ may be written in the form

$$(5b) \qquad f(x) = \int_0 [C(\alpha)\cos \alpha x + S(\alpha)\sin \alpha x]\, d\alpha$$

where

$$(5a) \qquad C(\alpha) = \frac{1}{\pi} \int f(\xi)\cos \alpha\xi d\xi, \quad S(\alpha) = \frac{1}{\pi} \int f(\xi)\sin \alpha\xi d\xi$$

Formula (5b) is the "harmonic analysis" of the function $f(x)$; — the representation of the function as "sums" of cosines and sines. If now

$f(x)$ is, in particular, even $[f(-x) = f(x)]$ or odd $[f(-x) = -f(x)]$, then $S(\alpha) = 0$ or $C(\alpha) = 0$. Thus one obtains the pair of formulas

(6a) $$C(\alpha) = \frac{2}{\pi}\int_0 f(\xi)\cos\alpha\xi d\xi$$

(6b) $$f(x) = \int_0 C(\alpha)\cos x\alpha d\alpha$$

(7a) $$S(\alpha) = \frac{2}{\pi}\int_0 f(\xi)\sin\alpha\xi d\xi$$

(7b) $$f(x) = \int_0 S(\alpha)\sin x\alpha d\alpha \quad .$$

By substitution, there results

(8) $$f(x) = \frac{2}{\pi}\int_0 d\alpha \cos x\alpha \int_0 f(\xi)\cos\alpha\xi d\xi$$

or

(9) $$f(x) = \frac{2}{\pi}\int_0 d\alpha \sin x\alpha \int_0 f(\xi)\sin\alpha\xi d\xi \quad .$$

Let λ and μ be any two numbers whose product is $2/\pi$. Multiplication of $C(\alpha)$ by $\lambda\pi/2$, transforms (6) into the symmetrical pair of formulas

(10) $$\psi(\alpha) = \lambda\int_0 f(\xi)\cos\alpha\xi d\xi, \qquad f(x) = \mu\int_0 \psi(\alpha)\cos x\alpha d\alpha \quad .$$

Similarly one obtains from (7)

(11) $$\chi(\alpha) = \lambda\int_0 f(\xi)\sin\alpha\xi d\xi, \qquad f(x) = \mu\int_0 \chi(\alpha)\sin x\alpha d\alpha \quad .$$

In what follows, we shall denote the passage from (6a) to (6b) as an inversion of (6a), and shall call the second the inverse (or also the converse [32]) of the first; similarly in the case of (7). By Theorem 11, there follows immediately

THEOREM 11a. The inversion of (6a) and (7a) is admissible provided the function $f(\xi)$, defined in $[0, \infty]$, satisfies 1) or 2) of Theorem 11 as $\xi \longrightarrow \infty$, and is of bounded variation in the neighborhood of x.

By inversion of §2, (4), we obtain for example

$$(12)\qquad \int_0 \frac{\cos x\alpha}{k^2+\alpha^2}\,d\alpha = \frac{\pi}{2k}e^{-kx}, \qquad \int_0 \frac{\alpha \sin x\alpha}{k^2+\alpha^2}\,d\alpha = \frac{\pi}{2}e^{-kx} .$$

If $f(x) = \frac{\sin x}{x}$, then by §4, (6), $C(\alpha) = 1$ for $0 < \alpha < 1$ and $= 0$ for $\alpha > 1$; hence actually

$$\int_0 C(\alpha)\cos \alpha x d\alpha = \int_0^1 \cos \alpha x d\alpha = f(x) .$$

3. If $f(x)$ is given in $[-\infty, \infty]$, the pair of relations (5) can also be interpreted as a pair of inversions, after introducing "complex" notation at any rate. Consider the transform of $f(x)$

$$(13)\qquad E(\alpha) = \frac{1}{2\pi}\int f(\xi)e(-\alpha\xi)\,d\xi ,$$

Then, cf. the proof of Theorem 11,

$$\int_\sigma^n E(\alpha)e(x\alpha)\,d\alpha + \int_{-n}^{-\sigma} E(\alpha)e(x\alpha)\,d\alpha = \int_\sigma^n [E(\alpha)e(x\alpha) + E(-\alpha)e(-x\alpha)]\,d\alpha$$

$$= \frac{1}{\pi}\int_\sigma^n d\alpha \int f(\xi)\cos \alpha(\xi - x)\,d\xi$$

and the following theorem ensues.

THEOREM 11b. If $f(\xi)$ satisfies 1) or 2) of Theorem 11 not only as $\xi \longrightarrow \infty$, but also as $\xi \longrightarrow -\infty$, then (13) may be converted to

$$(14)\qquad f(x) = \int E(\alpha)e(x\alpha)\,d\alpha .$$

This converse holds for points x in whose neighborhood $f(x)$ is of bounded variation provided one interprets the integral (14) in a suitable manner as a Cauchy principal-value.

4. Of interest is

THEOREM 12. If $f(\xi)$ behaves at infinity as specified (especially if $f(\xi)$ is absolutely integrable) and if the integral

$$\int E(\alpha)e(\alpha x)\, d\alpha$$

converges for almost all x in an interval (a, b) — perhaps as a Cauchy principal-value —, then the limit function is, for almost all x in (a, b) identical with $f(x)$.

PROOF. By the assumed property at infinity, we have

$$\int_{-n}^{n} E(\alpha)e(x\alpha)\, d\alpha = \frac{1}{\pi}\int f(\xi)\, \frac{\sin n(\xi - x)}{\xi - x}\, d\xi \quad .$$

Now apply Theorem 8.

§12. Trigonometric Integrals with e^{-x}.

The formulas $[\alpha > 0,\ 0 < \mu < 1]$

$$\int_0 x^{\mu-1} \begin{matrix}\cos\\ \sin\end{matrix}\ \alpha x\, dx = \frac{\Gamma(\mu)}{\alpha^{\mu}} \begin{matrix}\cos\\ \sin\end{matrix}\ \mu\frac{\pi}{2} \tag{1}$$

which can be consolidated in one as

$$\int_0 x^{\mu-1}e(-\alpha x)\, dx = \frac{\Gamma(\mu)}{\alpha^{\mu}}\, e(-\mu\frac{\pi}{2}) \quad , \tag{2}$$

originated with Euler [34]. Here $\Gamma(\mu)$ denotes the Euler gamma function

$$\Gamma(\mu) = \int_0 e^{-x}x^{\mu-1}\, dx \quad . \tag{3}$$

The formulas of Fresnel are important special cases of (1),

$$\int_0 \frac{\cos x}{\sqrt{x}}\, dx = \int_0 \frac{\sin x}{\sqrt{x}}\, dx = \frac{1}{\sqrt{2}}\Gamma(\tfrac{1}{2}) = \sqrt{\frac{\pi}{2}} \quad . \tag{4}$$

We shall discuss these separately later on. Formula (2) also results from $[k > 0,\ -\infty < \alpha < \infty,\ 0 < \mu < \infty]$

$$(5) \qquad \int_0 e^{-kx}x^{\mu-1}e(-\alpha x)\,dx = \frac{\Gamma(\mu)}{(k+i\alpha)^{\mu}}$$

by letting $k \longrightarrow 0$. Here

$$(k+i\alpha)^{\mu} = \sqrt{k^2+\alpha^2}^{\,\mu}\, e\left(\mu \arc\tan\frac{\alpha}{k}\right), \qquad -\frac{\pi}{2} < \arc\tan\frac{\alpha}{k} < \frac{\pi}{2} \;.$$

We shall, however, not prove (5) [35].

2. There results by inversion of (5) $[k > 0,\ \mu > 0]$

$$(6) \qquad \frac{\Gamma(\mu)}{2\pi}\int \frac{e(\alpha x)}{(k+i\alpha)^{\mu}}\,d\alpha = \begin{cases} e^{-kx}x^{\mu-1}, & x > 0 \\ 0\;, & x < 0\;. \end{cases}$$

Hence [36], by changing the signs of α and x,

$$(7) \qquad \frac{\Gamma(\mu)}{2\pi}\int \frac{e(\alpha x)}{(k-i\alpha)^{\mu}}\,d\alpha = \begin{cases} 0\;, & x > 0 \\ e^{kx}|x|^{\mu-1}, & x < 0\;. \end{cases}$$

Substituting $\alpha + q$ for α, one easily finds that both relations are valid also for such complex k for which $\Re(k) > 0$. Only then one must set

$$(8) \qquad (k+i\alpha)^{\mu} = |k+i\alpha|^{\mu} e\left(\mu \arc\tan\frac{\alpha + \Im(k)}{\Re(k)}\right).$$

This result can also be obtained by analytic continuation, and the formula is, in fact, valid for complex μ with $\Re(\mu) > 0$.

3. If in

$$(9) \qquad \int_0 e^{-kx}e(-\alpha x)\,dx = \frac{1}{k+i\alpha}\,, \qquad \Re(k) > 0$$

we replace α by $\alpha - \rho$ and $\alpha + \rho$ and then add, we obtain

$$2i\int_0 e^{-kx}e(-\alpha x)\cos\rho x\,dx = \frac{1}{\alpha-\rho-ik} + \frac{1}{\alpha+\rho-ik}\;.$$

Integration with respect to ρ between 0 and ρ gives, for $0 < \rho < \alpha$,

$$2i\int_0 e^{-kx}e(-\alpha x)\frac{\sin\rho x}{x}\,dx = \log\left(\frac{\alpha+\rho-ik}{\alpha-\rho-ik}\cdot\frac{\alpha-ik}{\alpha-ik}\right)$$

where the imaginary part of the logarithm is to be taken between $-\pi/2$ and $\pi/2$. Letting $k \longrightarrow 0$, and taking the real part we obtain for $0 < \alpha < \beta$ or $0 < \beta < \alpha$

$$\int_0 \frac{\sin \alpha x}{x} \sin \beta x \, dx = \frac{1}{2} \log \left| \frac{\alpha + \beta}{\alpha - \beta} \right| \tag{10}$$

Integrating (9) with respect to k gives $[a > 0, b > 0]$

$$\int_0 \frac{e^{-ax} - e^{-bx}}{x} e(-\alpha x) \, dx = \log \frac{b + i\alpha}{a + i\alpha} .$$

Hence follows [37]

$$\int_0 \frac{e^{-ax} - e^{-bx}}{x} \cos \alpha x \, dx = \log \sqrt{\frac{b^2 + \alpha^2}{a^2 + \alpha^2}} \tag{11}$$

$$\int_0 \frac{e^{-ax} - e^{-bx}}{x} \sin \alpha x \, dx = \operatorname{arc\,tan} \frac{b}{\alpha} - \operatorname{arc\,tan} \frac{a}{\alpha} . \tag{12}$$

For positive numbers a, x and μ, we have by (3)

$$\Gamma(\mu)\, (x + a)^{-\mu} = \int_0 e^{-(x+a)z} z^{\mu-1} \, dz . \tag{13}$$

Multiply by $e^{-kx} e(-\alpha x)$, $k > 0$, and integrate with respect to x between 0 and ∞. One can interchange the order of integration in the repeated right integral to obtain

$$\Gamma(\mu) \int_0 \frac{e^{-kx} e(-\alpha x)}{(x+a)^{\mu}} \, dx = \int_0 \frac{e^{-az} z^{\mu-1}}{z + k + i\alpha} \, dz . \tag{14}$$

If $\alpha \neq 0$, one can even consider the case where $k \longrightarrow 0$. Altogether, one obtains $[a > 0, \mu > 0, k \geq 0, \alpha \neq 0]$

$$\Gamma(\mu) \int_0 \frac{e^{-kx}}{(x+a)^{\mu}} \begin{Bmatrix} \cos \\ \sin \end{Bmatrix} \alpha x \, dx = \int_0 \frac{e^{-az}}{(z+k)^2 + \alpha^2} \begin{Bmatrix} z + k \\ \alpha \end{Bmatrix} dz . \tag{15}$$

4. The Fresnel integral

$$J = \sqrt{\frac{2}{\pi}} \int_0 \frac{\sin x}{\sqrt{x}} \, dx \tag{16}$$

can be evaluated "directly" by making use of §11, (11) with $\lambda = \mu$ [38]. Setting $f(x) = x^{-1/2}$, we obtain

$$(17) \qquad \chi(\alpha) = \sqrt{\frac{2}{\pi}} \int_0 \frac{\sin \alpha\xi}{\sqrt{\xi}} d\xi = \frac{1}{\sqrt{\alpha}} \chi(1) = f(\alpha)J .$$

Hence also $f(\alpha) = \chi(\alpha)J$, and therefore

$$(18) \qquad \chi(\alpha)\,(1 - J^2) = 0 .$$

By the decomposition

$$\int_0 \frac{\sin x}{\sqrt{x}} dx = \sum_{k=0}^{\infty} \int_{2k\pi}^{(2k+1)\pi} \sin x \left(\frac{1}{\sqrt{x}} - \frac{1}{\sqrt{x+\pi}}\right) dx ,$$

one recognizes that $J > 0$, and therefore because of (17), $\chi(\alpha) \neq 0$. Hence $(1 - J^2) = 0$ follows from (18), and because $J > 0$, it follows that $J = 1$. One can treat the cosine integral in a similar manner, only here the proof that $J > 0$ is somewhat more complicated. The proof rests on the fact that the function $x^{-1/2}$ is its one cosine — or sine-transform. Many other such "self reciprocal" functions exist [39].

§13. The Absolutely Integrable Functions. Their Faltung and Their Summation.

1. We shall now consider all those functions $f(x)$ which are defined and are absolutely integrable in $[-\infty, \infty]$. We denote the totality of such functions by $\mathfrak{F}_0$. Since we require of a function of $\mathfrak{F}_0$ only integrability in the infinite region, we shall disregard its behavior on a null set. In particular, we shall regard two functions of $\mathfrak{F}_0$ as identical, if the functions have the same value almost everywhere in $[-\infty, \infty]$. If $f(x)$ is a function of $\mathfrak{F}_0$, then the following also belong to $\mathfrak{F}_0$: $f(-x)$, $\overline{f(x)}$, $g(x)f(x)$ where $g(x)$ is bounded (integrable in the finite region), in particular the function $e(\lambda x)f(x)$ where λ is real, and in addition the function $f(x + \lambda)$. Moreover the class $\mathfrak{F}_0$ is linear.

2. We call a collection of functions linear if $c_1f_1 + c_2f_2$ is an element of the collection, where c_1 and c_2 are any two (complex) constants, and f_1 and f_2 are elements of the collection. In particular, for each constant c, cf is also an element of the collection, where f is an element of the collection.

Moreover, $\mathfrak{F}_0$ is closed in the following sense. If a sequence of functions $f_n(x)$, $n = 1, 2, 3, \ldots$ of $\mathfrak{F}_0$ is convergent in the integrated mean, i.e., if

$$(1) \qquad \lim_{\substack{m \to \infty \\ n \to \infty}} \int |f_m(x) - f_n(x)| \, dx = 0 \quad ,$$

then there is exactly one function $f(x)$ of $\mathfrak{F}_0$ towards which it converges in the sense that

$$(2) \qquad \lim_{n \to \infty} \int |f(x) - f_n(x)| \, dx = 0$$

cf. Appendix 11.

3. Let f_1 and f_2 be two given functions of $\mathfrak{F}_0$. For brevity, we set

$$(3) \qquad \int |f_i(x)| \, dx = C_i, \qquad i = 1, 2 \quad .$$

We now make full use of Fubini's theorem, cf. Appendix 7, 10. The function

$$(4) \qquad g(x, y) = f_1(y)f_2(x - y)$$

is a measurable function of the variables (x, y). For each point y in which $f_1(y)$ is finite

$$\int |g(x, y)| \, dx = |f_1(y)| \int |f_2(x - y)| \, dy = |f_1(y)| \cdot C_2 \quad .$$

Therefore

$$\int dy \int |g(x, y)| \, dx = C_2 \int |f_1(y)| \, dy = C_1C_2 \quad .$$

Hence the function $g(x, y)$ is absolutely integrable over the whole (x, y) plane. Therefore the integral

$$(5) \qquad \int g(x, y) \, dy = \int f_1(y)f_2(x - y) \, dy$$

exists for almost all x, and is again a function of $\mathfrak{F}_0$. Moreover it is independent of the order of f_1 and f_2 because

$$(6) \qquad \int f_1(y)f_2(x - y) \, dy = \int f_2(\eta)f_1(x - \eta) \, d\eta \quad .$$

If for any two functions in $[-\infty, \infty]$, _the integral (6) exists for almost all_ x, _then the function_

$$\frac{1}{2\pi}\int f_1(y)f_2(x-y)\,dy \tag{7}$$

<u>is called the Faltung of</u> f_1 <u>and</u> f_2 [40]. If $f_1(x)$ and $f_2(x)$ are functions of $\mathfrak{F}_0$, then their Faltung exists, and likewise belongs to $\mathfrak{F}_0$.

4. <u>To each function of</u> $\mathfrak{F}_0$, <u>we attach its transform</u> $E(\alpha)$

$$E(\alpha) = \frac{1}{2\pi}\int f(x)e(-\alpha x)\,dx \quad , \tag{8}$$

<u>and denote the totality of these functions</u> $E(\alpha)$ <u>by</u> $\mathfrak{J}_0$. In order that a function $E(\alpha)$ belong to $\mathfrak{J}_0$, it is necessary that it be bounded

$$2\pi|E(\alpha)| \leqq \int |f(x)|\,dx \quad , \tag{9}$$

that it be continuous (Theorem 2), and that it approach zero as $\alpha \longrightarrow \pm\infty$ (Theorem 1). The class $\mathfrak{J}_0$ is moreover linear: to $c_1f_1 + c_2f_2$ belong $c_1E_1 + c_2E_2$. If $E(\alpha)$ belongs to $\mathfrak{J}_0$, then the following do also: $E(-\alpha)$, $\overline{E(\alpha)}$, $E(\alpha+\lambda)$ and $e(\lambda\alpha)E(\alpha)$, λ real. They are the transforms of $f(-x)$, $\overline{f(-x)}$, $e(-\lambda x)f(x)$ and $f(x+\lambda)$. If functions of $\mathfrak{F}_0$ converge in the integrated mean, then owing to

$$2\pi|E(\alpha) - E_n(\alpha)| \leqq \int |f(x) - f_n(x)| \cdot |e(-\alpha x)|\,dx \leqq \int |f(x) - f_n(x)|\,dx \quad ,$$

their transforms are uniformly convergent in $-\infty < \alpha < \infty$.

THEOREM 13. The Faltung of functions of $\mathfrak{F}_0$ corresponds to the multiplication of their transforms.

PROOF. Since

$$\frac{1}{2\pi}\int f(x)e(-\alpha x)\,dx = \frac{1}{4\pi^2}\int dx \int f_1(y)e(-\alpha y)f_2(x-y)e(-\alpha(x-y))\,dy \quad ,$$

and since for fixed α, the functions $f_1(y)e(-\alpha y)$ and $f_2(\eta)e(-\alpha\eta)$ belong to $\mathfrak{F}_0$, we can interchange the order of integration to obtain

$$E(\alpha) = E_1(\alpha)E_2(\alpha) \quad . \tag{10}$$

Within the class $\mathfrak{J}_0$, the functions can therefore by multiplied arbitrarily. We also remark that the transform $E_1(\alpha)\overline{E_2(\alpha)}$ belongs to the function

(11) $$\frac{1}{2\pi}\int f_1(y)\overline{f_2(y-x)}\,dy = \frac{1}{2\pi}\int f_1(y+x)\overline{f_2(y)}\,dy \quad .$$

5. By definition, each function of $\mathfrak{F}_0$ is attached to exactly one function of $\mathfrak{J}_0$. It is an important fact, that the converse is also valid.

THEOREM 14. Two functions of $\mathfrak{F}_0$ which have the same transforms are identical [41].

PROOF. To $f(x) = f_1(x) - f_2(x)$ belongs $E(\alpha) = E_1(\alpha) - E_2(\alpha)$. Now let $E(\alpha) \equiv 0$. Then

$$\int E(\alpha)e(\alpha x)\,d\alpha \equiv 0 \quad .$$

Hence by Theorem 12, $f(x) \equiv 0$.

The one-to-one relation between functions of $\mathfrak{F}_0$ and $\mathfrak{J}_0$ will be expressed by the symbolic formula

(12) $$f(x) \sim \int e(\alpha x)E(\alpha)\,d\alpha \quad .$$

We call this formula, wholly independent of whether or not the integral converges, a representation of $f(x)$. We shall also speak of the function

(13) $$\int e(\alpha x)E(\alpha)\,d\alpha \quad ,$$

by which we shall mean the well determined function of $\mathfrak{F}_0$, whose transform agrees with the fiven function $E(\alpha)$ of $\mathfrak{J}_0$. We say that the representation of $f(x)$ converges if the integral (13) is convergent for almost all x. We then write, by Theorem 12

(14) $$f(x) = \int e(\alpha x)E(\alpha)\,d\alpha \quad .$$

6. Let $\varphi(\alpha)$ be a given continuous function in $[-\infty, \infty]$ with the following properties: 1) $\varphi(-\alpha) = \varphi(\alpha)$, 2) $\varphi(0) = 1$, 3) $\varphi(\alpha)$ is absolutely integrable, and 4) the function

(15) $$K(x) = \frac{1}{2\pi}\int e(\alpha x)\varphi(\alpha)\,d\alpha$$

is absolutely integrable (and therefore a function of $\mathfrak{F}_0$), bounded and at infinity equal to $o(|x|^{-1})$. — It is easily seen that $\varphi(\alpha)$ is the transform of $2\pi K(x)$, and more generally that $\varphi(\frac{\alpha}{n})$ is the transform of $2\pi nK(nx)$, i.e.,

$$\varphi(\tfrac{\alpha}{n}) = n \int e(-\alpha x)K(nx)\, dx \quad .$$

Hence by 2), it follows that

$$(16) \qquad \int K(\xi)\, d\xi = 1 \quad .$$

The functions $e^{-|\alpha|}$, $e^{-\alpha^2}$ and $\varphi(\alpha) = 1 - |\alpha|$ for $|\alpha| \leq 1$ and $= 0$ for $|\alpha| > 1$ are examples of functions satisfying the given conditions. The corresponding functions $K(x)$ are

$$\frac{1}{\pi}\frac{1}{1 + x^2}, \quad \frac{1}{2\sqrt{\pi}} e^{-\left(\frac{x}{2}\right)^2} \quad \text{and} \quad \frac{2}{\pi}\left(\frac{\sin\frac{x}{2}}{x}\right)^2 ,$$

cf. §12 (9); §15, (9) and §4, (14).

Now let $f(x)$ be any function of $\mathfrak{F}_0$, and $E(\alpha)$ its transform. The Faltung of $f(x)$ and $2\pi n K(nx)$ corresponds to the transform $\varphi(\frac{\alpha}{n})E(\alpha)$. If, now, $\varphi(\frac{\alpha}{n})$ is absolutely integrable and $E(\alpha)$ bounded, then the product is also absolutely integrable. Hence for almost all x

$$(17) \qquad \int \varphi(\tfrac{\alpha}{n})E(\alpha)e(\alpha x)\, d\alpha = n \int f(\xi)K[n(x - \xi)]\, d\xi \quad .$$

By Theorems 1 and 2, the left side is convergent and continuous for all x. The right side is likewise convergent and continuous for all x; this easily follows from the fact that $f(\xi)$ is absolutely integrable, and that $K(x)$, because of (15), is bounded and uniformly continuous for all x. In regard to the latter, cf. the remarks in 4. Two continuous functions which agree almost everywhere, agree point for point. The relation (17) is therefore valid for all points x. Therefore by (16) and Theorem 3, at each point x for which $f(x + 0)$ and $f(x - 0)$ exist, we have

$$(18) \qquad \frac{1}{2}[f(x + 0) + f(x - 0)] = \lim_{n \to \infty} \int \varphi(\tfrac{\alpha}{n})E(\alpha)e(\alpha x)\, d\alpha \quad .$$

By (1), we can also write for (18)

$$(19) \quad \frac{1}{2}[f(x + 0) + f(x - 0)] = \lim_{n \to \infty} \frac{1}{\pi}\int_0 \varphi(\tfrac{\alpha}{n})\, d\alpha \int f(\xi) \cos \alpha(x - \xi)\, d\xi \quad .$$

Letting now $n \longrightarrow \infty$ under the integral sign, formula §11, (1) results; but this extension of the limit is inadmissible because for the validity of the last formula, bounded variation in the neighborhood of x was assumed. However, we may view the situation in the following manner. Formula §11, (1) can still be maintained for absolutely integrable functions

$f(x)$ for points x at which the requirement of bounded variation is not fulfilled, provided that ordinary convergence of the integral be replaced by "summability" with the aid of a convergence producing factor $\varphi(\alpha)$. The factor can be of very general nature; in the two special cases $\varphi(\alpha) = e^{-|\alpha|}$ and $\varphi(\alpha) = e^{-\alpha^2}$, the integral (19) is called a <u>Sommerfeld integral</u> [42].

7. Theorem 13 can be employed for the calculation of definite integrals.

THEOREM 15. Let $f_1(x)$ and $f_2(x)$ be given functions of $\mathfrak{F}_0$. In order that the relation

$$2\pi \int E_1(\alpha)E_2(\alpha)e(\alpha x)\, d\alpha = \int f_1(y)f_2(x - y)\, dy \tag{20}$$

hold for all x, it is sufficient that one of the following conditions be fulfilled:

1) The function $E_1(\alpha)E_2(\alpha)$ be absolutely integrable and the right side of (20) represent a continuous function.

2) One of the functions $f_1(x)$ and $f_2(x)$ have an absolutely integrable derivative.

PROOF of 1). This half of the theorem has already been proved and used in 6. We repeat the proof briefly. By Theorem 12, (20) holds for almost all x. Since the functions on both sides of it are continuous, it holds for all x without exception.

PROOF of 2). Because of Theorem 11b, it is sufficient to prove that the right side of (20) is differentiable. We may assume, from the hypothesis, that $f_2(x)$ is the function which has an absolutely integrable derivative. We set

$$2\pi\varphi(x) = \int f_1(y)f_2'(x - y)\, dy$$

and form, for an arbitrary a, the integral

$$2\pi \int_a^x \varphi(\xi)\, d\xi = \int_a^x d\xi \int f_1(y)f_2'(\xi - y)\, dy \quad .$$

One can interchange the order of integration on the right, to obtain

$$\int f_1(y)\,dy \int_a^x f_2'(\xi - y)\,d\xi = \int f_1(y)f_2(x - y)\,dy - \int f_1(y)f_2(a - y)\,dy$$

$$= 2\pi f(x) - 2\pi f(a) .$$

Hence the Faltung $f(x)$ is the integral of a function $\varphi(x)$, Q.E.D.

If $f_1(x)$ and $f_2(x)$ both vanish for $x < 0$, then its Faltung also vanishes; for $x > 0$, it has the value

$$\frac{1}{2\pi}\int_0^x f_1(y)f_2(x - y)\,dy . \tag{21}$$

The first part of the above assertion is also valid if more functions are faltet one after the other. More generally, if for $\nu = 1, 2, \ldots, n$, $f_\nu(x) = 0$ for $x \leqq x_\nu$, then the Faltung $= 0$ for $x \leqq x_1 + x_2 + \cdots + x_n$. A similar remark is valid if the "$<$" sign is replaced by the "$>$" sign. All of these rules will be made use of in what follows.

8. From

$$\int \frac{\sin \rho\xi}{\xi}\, e(\sigma\xi)\,d\xi = 0 \quad \text{for} \quad |\sigma| > \rho > 0 ,$$

one obtains

$$\int \prod_1^n \frac{\sin \rho_\nu \xi}{\xi}\, e(\sigma\xi)\,d\xi = 0 \quad \text{for} \quad |\sigma| \geqq \rho_1 + \rho_2 + \cdots + \rho_n , \tag{22}$$

and in particular, cf. §4, 4

$$\int \left(\frac{\sin \rho\xi}{\xi}\right)^n e(\sigma\xi)\,d\xi = 0 \quad \text{for} \quad |\sigma| > n\rho > 0 \tag{23}$$

provided one faltet together the functions $f_\nu(x)$ of $\mathfrak{F}_0$ corresponding to the n functions

$$\frac{\sin \rho_\nu \alpha}{\alpha}$$

of $\mathfrak{F}_0$.

By §12, 6, we obtain [43], for $\Re(k_\nu) > 0$, $\mu_\nu > 0$, $\nu = 1, \ldots, \mu$

(24)
$$\frac{1}{2\pi}\int \frac{e(\alpha x)\, d\alpha}{(k_1+i\alpha)^{\mu_1} \cdots (k_n+i\alpha)^{\mu_n}} = 0 \quad \text{for} \quad x < 0 ,$$

and

(25)
$$\frac{1}{2\pi}\int \frac{e(\alpha x)\, d\alpha}{(k_1+i\alpha)^{\mu_1}(k_2+i\alpha)^{\mu_2}} = \frac{e^{-k_2 x}}{\Gamma(\mu_1)\Gamma(\mu_2)} \int_0^x e^{-(k_2-k_1)y} y^{\mu_1-1}(x-y)^{\mu_2-1}\, dy$$

for $x > 0$. By §12, (6) and §12, (7), we obtain, for $\Re(k) > 0$, $\mu > 0$, and $x > 0$

$$\frac{1}{2\pi}\int \frac{e(\alpha x)\, d\alpha}{(k^2+\alpha^2)^{\mu}} = \frac{e^{kx}}{\Gamma^2(\mu)} \int_x e^{-2ky} y^{\mu-1}(y-x)^{\mu-1}\, dy \quad [44] .$$

For $\mu = 1$, we obtain the already known formula of §11, (12), $\Re(\ell) > 0$,

(27)
$$\frac{1}{2\pi}\int \frac{e(\alpha x)\, d\alpha}{\ell^2 + a^2} = \frac{1}{2\ell}\, e^{-\ell|x|}, \quad -\infty < x < \infty .$$

More generally, we obtain, for integers $\mu > 0$,

(28)
$$\frac{1}{\pi}\int_0 \frac{\cos \alpha x\, d\alpha}{(\ell^2+\alpha^2)^{\mu}} = \frac{e^{-\ x}}{2^{2\mu-1}\Gamma(\mu)} \sum_{r=0}^{\mu-1} \frac{\Gamma(2\mu-r-1)}{\Gamma(r+1)\Gamma(\mu-r)} \frac{(2x)^r}{\ell^{2\mu-r-1}}$$

by computing the integral on the right of (26), for $x > 0$. If for $x < 0$, the function (27) is faltet n-times one after another with the function of §12, (6) for various values of k and μ [45], we obtain [$\Re(k_\nu) > 0$, $\Re(\ell) > 0$, $\mu_\nu > 0$, $x < 0$]

(29)
$$\frac{1}{2\pi}\int \frac{e(\alpha x)\, d\alpha}{(\ell^2+\alpha^2)(k+i\alpha)^{\mu_1}\cdots(k_n+i\alpha)^{\mu_n}} = \frac{e^{\ell x}}{2\ell} \prod_1^n \frac{1}{(\ell+k_\nu)^{\mu_\nu}} .$$

By §12 (6), and §12, (7), we obtain for $x > 0$ [46]

(30)
$$\frac{1}{2\pi}\int \frac{e(\alpha x)\, d\alpha}{(k+i\alpha)^{\rho}(k-i\alpha)^{\sigma}} = \frac{e^{kx}}{\Gamma(\rho)\Gamma(\sigma)} \int_x e^{-2ky} y^{\rho-1}(y-x)^{\sigma-1}\, dy .$$

If $\rho + \sigma > 1$, one can allow $x \longrightarrow 0$, and obtain

(31) $$\int \frac{d\alpha}{(k+i\alpha)^{\rho}(k-i\alpha)^{\sigma}} = 2\pi \frac{\Gamma(\rho+\sigma-1)}{\Gamma(\rho)\Gamma(\sigma)(2k)^{\rho+\sigma-1}} .$$

By the transformation $\alpha = k \tan x$ and the substitution $\sigma + \rho - 1 = p$ and $\sigma - \rho = q$, we obtain ($\rho > 0$, arbitrary q)

$$\int_{-\pi/2}^{\pi/2} \cos^{p-1}x \cos qx \, dx = \int_{-\pi/2}^{\pi/2} \cos^{p-1}x \, e(qx) \, dx = \frac{\pi}{2^{p-1}} \frac{\Gamma(p)}{\Gamma(\frac{p+1+q}{2})\Gamma(\frac{p+1-q}{2})}$$

From this, we obtain by inversion, for $p > 1$

$$\frac{1}{2^p} \int \frac{\Gamma(p)e(x\alpha)}{\Gamma(\frac{p+1+x}{2})\Gamma(\frac{p+1-x}{2})} = \begin{cases} \cos^{p-1}\alpha, & |\alpha| \leq \frac{\pi}{2} \\ 0 \quad , & |\alpha| > \frac{\pi}{2} \end{cases}$$

9. The following remark will be of use later on.

If the functions $f(\xi)$ and $E(\alpha)$ are interchanged in Theorem 12, and small changes are made in normalization, the following results. _If the function $E(\alpha)$ is absolutely integrable, and if the function_

$$f(x) = \int e(x\alpha)E(\alpha) \, d\alpha$$

belongs to $\mathfrak{F}_0$ (i.e., it is absolutely integrable), _then the function_

$$\frac{1}{2\pi} \int e(-\alpha x)f(x) \, dx$$

agrees with $E(\alpha)$ _for almost all_ α. _If, in particular,_ $E(\alpha)$ _is also continuous, then for all_ α

$$E(\alpha) = \frac{1}{2\pi} \int e(-\alpha x)f(x) \, dx \quad ,$$

and $E(\alpha)$, _as the transform of_ $f(x)$ _is a function of_ $\mathfrak{F}_0$.

Hence each function $\gamma(\alpha)$ defined in $[-\infty, \infty]$, which is two times differentiable, and which together with its both derivatives is absolutely integrable, is contained in the class $\mathfrak{F}_0$. In particular, therefore, each function $\gamma(\alpha)$ is in this class which is two times differentiable and vanishes outside of a finite interval. Since for the function

(32) $$K_{\gamma}(x) = \int e(x\alpha)\gamma(\alpha) \, d\alpha \quad ,$$

we obtain, by an appropriate application of §3, 4, the valuation

$$K_\gamma(x) = o(|x|^{-2}) \quad ,$$

it follows that $K_\gamma(x)$ is a function of $\mathfrak{F}_0$. We note again that

$$\gamma(\alpha) = \frac{1}{2\pi}\int e(-\alpha x)K_\gamma(x)\,dx \quad .$$

If $\gamma(\alpha)$ has r absolutely integrable derivatives, $r \geqq 2$, then by §3, 4

$$K_\gamma(x) = o(|x|^{-r}) \quad . \tag{33}$$

If for $r \geqq 1$, the functions

$$\alpha^\rho\gamma(\alpha), \qquad \rho = 0, 1, \ldots, r, \tag{34}$$

are absolutely integrable, then by application of §4, 2, (c) to the integral (32), we obtain the result without assuming differentiability that the function $K_\gamma(x)$ is r-times continuously differentiable, and

$$K_\gamma^{(\rho)}(x) = \int e(x\alpha)(i\alpha)^\rho\gamma(\alpha)\,d\alpha, \qquad \rho = 0, 1, \ldots, r \quad . \tag{35}$$

If in addition, the first two derivatives of the function (34) are absolutely integrable (which happens, for example, if $\gamma(\alpha)$ vanishes outside of a finite interval), then by the above, the functions (35) are likewise absolutely integrable.

§14. Trigonometric Integrals with Rational Functions

1. In order that a rational function come under §11, it is necessary and sufficient that the degree of the numerator be smaller than the degree of the denominator, and that no real poles exist. — By §12, (6) and §12, (7), the transform

$$\frac{1}{2\pi}\int \frac{e(-\alpha x)}{x - ik}\,dx \tag{1}$$

is known for any complex number k. It is, for $\Re(k) > 0$,

$$0 \text{ for } \alpha > 0 \quad \text{and} \quad ie^{k\alpha} \text{ for } \alpha < 0 \quad , \tag{2}$$

and for $\Re(k) < 0$,

$$-ie^{k\alpha} \text{ for } \alpha > 0 \quad \text{and} \quad 0 \text{ for } \alpha < 0 \quad . \tag{3}$$

By Faltung, or more simply by differentiation with respect to k, one can compute from this the transform of $(x - ik)^{-n}$, $n = 1, 2, \ldots$. By a partial fraction decomposition, one can obtain from this, in principle, the transform of every rational function. For example, one thus obtains for $\alpha > 0$, formula §11, (12), but this time more generally for $\Re(k) > 0$. By integration with respect to α, of the first expression therein, there results

$$(4) \qquad \frac{1}{\pi} \int_0 \frac{\sin \alpha x}{x(x^2+k^2)} \, dx = \frac{1}{2k^2} (1 - e^{\alpha k}) .$$

2. The partial fraction decomposition is burdensome. It is possible in certain special cases to proceed differently. For example, setting $k = re(\varphi)$, $-\frac{\pi}{2} < \varphi < \frac{\pi}{2}$, in §11, (12), and separating reals and imaginaries, we obtain

$$\frac{1}{\pi} \int_0 \frac{\cos \alpha x \, dx}{x^4 + 2r^2x^2 \cos 2\varphi + r^4} = \frac{1}{2r^3} e^{-r\alpha \cos \varphi} \frac{\sin(\varphi + r\alpha \sin \varphi)}{\sin 2\varphi} ,$$

$$\frac{1}{\pi} \int \frac{x^2 \cos \alpha x \, dx}{x^4 + 2r^2x^2 \cos 2\varphi + r^4} = \frac{1}{2r} e^{-r\alpha \cos \varphi} \frac{\sin(\varphi - r\alpha \sin \varphi)}{\sin 2\varphi} .$$

In particular, if $\varphi = \frac{\pi}{4}$, $[\alpha > 0, r > 0]$

$$\frac{1}{\pi} \int_0 \frac{\cos \alpha x \, dx}{x^4 + r^4} = \frac{\sqrt{2}}{4r^3} e^{-\frac{r\alpha}{\sqrt{2}}} \left[\cos\left(\frac{r\alpha}{\sqrt{2}}\right) + \sin\left(\frac{r\alpha}{\sqrt{2}}\right) \right] .$$

By differentiation with respect to α, it follows that

$$\frac{1}{\pi} \int_0 \frac{x \sin \alpha x \, dx}{x^4 + r^4} = \frac{1}{2r^2} e^{-\frac{r\alpha}{\sqrt{2}}} \sin\left(\frac{r\alpha}{\sqrt{2}}\right) .$$

Numerous other integrals can be evaluated in this manner [47].

3. The residue theory gives the most practical method to follow. If $f(x)$ is rational, $z = x + yi$, then

$$\int f(x) e(-\alpha x) \, dx$$

has the value

$$2\pi i \Sigma R^+, \qquad \text{for } \alpha < 0,$$

$$-2\pi i \Sigma R^-, \qquad \text{for } \alpha > 0.$$

Here ΣR^+ is the sum of the residues of the function

$$f(z)e(-\alpha z)$$

for those poles which lie in the half plane $y > 0$, and ΣR^- the sum for those poles which lie in the lower half plane. If $f(x)$ is even, then

$$\frac{1}{\pi}\int_0 f(x) \cos \alpha x \, dx = i\Sigma R^+ = - i\Sigma R^{-1}, \qquad \alpha > 0,$$

and if $f(x)$ is odd, then

$$\frac{1}{\pi}\int_0 f(x) \sin \alpha x \, dx = \Sigma R^+ = \Sigma R^-, \qquad \alpha > 0.$$

For example, let us select the simplest case; $f(z) = z - \lambda$, and $\Im(\lambda) \neq 0$. If $\Im(\lambda) > 0$, and if x is replaced by α, α by $- x$, we obtain

$$\int \frac{e(\alpha x)}{\alpha - \lambda}\, d\alpha = \begin{cases} 0 & \text{for } x < 0 \\ 2\pi i e(\lambda x) & \text{for } x > 0, \end{cases} \tag{5}$$

and if $\Im(\lambda) < 0$, we obtain similarly

$$\int \frac{e(\alpha x)}{\alpha - \lambda}\, d\alpha = \begin{cases} - 2\pi i e(\lambda x) & \text{for } x < 0 \\ 0 & \text{for } x > 0 . \end{cases} \tag{6}$$

These results are in agreement with the value (2) or (3) of 1. From this we shall make an application which will be useful to us later. Consider any function $f(x)$ of $\mathfrak{F}_0$ and denote its transform by $\varphi(\alpha)$. Since the functions on the right of (5) and (6) likewise belong to $\mathfrak{F}_0$, there is, for $\Im(\lambda) \neq 0$, a function of $\mathfrak{F}_0$ which we denote by $g(x)$, whose transform agrees with

$$\frac{\varphi(\alpha)}{i(\alpha-\lambda)} . \tag{7}$$

It has the value

$$e(\lambda x)\int^x e(- \lambda x_1)f(x_1)\, dx_1, \quad \text{or} \quad - e(\lambda x)\int_x e(- \lambda x_1)f(x_1)\, dx_1 ,$$

according as

$$\Im(\lambda) > 0 \quad \text{or} \quad \Im(\lambda) < 0 .$$

This process can be repeated. For each integer $p > 0$, there is a function $g_p(x)$ of $\mathfrak{F}_0$ whose transform agrees with

$$\frac{\varphi(\alpha)}{\{i(\alpha-\lambda)\}^p} \quad . \tag{8}$$

Hence

$$g_p(x) = e(\lambda x) f_p(x) \quad , \tag{9}$$

where $f_p(x)$ has the value

$$\int^x dx_1 \int^{x_1} dx_2 \cdots \int^{x_{p-2}} dx_{p-1} \int^{x_{p-1}} e(-\lambda x_p) f(x_p)\, dx_p \quad , \tag{10}$$

or

$$(-1)^p \int_x dx_1 \int_{x_1} dx_2 \cdots \int_{x_{p-2}} dx_{p-1} \int_{x_{p-1}} e(-\lambda x_p) f(x_p)\, dx_p \quad , \tag{11}$$

according as $\Im(\lambda) > 0$ or $\Im(\lambda) < 0$.

4. For $k > 0$, a partial fraction decomposition yields

$$\frac{2}{i}\int_0 \frac{e^{-kx}}{1+x^2}\, dx = \int_0 \frac{e^{-y}}{y+ki}\, dy - \int_0 \frac{e^{-y}}{y-ki}\, dy \quad , \tag{13}$$

a result which is also evidently valid for $\Re(k) > 0$. We now set $k = \epsilon + \alpha i$, $\alpha > 0$ and $\epsilon > 0$, and write for the first integral on the right

$$\int_0^{2\alpha} \frac{e^{-y}}{y - \alpha + \epsilon i}\, dy + \int_{2\alpha} \frac{e^{-y}}{y - \alpha + \epsilon i}\, dy \quad .$$

Separating the real part on both sides of (13), and allowing $\epsilon \longrightarrow 0$, then there results for $\alpha > 0$, cf. §5, 5

$$2\int_0 \frac{\sin \alpha x}{1+x^2}\, dx = \int_0 \frac{e^{-y}}{y+\alpha}\, dy - \int \frac{e^{-y}}{y-\alpha}\, dy \quad ,$$

where for the second integral on the right, we have to take the Cauchy principle value. If now we introduce the function

$$\ell i\, y = \int_{-\log y} \frac{e^{-\xi}}{\xi}\, d\xi$$

(logarithmic integral of y), it is possible to write for this [48]

(14) $$\int_0 \frac{\sin \alpha x}{1 + x^2}\, dx = \frac{1}{2}\left[e^{-\alpha} \ell i\ e^{\alpha} - e^{\alpha} \ell i\ e^{-\alpha}\right] .$$

Differentiating, we obtain

(15) $$\int_0 \frac{x \cos \alpha x}{1 + x^2}\, dx = -\frac{1}{2}\left[e^{-\alpha} \ell i\ e^{\alpha} + e^{\alpha} \ell i\ e^{-\alpha}\right] .$$

By inversion, we obtain the related formulas

$$\int_0 e^{x}\ \ell i\ e^{-x} \cos \alpha x\, dx = \frac{\log \alpha - \alpha \frac{\pi}{2}}{1 + \alpha^2}$$

$$\int_0 e^{-x} \ell i\ e^{x} \cos \alpha x\, dx = \frac{-\log \alpha - \alpha \frac{\pi}{2}}{1 + \alpha^2}$$

§15. Trigonometric Integrals with e^{-x^2}

We start from the equality

(1) $$\int_0 e^{-x^2}\, dx = \frac{1}{2}\sqrt{\pi} .$$

Its simplest direct proof proceeds in the following well known manner. We denote the integral by J. Then

$$J^2 = \int_0 e^{-y^2}\, dy \int_0 e^{-x^2}\, dx = \int_0\int_0 e^{-(x^2+y^2)}\, dx\, dy ,$$

and by the introduction of polar coordinates

$$J^2 = \int_0 e^{-r^2} r\, dr \int_0^{\pi/2} d\varphi = \frac{\pi}{4} .$$

Letting $x = y\sqrt{\lambda}$, $\lambda > 0$, and then differentiating repeatedly with respect to λ, we obtain

(2) $$\int_0 e^{-\lambda y^2}\, dy = \frac{1}{2}\sqrt{\pi}\, \lambda^{-1/2}$$

$$\int_0 y^{2n} e^{-\lambda y^2}\, dy = \frac{1}{2}\sqrt{\pi}\, \frac{1\cdot 3\cdot \ \ldots\ \cdot(2n-1)}{2^n}\, \lambda^{-\frac{2n+1}{2}}$$

We shall now calculate the integral

(3) $$\int_0 e^{-x^2} \cos 2bx\, dx$$

by a procedure which can also be used in many more difficult cases. We make use of the series expansion

$$\cos 2bx = \sum_0^\infty (-1)^n \frac{x^{2n}(2b)^{2n}}{(2n)!} .$$

Inserting it in (3), and then integrating termwise, one obtains by (2)

$$\frac{1}{2}\sqrt{\pi}\sum_0^\infty (-1)^n \frac{1 \cdot 3 \ldots (2n-1)}{1 \cdot 2 \ldots \cdot 2n} \frac{2^{2n}}{2^n} b^{2n} = \frac{1}{2}\sqrt{\pi}\sum_0^\infty (-1)^n \frac{b^{2n}}{1 \cdot 2 \ldots n}$$

$$= \frac{1}{2}\sqrt{\pi}\, e^{-b^2} .$$

It is still necessary to justify the interchange just used of series summation and integration over an infinite interval. Set

$$\cos 2bx = \sum_0^n (-1)^\nu \frac{x^{2\nu}(2b)^{2\nu}}{(2\nu)!} + R_n(x) .$$

Then by the above

(4) $$\int_0 e^{-x^2} \cos 2bx\, dx = \frac{1}{2}\sqrt{\pi}\sum_{\nu=0}^n (-1)^\nu \frac{b^{2\nu}}{\nu!} + \int_0 R_n(x)e^{-x^2}\, dx .$$

But now by Taylor's theorem, for a suitable θ between 0 and 1,

$$|R_n(x)| = \left|\frac{(-1)^{n+1}(2bx)^{2n+2}\cos(\theta 2bx)}{1 \cdot 2 \cdot \ldots \cdot (2n+2)}\right| \leqq \frac{x^{2n+2}(2b)^{2n+2}}{1 \cdot 2 \ldots \cdot (2n+2)} .$$

Hence by (2)

$$\left|\int_0 R_n(x)e^{-x^2}\, dx\right| \leqq \frac{1}{2}\sqrt{\pi}\frac{b^{2n+2}}{(n+1)!} .$$

Therefore the remainder term approaches zero as $n \longrightarrow \infty$, and we have [50]

(5) $$\int_0 e^{-x^2} \cos 2bx\, dx = \frac{1}{2}\sqrt{\pi}\, e^{-b^2} .$$

Differentiating this equation repeatedly with respect to b, and making use of (2), there results [51] for $k = (0), 1, 2, 3, \dots$

$$\int_0 x^{2k} e^{-x^2} \cos 2bx \, dx = e^{-b^2} \int_0 \frac{(x+bi)^{2k} + (x-bi)^{2k}}{2} e^{-x^2} \, dx \; .$$

(6)

$$\int_0 x^{2k-1} e^{-x^2} \sin 2bx \, dx = e^{-b^2} \int_0 \frac{(x+bi)^{2k-1} - (x-bi)^{2k-1}}{2i} e^{-x^2} \, dx \; .$$

There is also a relation for the integral

$$\varphi(b) = \int_0 e^{-x^2} \sin 2bx \, dx \; ,$$

namely

$$\varphi'(b) + 2b\varphi(b) = -\int_0 \frac{d}{dx}\left[e^{-x^2} \cos 2bx \right] dx = 1 \; .$$

Therefore

$$\varphi(b) = e^{-b^2} \int_0^b e^{b^2} db + ce^{-b^2} \; .$$

From the fact that $\varphi(0) = 0$, we have $c = 0$, and hence [52]

$$(7) \qquad \int_0 e^{-x^2} \sin 2bx \, dx = e^{-b^2} \int_0^b e^{\beta^2} \, d\beta \; .$$

By a simple transformation of the variable, one obtains from (5), for $\lambda > 0$

$$(8) \qquad \int_0 e^{-\lambda x^2} \cos \alpha x \, dx = \frac{1}{2}\sqrt{\frac{\pi}{\lambda}} \, e^{-\frac{\alpha^2}{4\lambda}} \; ,$$

and from it

$$(9) \qquad \int e^{-\lambda x^2} e(-\alpha x) \, dx = \sqrt{\frac{\pi}{\lambda}} \, e^{-\frac{\alpha^2}{4\lambda}} \; .$$

By §12, (3), for $k \geq 0$ and $\mu > -1$

$$\frac{1}{(x^2+k^2)^{\mu+1}} = \frac{1}{\Gamma(\mu+1)} \int_0 y^{\mu} e^{-(x^2+k^2)y} \, dy \; .$$

From this follows [53]

(10) $$\int_0 \frac{\cos \alpha x \, dx}{(x^2+k^2)^{\mu+1}} = \frac{\sqrt{\pi}}{2\Gamma(\mu+1)} \int_0 y^{\mu - \frac{1}{2}} e^{-\left(k^2 y + \frac{\alpha^2}{4y}\right)} dy \; .$$

This gives, for $\mu = 0$

(11) $$\frac{1}{2}\sqrt{\pi} \int_0 y^{-1/2} e^{-\left(k^2 y + \frac{\alpha^2}{4y}\right)} dy = \frac{\pi}{2k} e^{-|\alpha| k} \; ,$$

and (by substituting $y = x^2$, $k^2 = p$, $\alpha^2 = 4q$)

(12) $$\int_0 e^{- px^2 - \frac{q}{x^2}} dx = \frac{1}{2}\sqrt{\frac{\pi}{p}}\, e^{-2\sqrt{pq}}$$

for $p > 0$ and $q > 0$. The integral on the left is, for fixed $p > 0$, convergent and defined for all complex q with positive real part, and as can easily be shown, is an analytic function in q. Therefore by the principle of analytic continuation, (12) is also valid for $q = \rho + i\alpha$ with $\rho > 0$. Hence if one now sets $p = 1$, and as can be justified, allows $\rho \longrightarrow 0$, one obtains for $\alpha > 0$

(13) $$\int_0 e^{-x^2} e\left(-\frac{\alpha}{x^2}\right) dx = \frac{1}{2}\sqrt{\pi}\, e^{-\sqrt{2\alpha}}\, e(-\sqrt{2\alpha}) \; .$$

One obtains similarly [54]

(14) $$\int_0 e^{-\frac{1}{x^2}} e(-\alpha x^2)\, dx = \frac{1}{2}\sqrt{\frac{\pi}{2\alpha}}\,(1 - i)\, e^{-\sqrt{2\alpha}(1+i)} \; .$$

§16. Bessel Functions

In the theory of cylindrical functions, there is a large number of trigonometric integrals, the most important of which will be stated here. We refer to the basic work of G. N. Watson, Theory of Bessel Functions, Cambridge 1922: quoted as "Watson", for notations and proofs.

For the Bessel function of any complex order, the following relation is valid, $\Re(\nu) > -\frac{1}{2}$ [55]

(1) $$J_\nu(x) = \frac{2(x/2)^\nu}{\sqrt{\pi}\,\Gamma(\nu+1/2)} \int_0^1 (1 - t^2)^{\nu-1/2} \cos xt \, dt \; .$$

One obtains by inversion, $\Re(\nu) > -1/2$,

$$(2)\quad \int_0 \left(\frac{x}{2}\right)^{-\nu} J_\nu(x)\cos\alpha x\,dx = \begin{cases} \frac{\sqrt{\pi}}{\Gamma(\nu+1/2)}(1-\alpha^2)^{\nu-1/2}, & 0<\alpha<1 \\ 0, & 1<\alpha<\infty, \end{cases}$$

or writing it somewhat differently $[a > 0,\ b > 0,\ \Re(\nu) > -1/2]$

$$(3)\quad \int_0 \frac{J_\nu(ax)}{x^\nu}\cos bx\,dx = \begin{cases} \frac{\sqrt{\pi}}{2^\nu\Gamma(\nu+1/2)a^\nu}(a^2-b^2)^{\nu-1/2}; & b<a \\ 0, & b>a. \end{cases}$$

For $\Re(\nu) > 0,\ a > 0,\ b > 0,$ we have [55a]

$$(4)\quad \int_0 J_\nu(ax)e(bx)\,dx = \begin{cases} \frac{1}{\sqrt{a^2-b^2}}\, e(\nu \arcsin b/a), & b<a \\ \frac{a^\nu e(\pi/2+\nu\,\pi/2)}{\sqrt{b^2-a^2}}\left(b+\sqrt{b^2-a^2}\right)^{-\nu}, & b>a, \end{cases}$$

and

$$(5)\quad \int_0 \frac{J_\nu(ax)}{x}e(bx)\,dx = \begin{cases} \frac{1}{\nu}\, e(\nu \arcsin b/a), & b<a \\ \frac{a^\nu e(\nu\pi/2)}{\nu}\left(b+\sqrt{b^2-a^2}\right)^{-\nu}, & b>a. \end{cases}$$

Important special cases obtained from the above ("Weber Discontinuous Integrals") are

$$(6)\quad \int_0 J_0(ax)\cos bx\,dx = \frac{1}{\sqrt{a^2-b^2}} \quad \text{or} \quad 0,$$

$$(7)\quad \int_0 J_0(ax)\sin bx\,dx = 0 \quad \text{or} \quad \frac{1}{\sqrt{b^2-a^2}},$$

according as $b < a$ or $b > a$.

For integers n, one obtains [56]

$$J_{2n+1/2}(x) = (-1)^n \sqrt{\frac{2x}{\pi}} \int_0^1 P_{2n}(t)\cos xt\,dt$$

$$J_{2n+3/2}(x) = (-1)^n \sqrt{\frac{2x}{\pi}} \int_0^1 P_{2n+1}(t) \sin xt \, dt ,$$

where $P_m(t)$ are the Legendre polynomials.

More general than (3) are the following formulas [$\Re(\nu) > -1/2$, z any complex number] [57]

$$(8) \qquad \int_0^{\infty} \frac{J_\nu\left\{a\sqrt{x^2+z^2}\right\}}{\left(\sqrt{x^2+z^2}\right)^\nu} \cos bx \, dx = \sqrt{\frac{\pi}{2}}\, a^{-\nu} J_{\nu-1/2}\left\{z\sqrt{a^2+b^2}\right\} \cdot \left(\frac{\sqrt{a^2-b^2}}{z}\right)^{\nu-1/2} ,$$

for $b < a$, and $= 0$ for $b > a$. The special case $\nu = 1/2$ gives [$a > 0$, $b > 0$]

$$\frac{2}{\pi}\int_0^{\infty} \frac{\sin\left(a\sqrt{x^2+z^2}\right)}{\sqrt{x^2+z^2}} \cos bx \, dx = \begin{cases} J_0\left(z\sqrt{a^2-b^2}\right) , & b < a \\ 0 , & b > a. \end{cases}$$

The function $\alpha^{-\nu}J_\nu(\alpha)$ is an even function of α. By (2) and §4, (11), one obtains by Faltung, for the expression

$$(9) \qquad \int \frac{\sin \rho(\alpha - t)}{\alpha - t} \, \frac{J_\nu(\alpha)}{\alpha^\nu} \, d\alpha ,$$

the value [58]

$$(10) \qquad \frac{2^{1-\nu}\sqrt{\pi}}{\Gamma(\nu+1/2)} \int_0^\rho \cos ty \cdot (1-y^2)^{\nu-1/2} \, dy \quad \text{or} \quad \pi \frac{J_\nu(t)}{t^\nu} ,$$

according as $0 < \rho < 1$ or $1 < \rho$. Here $\Re(\nu) > -1/2$ and t is any real number.

From the formula [$\Re(\mu+\nu) > -1$, z any number] [59]

$$(11) \qquad J_{\mu+\xi}(z) J_{\nu-\xi}(z) = \frac{1}{2\pi}\int_{-\pi}^{\pi} J_{\mu+\nu}(2z \cos Q/2) e(\overline{\nu-\mu}\, Q/2) e(-\xi Q) \, dQ ,$$

one obtains by inversion that

$$(12) \quad \int J_{\mu+\xi}(z)J_{\nu-\xi}(z)e(t\xi)\,d\xi = \begin{cases} J_{\mu+\nu}(2z\cos t/2)e(\overline{\nu-\mu}\;t/2), & \text{for } |t| < \pi \\ 0, & \text{for } |t| > \pi. \end{cases}$$

Hence in particular [60]

$$(13) \quad \int J_{\mu+\xi}(x)J_{\nu-\xi}(x)\,d\xi = J_{\mu+\nu}(2x) \; .$$

There are two integrals for the Hankel function $H_\nu^{(1)}(x)$, $[-1/2 < \Re(\nu) < 1/2,\ x > 0]$ [61]

$$(14) \quad H_\nu^{(1)}(x) = \frac{\Gamma(1/2-\nu)\cdot(x/2)^\nu}{i\sqrt{\pi^3}}\,[1 + e(-2\nu\pi)]\int_1 (t^2-1)^{\nu-1/2}e(xt)\,dt \; ,$$

$$(15) \quad H_\nu^{(1)}(x) = \frac{2(x/2)^{-\nu}}{i\sqrt{\pi}\,\Gamma(1/2-\nu)}\int_1 \frac{e(xt)}{(t^2-1)^{\nu+1/2}}\,dt \; .$$

The corresponding integrals for $H^{(2)}(x)$ are obtained by interchanging i and $-i$.

This implies [61]

$$(16) \quad J_\nu(x) = \frac{2(x/2)^{-\nu}}{\sqrt{\pi}\,\Gamma(1/2-\nu)}\int_1 \frac{\sin xt}{(t^2-1)^{\nu+1/2}}\,dt \; ,$$

$$(17) \quad Y_\nu(x) = -\frac{2(x/2)^{-\nu}}{\sqrt{\pi}\,\Gamma(1/2-\nu)}\int_1 \frac{\cos xt}{(t^2-1)^{\nu+1/2}}\,dt \; .$$

For the function

$$K_\nu(z) = \frac{1}{2}\,\pi i e\left(\frac{\nu\pi}{2}\right)\cdot H_\nu^{(1)}(iz) \; ,$$

we have $[\Re(\nu) > -1/2]$ [62]

$$(18) \quad K_\nu(x) = \frac{2^\nu\Gamma(\nu+1/2)}{\sqrt{\pi}\,x^\nu}\int_0 \frac{\cos xu}{(u^2+1)^{\nu+1/2}}\,du \; .$$

For example, there results by inversion [63]

$$(19) \quad \int_0 K_0(t)\cos\alpha t\,dt = \frac{\pi}{2\sqrt{1+\alpha^2}} \; .$$

In addition we have the formula

(20) $$\int_0 K_o(t) \sin \alpha t \, dt = \frac{\text{arc sinh } x}{\sqrt{1 + \alpha^2}}$$

§17. Evaluation of Certain Repeated Integrals

Let $\Phi(xyz)$ be a function defined and absolutely integrable in the whole (xyz)-space. Let

(1) $$E(\alpha) = \iiint \Phi(xyz) e\,[\alpha(x + y + z)]\, dx\, dy\, dz$$

be a function of $\mathfrak{T}_o$; indeed let it be the transform of a function $\Psi(\xi)$ of $\mathfrak{F}_o$. Further, let $D(\xi)$ be a given function in an interval $\lambda \leq \xi \leq \mu$ which is itself two times differentiable, and which, together with its first derivative, vanishes at the end points of the interval,

(2) $$D(\lambda) = D'(\lambda) = D(\mu) = D'(\mu) = 0 .$$

Outside of (λ, μ), we extend it by zero values, and denote this new function by $F(\xi)$. Call its transform $E_o(\alpha)$. By partial integration, two times, of

$$\frac{1}{2\pi}\int F(\xi) e(-\alpha\xi)\, d\xi = \frac{1}{2\pi}\int_\lambda^\mu F(\xi) e(-\alpha\xi)\, d\xi ,$$

it follows that $E_o(\alpha) = O(|\alpha|^{-2})$. Hence $E_o(\alpha)$ is absolutely integrable in $[-\infty, \infty]$.

We shall now compute the integral

(3) $$J = \iiint \Phi(xyz) F(x + y + z)\, dx\, dy\, dz .$$

Substitute

(4) $$F(x + y + z) = \int E_o(\alpha) e[\alpha(x + y + z)]\, d\alpha .$$

Since not only $\Phi(xyz)$ but also $E_o(\alpha)$ is absolutely integrable, one can interchange the order of integration and obtain

(5) $$J = \int E_o(\alpha) E(\alpha)\, d\alpha .$$

Hence by Theorem 15

$$J = \frac{1}{2\pi}\int F(y)\,\Psi(-y)\,dy .$$

THEOREM 16 [64]. Let the function $\Phi(xyz)$ be defined and absolutely integrable in the whole (xyz)-space, and let the function

$$E(\alpha) = \iiint \Phi(xyz)e[\alpha(x+y+z)]\,dx\,dy\,dz$$

be the transform of an absolutely integrable function $\Psi(z)$, i.e.,

$$E(\alpha) = \frac{1}{2\pi}\int \Psi(\xi)e(-\alpha\xi)\,d\xi, \qquad \Psi(\xi) = \int E(\alpha)e(\xi\alpha)\,d\alpha .$$

For each function $D(\xi)$ defined and bounded in an interval $\lambda \leqq \xi \leqq \mu$, we have

$$(6)\qquad \iiint\limits_{\lambda<x+y+z<\mu} \Phi(xyz)D(x+y+z)\,dx\,dy\,dz = \frac{1}{2\pi}\int_{\lambda}^{\mu} \Psi(-y)D(y)\,dy .$$

For the special case $D(\xi) = 1$, we obtain

$$(7)\qquad \iiint\limits_{\lambda<x+y+z<\mu} \Phi(xyz)\,dx\,dy\,dz = \frac{1}{2\pi}\int_{\lambda}^{\mu} \Psi(-y)\,dy .$$

PROOF. We have already proved the theorem for the case where $D(\xi)$ is two times differentiable in (λ, μ) and the boundary conditions (2) hold. If $D(\xi)$ is bounded in (λ, μ), then it is possible to specify a set of functions which has the above special structure, is uniformly bounded, $|D_n(\xi)| \leqq G$, and converges to $D(\xi)$ for almost all ξ in (λ, μ). We shall not however prove this assertion. The relation (6) holds for each such function $D_n(\xi)$. By a general convergence theorem [Appendix 7, 8], one can then allow $n \longrightarrow \infty$ under the integral sign on both sides of (6). Hence (6) is also valid for this limit function $D(\xi)$, Q.E.D.

EXAMPLES. In the examples which follow, the space integral is to extend over the parallel layer

$$\lambda < x + y + z < \mu .$$

1. Let [$a > 0$, $b > 0$, $c > 0$, or more generally $\Re(a) > 0$, $\Re(b) > 0$, $\Re(c) > 0$]

$$\Phi(xyz) = \left\{ (a^2 + x^2)(b^2 + y^2)(c^2 + z^2) \right\}^{-1} .$$

Then by §13, (27)

$$E(\alpha) = \int \frac{e(\alpha x)\, dx}{a^2 + x^2} \int \frac{e(\alpha y)\, dy}{b^2 + y^2} \int \frac{e(\alpha z)\, dz}{c^2 + z^2} = \frac{\pi^3}{abc} e^{-|\alpha|(a+b+c)} .$$

Therefore

$$\Psi(-y) = \frac{\pi^3}{abc} \int e^{-|\alpha|(a+b+c)} e(-y\alpha)\, d\alpha = \frac{2\pi^3}{abc} \frac{a + b + c}{(a+b+c)^2 + y^2} .$$

Hence

$$\iiint\limits_{\lambda<x+y+z<\mu} \frac{dx\, dy\, dz}{(a^2+x^2)(b^2+y^2)(c^2+z^2)} = \frac{\pi^2}{abc} \left\{ \arctan \frac{\mu}{a+b+c} - \arctan \frac{\lambda}{a+b+c} \right\},$$

$$\iiint\limits_{\lambda<x+y+z<\mu} \frac{(a+b+c)^2 + (x+y+z)^2}{(a^2+x^2)(b^2+y^2)(c^2+z^2)}\, dx\, dy\, dz = \frac{\pi^2}{abc} (a + b + c)(\mu - \lambda) .$$

2. Let [$a > 0$, $b > 0$, $c > 0$]

$$\Phi(xyz) = e^{-a^2x^2-b^2y^2-c^2z^2} .$$

Then by §15, (9)

$$\iiint\limits_{\lambda<x+y+z<\mu} e^{-a^2x^2-b^2y^2-c^2z^2} D(x + y + z)\, dx\, dy\, dz = \frac{\pi\rho}{abc} \int_\lambda^\mu e^{-\rho^2y^2} D(y)\, dy ,$$

where $\dfrac{1}{\rho^2} = \dfrac{1}{a^2} + \dfrac{1}{b^2} + \dfrac{1}{c^2}$.

3. Let [$k > 0$, $a > 0$, $b > 0$, $c > 0$]

$$\begin{aligned} \Phi(xyz) &= e^{-k(x+y+z)} x^{a-1} y^{b-1} z^{c-1}, \qquad && \text{for } x > 0,\ y > 0,\ z > 0 \\ &= 0 , && \text{for other values.} \end{aligned}$$

Then for $\mu > \lambda \geqq 0$

$$\iiint\limits_{\lambda<x+y+z<\mu} e^{-k(x+y+z)}x^{a-1}y^{b-1}z^{c-1}D(x + y + z)\ dz$$

$$= \frac{\Gamma(a)\Gamma(b)\Gamma(c)}{\Gamma(a+b+c)} \int_{\lambda}^{\mu} e^{-ky}y^{a+b+c-1}D(y)\ dy \ .$$

Since the actual region of integration on the left is also a bounded point set, one can allow $k \longrightarrow 0$, and obtain

$$\iiint\limits_{\lambda<x+y+z<\mu} x^{a-1}y^{b-1}z^{c-1}D(x + y + z)\ dx\ dy\ dz$$

$$= \frac{\Gamma(a)\Gamma(b)\Gamma(c)}{\Gamma(a+b+c)} \int_{\lambda}^{\mu} y^{a+b+c-1}D(y)\ dy \ .$$

CHAPTER IV

STIELTJES INTEGRALS

§18. The Function Class $\mathfrak{P}$

1. We recall the concept of the Stieltjes Integral [65]. Let $\chi(\alpha)$ be a continuous function and $\psi(\alpha)$ be a monotonically increasing function in a finite interval $a \leq \alpha \leq b$. The last function need not be continuous, but it should have a well determined value at each point. Take any arbitrary set of numbers

$$a = \alpha_0 < \alpha_1 < \alpha_2 < \cdots < \alpha_{n-1} < \alpha_n = b \quad ,$$

and form the sum

$$\sum_{\nu=0}^{n-1} \chi(\beta_\nu)\Big[\psi(\alpha_{\nu+1}) - \psi(\alpha_\nu)\Big], \qquad \alpha_\nu \leqq \beta_\nu \leqq \alpha_{\nu+1}.$$

If the interval $(\alpha_\nu, \alpha_{\nu+1})$ becomes sufficiently small, then this sum differs from a well determined limit by an arbitrarily small value. One denotes the limit by

$$\int_a^b \chi(\alpha)d\psi(\alpha) \tag{1}$$

and calls it a Stieltjes integral. If $\psi(\alpha)$ is differentiable, then

$$\int_a^b \chi(\alpha)d\psi(\alpha) = \int_a^b \chi(\alpha)\psi'(\alpha)\, d\alpha \ , \tag{2}$$

and if $\psi_1(\alpha)$ and $\psi_2(\alpha)$ differ only by a constant, then

$$\int_a^b \chi(\alpha)d\psi_1(\alpha) = \int_a^b \chi(\alpha)d\psi_2(\alpha) \ . \tag{3}$$

The usual rules of calculation apply; in particular

$$\int_a^b d\psi(\alpha) = \psi(b) - \psi(a) ,$$

$$(4) \qquad \left|\int_a^b \chi(\alpha)d\psi(\alpha)\right| \leq \int_a^b |\chi(\alpha)|d\psi(\alpha) \leq \text{Max } |\chi(\alpha)| \cdot [\psi(b) - \psi(a)] .$$

If $\chi(\alpha)$ is differentiable, then one obtains by partial integration

$$(5) \qquad \int_a^b \chi(\alpha)d\psi(\alpha) = \chi(b)\psi(b) - \chi(a)\psi(a) - \int_a^b \chi'(\alpha)\psi(\alpha)\, d\alpha .$$

If $\chi(\xi, \alpha)$ is continuous in the rectangle $a \leq \alpha \leq b$, $\xi_0 \leq \xi \leq \xi_1$, then, as can be verified without difficulty, the following rule is valid

$$(6) \qquad \int_{\xi_0}^{\xi_1} d\xi \int_a^b \chi(\xi, \alpha)d\psi(\alpha) = \int_a^b \left[\int_{\xi_0}^{\xi_1} \chi(\xi, \alpha)d\xi\right] d\psi(\alpha).$$

It states that the order of integration with respect to α and ξ may be interchanged.

2. In addition, we define

$$(7) \qquad \int_a \chi(\alpha)d\psi(\alpha) = \lim_{b \to \infty} \int_a^b \chi(\alpha)d\psi(\alpha) ,$$

insofar as the right hand limit exists, and analogously for the lower limit $-\infty$. The same rules for calculating proper integrals comes over to improper ones, and the partial integration formula is valid provided $\chi(\alpha)\psi(\alpha)$ approaches a limit as $\alpha \to \infty$ or $\alpha \to -\infty$. It is sufficient for the existence of the integral (7), that $\chi(\alpha)$ and $\psi(\alpha)$ be bounded. For example, if $\psi(\alpha)$ is monotonic and bounded, then the limit

$$\psi(\infty) = \lim_{\alpha \to \infty} \psi(\alpha)$$

exists, and by (4), for $B > A$,

$$(8) \qquad \left|\int_A^B \chi(\alpha)d\psi(\alpha)\right| \leq G[\psi(\infty) - \psi(A)] ,$$

where G denotes a bound of $\chi(\alpha)$. Afterwards B can also be replaced in (8) by $+\infty$.

3. If two functions $\psi_1(\alpha)$ and $\psi_2(\alpha)$ defined in $[-\infty, \infty]$, deviate from one another only at points of discontinuities, then if the integral

$$\int \chi(\alpha)\, d\psi(\alpha) \tag{9}$$

converges for $\psi(\alpha) = \psi_1(\alpha)$, it also converges for $\psi(\alpha) = \psi_2(\alpha)$, and the value in both cases is the same. We can therefore always assume for the second function $\psi(\alpha)$ in the integral (9) that its value is normalized at each point by the relation

$$\psi(\alpha) = \frac{1}{2}\,[\psi(\alpha + 0) + \psi(\alpha - 0)] \quad .$$

4. <u>By a distribution function, we mean a function which is defined in</u> $[-\infty, \infty]$, <u>is bounded and monotonically increasing, and for which</u>

$$V(\alpha) = \frac{1}{2}\,[V(\alpha + 0) + V(\alpha - 0)]$$

everywhere.[1] We call two distributions <u>equivalent</u> and write

$$V_1(\alpha) \asymp V_2(\alpha) \quad ,$$

if these functions differ only by a constant

$$V_1(\alpha) = V_2(\alpha) + c \quad .$$

We shall frequently make use of the inequality

$$\left|\int_b e(\alpha x)\, dV(\alpha)\right| \leq V(\infty) - V(b), \quad \left|\int^a e(x\alpha)\, dV(\alpha)\right| \leq V(a) - V(-\infty) \quad . \tag{10}$$

For each distribution function $V(\alpha)$, the integral

$$f(x) = \int e(x\alpha)\, dV(\alpha) \tag{11}$$

exists for all x. Two equivalent distribution functions $V(\alpha)$ yield the

---

[1] On the other hand we do not make the usual assumption

$$V(\infty) - V(-\infty) = 1$$

because we cannot use it.

same function $f(x)$. <u>We denote the totality of the thus defined functions (11) as the function class</u> $\mathfrak{P}$. We shall give, in §21, a "direct" characterization of these functions.

If f_1 and f_2 are two functions of $\mathfrak{P}$, and c_1 and c_2 positive constants, then $c_1f_1 + c_2f_2$ is also a function of $\mathfrak{P}$. To $\mathfrak{P}$ belongs each function of $\mathfrak{F}_0$ whose transform $E(\alpha)$ is positive (more exactly: not negative), and absolutely integrable. In fact, we only have to set

$$V(\alpha) = \int^{\alpha} E(\beta)\, d\beta \quad .$$

Each function of $\mathfrak{P}$ is bounded

$$|f(x)| \leq f(0) = V(\infty) - V(-\infty) \quad , \tag{12}$$

and "hermitian", i.e.,

$$\overline{f(-x)} = f(x) \quad . \tag{13}$$

If $V(\alpha)$, in a certain normalization of the additive constants, is odd, then

$$\int e(x\alpha)\, dV(\alpha) = \int \cos\, (x\alpha)\, dV(\alpha) \quad .$$

Therefore $f(x)$ is a real even function, and hence

$$f(0) = 2V(\infty) \tag{14}$$

Conversely, if $f(x)$ is real, then by the formula (18) proved below, one finds that in the normalization $V(0) = 0$, the function $V(\alpha)$ is odd, and therefore, in particular, satisfies (14).

5. If one sets

$$f_n(x) = \int_{-n}^{n} e(x\alpha)\, dV(\alpha) \quad , \tag{15}$$

then

$$|f(x) - f_n(x)| \leq [V(\infty) - V(n)] + [V(-n) - V(-\infty)] \quad .$$

Hence in $[-\infty, \infty]$, the sequence (15) converges uniformly to $f(x)$ as $n \longrightarrow \infty$. Moreover, since, as is easily seen, $f_n(x)$ is uniformly continuous, therefore $f(x)$ is also uniformly continuous.

6. Let $g(\xi)$ be a continuous function of $\mathfrak{F}_0$. Set

$$R(\alpha, \rho) = \int_{-\rho}^{\rho} g(\xi)e(\xi\alpha)\, d\xi \quad .$$

Then by (6)

$$\int_{-\rho}^{\rho} g(\xi)f_n(\xi)\, d\xi = \int_{-n}^{n} R(\alpha, \rho)\, dV(\alpha)$$

is valid for the function (15). Since $f_n(x)$ converges uniformly to $f(x)$ and since $R(\alpha, \rho)$ is bounded

$$|R(\alpha, \rho)| \leqq \int |g(\xi)|\, d\xi = c \quad ,$$

it is permissible to let n go to ∞, and

$$\int_{-\rho}^{\rho} g(\xi)f(\xi)\, d\xi = \int R(\alpha, \rho)\, dV(\alpha) \quad .$$

As $\rho \longrightarrow \infty$, $R(\alpha, \rho)$ converges uniformly to

$$\int g(\xi)e(\xi\alpha)\, d\xi$$

in each finite α-interval. For each $a > 0$, one easily finds, therefore

$$\int_{-a}^{a} R(\alpha, \rho)\, dV(\alpha) \longrightarrow \int_{-a}^{a} \left[\int g(\xi)e(\xi\alpha)\, d\xi\right] dV(\alpha) \quad .$$

On the other hand, it follows from (8) that

$$\left| \int^{-a} + \int_{a} R(\alpha, \rho)\, dV(\alpha) \right| \leqq C\left\{[V(\infty) - V(a)] + [V(-a) - V(-\infty)]\right\} \quad ,$$

and the right side becomes arbitrarily small for sufficiently large a. Hence it follows from all this that

$$(16) \qquad \int g(\xi)f(\xi)\, d\xi = \int\left[\int g(\xi)e(\xi\alpha)\, d\xi\right] dV(\alpha) \quad .$$

Now replacing $g(\xi)$ by $g(x - \xi)$ and denoting the transform of $g(\xi)$ by $\Phi(\alpha)$,

$$g(\xi) \sim \int e(\xi\alpha)\Phi(\alpha)\, d\alpha \quad ,$$

we obtain the Faltung rule

$$\frac{1}{2\pi}\int g(x - \xi)f(\xi)\,d\xi = \int e(x\alpha)\Phi(\alpha)\,dV(\alpha) \quad . \tag{17}$$

7. THEOREM 17 [66]. Formula (11) has a converse and it is

$$V(\rho) - V(0) = \lim_{\omega \to \infty} \frac{1}{2\pi} \int_{-\omega}^{\omega} f(x)\,\frac{e(-\rho x) - 1}{-ix}\,dx \quad . \tag{18}$$

PROOF. For fixed ρ, set

$$Q(\alpha) = V(\alpha + \rho) - V(\alpha) \quad . \tag{19}$$

Since $V(\alpha)$ approaches a limit as $\alpha \longrightarrow \pm\infty$, therefore

$$Q(\alpha) \longrightarrow 0 \quad \text{as} \quad \alpha \longrightarrow \pm\infty \quad . \tag{20}$$

Furthermore $Q(\alpha)$ is absolutely integrable, since the integral

$$\int_a^b |Q(\alpha)|\,d\alpha$$

has the value

$$\int_a^b Q(\alpha)\,d\alpha = \int_0^\rho [V(b + \alpha) - V(a + \alpha)]\,d\alpha$$

and, therefore, converges to zero provided a and b both go to $+\infty$ (or $-\infty$).

We have,

$$[e(-\rho x)-1]f(x) = \int e[(\alpha-\rho)x]\,dV(\alpha) - \int e(\alpha x)\,dV(\alpha)$$

$$= \int e(\alpha x)\,dV(\alpha+\rho) - \int e(\alpha x)\,dV(\alpha)$$

$$= \lim_{n \to \infty}\left[e(nx)Q(n) - e(-nx)Q(-n) - \int_{-n}^{n} ixe(\alpha x)Q(\alpha)\,d\alpha \right],$$

and therefore

$$f_\rho(x) \equiv f(x)\,\frac{e(-\rho x) - 1}{-ix} = \int e(\alpha x)Q(\alpha)\,d\alpha \quad . \tag{21}$$

Now using Theorem 11b, we have

$$Q(0) = V(\rho) - V(0) = \lim_{\omega\to\infty} \frac{1}{2\pi}\int_{-\omega}^{\omega} f_\rho(x) \; ,$$

Q.E.D.

We conclude immediately from Theorem 17, the important

THEOREM 18. Two functions of $\mathfrak{P}$ are (then and) only then identical, if their distribution functions are equivalent.

8. The following theorem is significant.

THEOREM 19. Let

$$f(x) = \int e(x\alpha)\,dV(\alpha), \quad \text{and} \quad g(x) = \int e(x\alpha)\,dW(\alpha)$$

be functions of $\mathfrak{P}$, and let $V(\alpha)$ be continuous. The Stieltjes integral

$$U(\beta) = \int V(\beta - \alpha)\,dW(\alpha) \tag{22}$$

is a continuous distribution function, and the corresponding function $F(x)$ of $\mathfrak{P}$ has the value $f(x)g(x)$.

PROOF. By 2., the integral (22) is convergent and bounded for all β. If $\beta_2 > \beta_1$, then $V(\beta_2 - \alpha) \geq V(\beta_1 - \alpha)$ and therefore $U(\beta)$ is monotonically increasing. The continuity of $U(\beta)$ follows easily from the combination of facts that $V(\gamma)$ is continuous and that the limit

$$U(\beta) = \lim_{a\to\infty}\int_{-a}^{a} V(\beta - \alpha)\,dW(\alpha)$$

takes place uniformly in each β-interval.

We now consider alongside the functions (19) and (21), the corresponding functions

$$P(\beta) = U(\beta + \rho) - U(\beta) = \int Q(\beta - \alpha)\, dW(\alpha) \quad , \tag{23}$$

$$F_\rho(x) \equiv F(x)\, \frac{e(-\rho x) - 1}{- ix} = \int e(\beta x) P(\beta)\, d\beta \quad . \tag{24}$$

If the integral (23) is substituted for $P(\beta)$ in (24), and the order of integration with respect to α and β then interchanged (an interchange permissible by the same considerations which were employed in (16), and indeed now justified because of the continuity and absolute integrability of $Q(\gamma)$), we obtain

$$F_\rho(x) = \int \left[\int e(x\beta) Q(\beta - \alpha)\, d\beta \right] dW(\alpha)$$

$$= \int e(x\alpha) f_\rho(x)\, dW(\alpha) = f_\rho(x) g(x) \quad .$$

Returning to the definition of f_ρ and F_ρ, there results

$$F(x) = f(x) g(x) \quad . \tag{25}$$

9. If the assumption that $V(\alpha)$ is continuous is dropped, then Theorem 19 and its proof are still valid, but an expansion of the concept of the Stieltjes integral (1) is then required to include the case where $\chi(\alpha)$ is also discontinuous. Although we shall not make this expansion, we shall in §19, 5 prove by other means that the product of any two functions of $\mathfrak{P}$ is again a function of $\mathfrak{P}$.

10. The definition and properties of the Stieltjes integral (1) carries over immediately to the case where $\psi(\alpha)$ is not monotonic, but more generally is of bounded variation. We could generalize the class of functions $\mathfrak{P}$ to the class of integrals

$$f(x) = \int e(x\alpha)\, d\Phi(\alpha)$$

in which the underlying function $\Phi(\alpha)$ can be interpreted as the difference of two distribution functions. Many properties of the $\mathfrak{P}$ functions would be preserved, as for example Theorems 17 - 19. However the investigations in the paragraphs which follow are substantially based on the $\mathfrak{P}$ functions, so that we shall hereafter concentrate on such functions only.

§19. Sequences of Functions of $\mathfrak{P}$

1. Let it be known of a sequence of distribution functions

$$V_1(\alpha),\ V_2(\alpha),\ V_3(\alpha),\ \ldots \tag{1}$$

that it is convergent on a point set

$$\alpha_1,\ \alpha_2,\ \alpha_3,\ \ldots \tag{2}$$

which is everywhere dense,

$$\lim_{n\to\infty} V_n(\alpha_\nu) = \tau_\nu, \qquad (\nu = 1, 2, 3, \ldots),$$

and that it is uniformly bounded

$$|V_n(\alpha)| \leqq M, \qquad (n = 1, 2, 3, \ldots). \tag{3}$$

The totality of values of τ_ν is of itself also monotonically increasing, i.e., if $\alpha_\mu < \alpha_\nu$, then $\tau_\mu \leqq \tau_\nu$. Since the α_ν are by assumption everywhere dense, the τ_ν determine uniquely, as one finds without difficulty, a distribution function $V(\alpha)$ which has the following property. For each α, $V(\alpha + 0)$ is the lower bound of the numbers τ_ν for such ν for which α_ν is larger than α; and correspondingly $V(\alpha - 0)$ is the upper bound of the numbers τ_ν for such ν for which α_ν is smaller than α. It is also evident that

$$|V(\alpha)| \leqq M . \tag{4}$$

We shall show that the relation

$$V(\alpha) = \lim_{n\to\infty} V_n(\alpha) \tag{5}$$

holds at every point of continuity of $V(\alpha)$. Pick an arbitrary α. For $\alpha < \alpha_\nu$

$$V_n(\alpha) \leqq V_n(\alpha_\nu) ;$$

therefore

$$\overline{\lim_{n\to\infty}}\, V_n(\alpha) \leqq \overline{\lim_{n\to\infty}}\, V_n(\alpha_\nu) = \tau_\nu .$$

The left side of the inequality is independent of ν; on the right side any τ_ν can stand for which $\alpha < \alpha_\nu$. But since $V(\alpha + 0)$ is the lower bound of such τ_ν, we have

$$\overline{\lim_{n\to\infty}}\, V_n(\alpha) \leqq V(\alpha + 0) .$$

Similarly, one obtains

$$V(\alpha - 0) \leq \underline{\lim_{n \to \infty}} V(\alpha) .$$

Hence

$$V(\alpha - 0) \leq \underline{\lim_{n \to \infty}} V_n(\alpha) \leq \overline{\lim_{n \to \infty}} V_n(\alpha) \leq V(\alpha + 0) .$$

But at a continuity point α, $V(\alpha - 0) = V(\alpha + 0)$. Hence all four terms of the inequality are equal to one another, a result which is synonymous with (5).

<u>If we are given distribution functions</u> $V_n(\alpha)$, $n = 1, 2, \ldots$, <u>and</u> $V(\alpha)$, <u>and if relation (5) holds at all points at which</u> $V(\alpha)$ <u>is continuous, then the sequence</u> $V_n(\alpha)$ <u>is said to converge to</u> $V(\alpha)$. Since two distribution functions which agree as their common points of continuity, also agree at all other points, it follows that the limit function of a convergent sequence $V_n(\alpha)$ is unique.

THEOREM 20. If the distribution functions $V_n(\alpha)$, $n = 1, 2, 3, \ldots$, converge to a distribution function $V_0(\alpha)$, and if

$$\lim_{n \to \infty} V_n(\pm\infty) = V_0(\pm\infty) \tag{6}$$

then the functions $f_n(x)$ of $\mathfrak{P}$ belonging to them converge at each point x to the function $f_0(x)$ of $\mathfrak{P}$ which belongs to $V_0(\alpha)$ [67].

PROOF. For fixed $a > 0$ and for $n = 0, 1, 2, \ldots$, set

$$f_n(x) = \int_{-a}^{a} e(x\alpha)\, dV_n(\alpha) + \int^{-a} + \int_{a} = f_n(x, a) + Q_n(x, a) + P_n(x, a) ,$$

where a is selected in such a way that $V_0(\alpha)$ is continuous at $\alpha = +a$ and $\alpha = -a$. Therefore

$$\lim_{n \to \infty} V_n(a) = V_0(a) , \qquad \lim_{n \to \infty} V_n(-a) = V_0(-a) . \tag{7}$$

Hence as $n \longrightarrow \infty$

$$\overline{\lim} |f_0(x) - f_n(x)| \leq A(x, a) + B(x, a) + C(x, a) ,$$

where

$$A(x, a) = \overline{\lim} |f_o(x, a) - f_n(x, a)|, \; B(x, a) = \overline{\lim} |P_o - P_n|,$$

$$C(x, a) = \overline{\lim} |Q_o - Q_n| \quad .$$

On the one hand, we have

$$f_n(x, a) = e(xa)V_n(a) - e(-xa)V_n(-a) - ix \int_{-a}^{a} e(x\alpha)V_n(\alpha)\, d\alpha \quad ,$$

and because of (7) and the convergence of $V_n(\alpha)$ to $V_o(\alpha)$, it follows that (cf. Appendix 7, 8)

$$A(x, a) = 0 \quad .$$

On the other hand

$$|B(x, a)| \leqq |P_o| + \overline{\lim} |P_n| \leqq V_o(\infty) - V_o(a) + \overline{\lim} [V_n(\infty) - V_n(a)] \quad ,$$

and because of (6) and (7), it follows that

$$|B(x, a)| \leqq 2 [V_o(\infty) - V_o(a)] \quad .$$

Therefore, for a given $\epsilon > 0$, we can choose an a so large that $|B(x, a)| \leqq \epsilon$. A corresponding assertion is valid for $C(x, a)$. Hence for each ϵ

$$\overline{\lim} |f_o(x) - f_n(x)| \leqq 2\epsilon \quad .$$

From this it follows that

$$f_o(x) = \lim_{n \to \infty} f_n(x)$$

Q.E.D.

2. By assumption (6), it follows very easily that the $V_n(\alpha)$ are uniformly bounded

$$(8) \qquad |V_n(\alpha)| \leqq M \, , \qquad (n = 1, 2, 3, \ldots) \, .$$

But one must not replace assumption (6) in Theorem 20, by the weaker assumption (8), as can be seen from the following counter example: $V_n(\alpha) = 0$ for $\alpha < n$ and $= 1$, for $\alpha > n$. Hence $f_n(x) = e(nx)$, $V_o(\alpha) \equiv 0$, and therefore $f_o(x) = 0$. Note that in this example the sequence $f_n(x)$ not only fails to converge to $f_o(x)$ but it fails to converge at all. Now, this is not an accidental phenomenon, and the following assertion can be made in fact.

3. If the $V_n(\alpha)$ are uniformly bounded, and converge to $V_o(\alpha)$, and if the $f_n(x)$ converge to a continuous function $F(x)$, then

$$F(x) = f_o(x) \quad .$$

We proceed to prove this statement. By §18, 7, we have for fixed ρ,

$$f_n(x)\, \frac{e(-\rho x) - 1}{-ix} = \int e(x\alpha)\, dW_n(\alpha), \qquad (n = 0, 1, 2, \dots),$$

where the functions

$$W_n(\alpha) = \int_0^\alpha [V_n(\beta + \rho) - V_n(\beta)]\, d\beta$$

are again distribution functions. By "iteration", one finds for the functions

$$g_n(x) = f_n(x) \left(\frac{e(-\rho x) - 1}{-ix} \right)^2 \quad ,$$

the relation

$$g_n(x) = \int e(x\alpha) E_n(\alpha)\, d\alpha \quad , \tag{9}$$

where the functions

$$E_n(\alpha) = W_n(\alpha + \rho) - W_n(\alpha)$$

are absolutely integrable. By (8), it follows that, cf. §18, (12),

$$|f_n(x)| \leq 2M \quad . \tag{10}$$

Hence the functions $g_n(x)$ belong to $\mathfrak{F}_o$, and since the $E_n(\alpha)$ are also absolutely integrable, (9) states, cf. §13, 9, that $E_n(\alpha)$ is the transform of $g_n(x)$. Because of the convergence of $f_n(x)$ to $F(x)$, the convergence of $g_n(x)$ to

$$G(x) = F(x) \left(\frac{e(-\rho x) - 1}{-ix} \right)^2$$

follows. Since by (10), the $g_n(x)$ are uniformly bounded, therefore for $a > 0$ (cf. Appendix 10)

$$\int_{-a}^{a} |g_n(x) - G(x)|\, dx \longrightarrow 0 \quad .$$

And because

$$|g_n(x)| \leqq \frac{4M}{x^2} , \qquad |G(x)| \leqq \frac{4M}{x^2} ,$$

one also easily finds that

$$\int |g_n(x) - G(x)| \, dx \longrightarrow 0 \quad .$$

Hence for the transform $\Phi(\alpha)$ of $G(x)$, we have

$$\Phi(\alpha) = \lim_{n \to \infty} E_n(\alpha) \quad .$$

But by the convergence of $V_n(\alpha)$ to $V_o(\alpha)$ and its uniform boundedness, it follows that

$$E_o(\alpha) = \lim_{n \to \infty} E_n(\alpha) \quad .$$

From this, we obtain successively: $E_o(\alpha) = \Phi(\alpha)$, $g_o(x) = G(x)$, $f_o(x) = F(x)$.

4. THEOREM 21 [68]. If the functions $f_n(x)$ of $\mathfrak{P}$, $n = 1, 2, 3, \ldots$, are uniformly bounded

$$|f_n(x)| \leqq M \quad , \tag{11}$$

and converge, for all x, to a continuous limit function $f(x)$, then the following assertions are valid:

1) The function $f(x)$ likewise belongs to $\mathfrak{P}$. By (11), one can normalize, after §18, (12), the additive constants in the distribution functions $V_n(\alpha)$ of $f_n(x)$ in such a way that the $V_n(\alpha)$ are also uniformly bounded

$$|V_n(\alpha)| \leqq N \quad . \tag{12}$$

2) There is at least one such normalization in which the $V_n(\alpha)$ are convergent, and

3) If in any such normalization the $V_n(\alpha)$ converge, then its limit function is equivalent to the distribution function of $f(x)$.

PROOF of 1). Let the $V_n(\alpha)$ be given in a normalization (12). We take on the α-axis any countable set of everywhere dense numbers (2), and determine by the well known diagonal argument a subsequence

(13)
$$V_{n_1}(\alpha),\ V_{n_2}(\alpha),\ V_{n_3}(\alpha),\ \dots$$

which converges at all these points. By 1, there is a distribution function $V(\alpha)$ such that

(14)
$$V(\alpha) = \lim_{k\to\infty} V_{n_k}(\alpha) \quad .$$

Since by assumption, the set $f_{n_k}(x)$ as a subset of $f_n(x)$, converges to $f(x)$, it follows by the proof in 3 that $f(x)$ is a function of $\mathfrak{P}$, and $V(\alpha)$ is its distribution function.

PROOF of 2). We take a fixed point α_0 at which $V(\alpha)$ is continuous. We can add to the $V_n(\alpha)$ constants a_n so that

$$V(\alpha_0) = \lim_{n\to\infty} V_n(\alpha_0)$$

and the new functions $V_n(\alpha)$ continue to be uniformly bounded. We claim that they are also convergent to $V(\alpha)$. If not, there would be a point $\alpha = \beta$ at which $V(\alpha)$ is continuous and at which the set $V_n(\beta)$ would not converge to $V(\beta)$. We could then, using the arguments employed in the proof of 1), specify a subsequence

(16)
$$V_{m_1}(\alpha),\ V_{m_2}(\alpha),\ V_{m_3}(\alpha),\ \dots$$

which would converge to a distribution function $U(\alpha)$, but without converging to $V(\beta)$ at the point $\alpha = \beta$. Again referring to the proof of 1), since $U(\alpha)$ were a distribution function of $f(x)$, there would be a constant c such that

(17)
$$U(\alpha) = V(\alpha) + c$$

therefore, $U(\alpha)$ would be continuous at $\alpha = \alpha_0$, and hence

(18)
$$U(\alpha_0) = \lim_{k\to\infty} V_{m_k}(\alpha_0) \quad .$$

By (15) and (18), $U(\alpha_0) = V(\alpha_0)$; on the other hand by (17)

$$U(\alpha_0) = V(\alpha_0) + c \quad .$$

Therefore $c = 0$, i.e., $U(\alpha) = V(\alpha)$. Furthermore, $U(\alpha)$ is also continuous as $\alpha = \beta$, and thus the sequence (16) is convergent to $U(\beta)$ at $\alpha = \beta$. But because of $U(\beta) = V(\beta)$ this is a contradiction to the assumption that the subsequence (16) does not converge to $V(\beta)$ at $\alpha = \beta$.

PROOF of 3). This proof follows immediately from 3.

5. THEOREM 22. The product of two functions of $\mathfrak{P}$ is a function of $\mathfrak{P}$.

PROOF. Let $f(x)$ and $g(x)$ be any two functions of $\mathfrak{P}$. If $\varphi(x)$ is some function of $\mathfrak{P}$ which has a continuous distribution function, then by Theorem 19, $\varphi(x)f(x)$ is also a function of $\mathfrak{P}$ with continuous distribution function. Repeated application of Theorem 19 results in

$$\varphi(x)f(x)g(x)$$

also belonging to $\mathfrak{P}$. Now by §15, (4)

$$e^{-\frac{x^2}{4n^2}} = \int e(x\alpha)\, dV(\alpha) \quad ,$$

where

$$V(\alpha) = \frac{n}{\sqrt{\pi}} \int_0^{\alpha} e^{-\frac{\beta^2}{n^2}}\, d\beta \quad ,$$

and therefore the function

$$e^{-\frac{x^2}{4n^2}} f(x)g(x)$$

is contained in $\mathfrak{P}$. But now consider this function for variable values of n. As $n \longrightarrow \infty$, this set converges to $f(x)g(x)$. It is moreover uniformly bounded; hence by Theorem 21, $f(x)g(x)$ is also a function of $\mathfrak{P}$.

§20. Positive-Definite Functions

1. We call a function $f(x)$ positive-definite [69] if 1) it is continuous in the finite region, and is bounded in $[-\infty, \infty]$, 2) it is "hermitian", i.e.,

$$\overline{f(-x)} = f(x) \tag{1}$$

and 3) it satisfies the following conditions: For any points $x_1, x_2, \ldots, x_m$, $(m = 1, 2, 3, \ldots)$, and any numbers $\rho_1, \rho_2, \ldots, \rho_m$

$$\sum_{\mu=1}^{m} \sum_{\nu=1}^{m} f(x_\mu - x_\nu)\rho_\mu\overline{\rho_\nu} \geqq 0 \quad . \tag{2}$$

2. Each function of $\mathfrak{P}$ is positive-definite. Property 1) is

already known and 2) is easily verified. In regard to 3), the left side of (2) has the value

$$\int \mathfrak{P}(\alpha)\, dV(\alpha)$$

where

$$\mathfrak{P}(\alpha) = \sum_{\mu=1}^{m} \sum_{\nu=1}^{m} e([x_\mu - x_\nu]\alpha)\rho_\mu \overline{\rho_\nu} = \left| \sum_{\mu=1}^{m} e(x_\mu \alpha)\rho_\mu \right|^2 \geqq 0 \quad ,$$

and is therefore actually $\geqq 0$.

3. Conversely, we shall now show that each positive-definite function $f(x)$ belongs to $\mathfrak{P}$. We shall first prove this for all positive-definite functions $f(x)$ which happen to belong to $\mathfrak{F}_0$,

$$f(x) \sim \int e(x\alpha)e(\alpha)\, d\alpha \quad .$$

In addition to $f(\xi)$, let $g(\xi)$ be a given continuous function in $[-\infty, \infty]$. For each $A > 0$, the expression

$$\int_{-A}^{A} \int_{-A}^{A} f(x - y)g(-x)\overline{g(-y)}\, dx\, dy$$

is non-negative. In fact, by the definition of the Riemann integral, it is the limit, as $n \longrightarrow \infty$, of

$$h^2 \sum_{\mu=-n}^{n-1} \sum_{\nu=-n}^{n-1} f(\mu h - \nu h)g(-\mu h)g(-\nu h), \qquad \left(h = \frac{A}{n} \right),$$

and because of the assumption of the positive-definiteness of $f(\xi)$, this double sum is $\geqq 0$. If further $g(\xi)$ is also absolutely integrable,

$$g(\xi) \sim \int e(x\alpha)\Gamma(\alpha)\, d\alpha \quad ,$$

then by allowing $A \longrightarrow \infty$, we obtain

$$\int\int f(x - y)g(-x)\overline{g(-y)}\, dx\, dy \geqq 0 \quad . \tag{3}$$

Denoting by $F(\xi)$ the Faltung of the functions $f(\xi)$, $g(\xi)$ and $\overline{g(-\xi)}$, all three of which belong to $\mathfrak{F}_0$, we have

$$(2\pi)^2 F(\xi) = \int\int f(\xi - x - y) g(x) \overline{g(-y)}\, dx\, dy$$

$$= \int\int f(\xi + x - y) g(-x) \overline{g(-y)}\, dx\, dy \quad .$$

The transform of $F(\xi)$ has the value $|\Gamma(\alpha)|^2 E(\alpha)$. If $\Gamma(\alpha)$ is in addition also absolutely integrable, then since $F(\xi)$ is a continuous function, we have by Theorem 15

$$F(\xi) = \int e(\xi\alpha) |\Gamma(\alpha)|^2 E(\alpha)\, d\alpha \quad ,$$

and in particular

$$F(0) = \int |\Gamma(\alpha)|^2 E(\alpha)\, d\alpha \quad .$$

But since, by (3), $F(0) \geqq 0$, we conclude as follows. The transform $E(\alpha)$ of $f(x)$ is such that for each absolutely integrable function $\Gamma(\alpha)$ of $\mathfrak{T}_0$

$$(4) \qquad \int |\Gamma(\alpha)|^2 E(\alpha)\, d\alpha \geqq 0 \quad .$$

The transform of $\overline{f(-x)}$ is $\overline{E(\alpha)}$. Since $\overline{f(-x)} = f(x)$, therefore $\overline{E(\alpha)} = E(\alpha)$, and hence $E(\alpha)$ is real. Moreover, it follows by (4) that $E(\alpha) \geqq 0$ for all x. If it were not so, there would be an interval $\alpha_0 > \alpha > \alpha_1$ in which $E(\alpha)$ were < 0. A function $\gamma(\alpha)$ which is two times differentiable and which vanishes outside of (α_0, α_1) is a function of $\mathfrak{T}_0$. But for such a function

$$\int |\gamma(\alpha)|^2 E(\alpha)\, d\alpha < 0$$

in contradiction to (4). - We shall now show that $E(\alpha)$ is absolutely integrable. Consider the function

$$f_n(x) = \frac{1}{\pi n} \int f(\xi) \left(\frac{\sin n(x-\xi)}{x - \xi} \right)^2 d\xi = \int_{-2n}^{2n} e(x\alpha) \left(1 - \frac{|\alpha|}{2n} \right) E(\alpha)\, d\alpha \quad .$$

Since $f(\xi)$ is bounded, there is a constant A such that $|f_n(0)| \leqq A$. Hence, because $E(\alpha) \geqq 0$, it follows for $2n > a$, that

$$\int_{-a}^{a} \left(1 - \frac{|\alpha|}{2n}\right) E(\alpha)\, d\alpha \leqq A \quad ,$$

and by allowing $n \longrightarrow \infty$, that

$$\int_{-a}^{a} E(\alpha)\, d\alpha \leqq A \quad .$$

One now replaces a by ∞, Q.E.D. But if the transform $E(\alpha)$ of $f(x)$ is positive and absolutely integrable, then $f(x)$ is a function of $\mathfrak{P}$, cf. §18, 4.

4. Let $f(x)$ be any positive-definite function, and let $\gamma(\alpha)$ be positive and absolutely integrable. Consider the product

$$F(x) = f(x) \int e(x\alpha)\gamma(\alpha)\, d\alpha \quad .$$

$F(x)$ is likewise positive-definite. In fact, the boundedness and continuity of $F(x)$ and the relation $\overline{F(-x)} = F(x)$ are at once verified. Moreover

$$\sum_{\mu,\nu=1}^{n} F(x_\mu - x_\nu)\rho_\mu\overline{\rho_\nu} = \int \left[\sum_{\mu,\nu} f(x_\mu - x_\nu) e(x_\mu\alpha)\rho_\mu \cdot \overline{e(x_\nu\alpha)\rho_\nu} \right] \gamma(\alpha)\, d\alpha \quad ,$$

and since the cornered bracket is $\geqq 0$, the whole expression is therefore also $\geqq 0$. In particular, the functions

$$f_n(x) = f(x)e^{-\frac{x^2}{n}}, \qquad (n > 0) \quad ,$$

belong to the functions $F(x)$. These functions are therefore positive-definite, and since they belong to $\mathfrak{F}_0$, they are, as already shown, functions of $\mathfrak{P}$. Furthermore, they are uniformly bounded, and converge to $f(x)$ as $n \longrightarrow \infty$. By Theorem 21, therefore, $f(x)$ is also a function of $\mathfrak{P}$. We have thus proved

THEOREM 23 [69]. In order that a function belong to class $\mathfrak{P}$, it is necessary and sufficient that it be positive-definite.

5. One can state exactly for which ρ the functions

$$(5) \qquad f_\rho(x) = e^{-|x|^\rho}, \qquad (0 < \rho < \infty),$$

belong to $\mathfrak{P}$, and for which they do not. For $\rho = 1$ and $\rho = 2$, the transforms of (5) are positive and absolutely integrable. Hence the functions themselves are in $\mathfrak{P}$. We shall show below that they will also belong to $\mathfrak{P}$ if $0 < \rho < 1$ [70], but will not if $2 < \rho < \infty$ [71]. They are likewise positive-definite if $1 < \rho < 2$, but the proof of this assertion is somewhat troublesome and we shall, therefore, not give it [72].

Let us denote the transform of (5) by $E_\rho(\alpha)$. Therefore

$$\pi E_\rho(\alpha) = \frac{1}{2}\int e(-x\alpha) f_\rho(x)\,dx = \int_0 \cos(x\alpha) \cdot e^{-x^\rho}\,dx .$$

We obtain by partial integration for $0 < \rho < 1$,

$$\pi\,\alpha\,E_\rho(\alpha) = \int_0 \sin(x\alpha) \cdot \left[\rho x^{\rho-1} e^{-x^\rho}\right] dx .$$

The factor of $\sin x\alpha$ is positive and monotonically decreasing. Since

$$\int_0 \sin(x\alpha) \cdot g(x)\,dx$$

$$= \sum_{n=0}^{\infty} \int_0^{\pi/\alpha} \sin(x\alpha) \cdot \left[g\left(\frac{2n\pi}{\alpha} + x\right) - g\left(\frac{(2n+1)\pi}{\alpha} + x\right) \right] dx ,$$

it follows that $E_\rho(\alpha)$ is non-negative for $\alpha > 0$. As in 3., the absolute integrability of $E_\rho(\alpha)$ now follows because of the boundedness and continuity of $f(x)$. Hence $f_\rho(x)$ is a function of $\mathfrak{P}$ for $0 < \rho < 1$.

Let $E(\alpha)$ be the transform of a function $f(x)$ of $\mathfrak{F}_0$. If $f(x)$ is p-times differentiable, and if the derivatives up to the pth order also belong to $\mathfrak{F}_0$, then the transform of $f^{(p)}(x)$ is $(i\alpha)^p E(\alpha)$. One recognizes from this the following. If a function $f(x)$ of $\mathfrak{F}_0$ has $2n$ absolutely integrable derivatives and if the function $f^{(2n)}(x)$ is continuous and bounded, then the function $f(x)$ is positive-definite if and only if the function $(-1)^n f^{(2n)}(x)$ so is. For $\rho > 2$ (5) has two derivatives which are absolutely integrable, continuous and bounded. Because $f''_\rho(0) = 0$, $-f''_\rho(x)$ is not a function of $\mathfrak{P}$, since such a function has a value > 0 for $x = 0$, unless it vanishes identically. Hence for $\rho > 2$, (5) is not a function of $\mathfrak{P}$.

§21. Spectral Decomposition of Positive-Definite Functions. An Application to Almost Periodic Functions

1. A distribution function $V(\alpha)$ has at most a countable number of discontinuous points, but these can pile up in any manner on the α-axis. We denote them in some sequence by

$$\lambda_0, \lambda_1, \lambda_2, \dots \tag{1}$$

and the corresponding function jumps by a_ν. Thus

$$a_\nu = V(\lambda_\nu + 0) - V(\lambda_\nu - 0) \quad , \tag{2}$$

and

$$\sum_\nu a_\nu \leqq V(\infty) - V(-\infty) \quad . \tag{3}$$

It is possible to write $V(\alpha)$ as the sum of two distribution functions

$$V(\alpha) = S(\alpha) + D(\alpha) \tag{4}$$

so that $S(\alpha)$ is continuous and $D(\alpha)$ consists only of the jumps of $V(\alpha)$. Indeed, the function $D(\alpha)$ is defined in an unique manner by the property that at each continuity point α of $V(\alpha)$, it is equal to the sum of the jumps of $V(\alpha)$ lying to the left of α,

$$D(\alpha) = \sum_{\lambda_\nu < \alpha} a_\nu \quad . \tag{5}$$

Therefore

$$D(\lambda_\nu + 0) - D(\lambda_\nu - 0) = V(\lambda_\nu + 0) - V(\lambda_\nu - 0) \quad .$$

If $V(\alpha)$ is continuous, then $D(\alpha) \equiv 0$. The other extreme occurs when $S(\alpha) = \text{constant}$.

2. We write the function

$$f(x) = \int e(x\alpha)\, dV(\alpha)$$

in the form

$$f(x) = g(x) + h(x) \quad ,$$

where

$$g(x) = \int e(x\alpha)\, dD(\alpha) \tag{6}$$

and

$$h(x) = \int e(x\alpha)\, dS(\alpha) \quad . \tag{7}$$

The elements $g(x)$ and $h(x)$ are themselves functions of $\mathfrak{P}$. For each continuous and bounded function $\chi(\alpha)$,

$$\int \chi(\alpha)\, dD(\alpha) = \sum_\nu \chi(\lambda_\nu) a_\nu \quad .$$

Hence in particular

$$g(x) = \sum_\nu a_\nu e(\lambda_\nu x) \tag{8}$$

Conversely each exponential series of the form (8) in which the λ_ν are real numbers and the a_ν positive numbers with convergent sums

$$\sum_\nu a_\nu \quad ,$$

is a function of the form (6). $D(\alpha)$ needs only to be determined as follows. If α is different from all λ_ν, then (5) holds; for the remaining points

$$D(\alpha) = \frac{1}{2}\, [D(\alpha + 0) + D(\alpha - 0)] \quad .$$

3. If $S(\alpha)$ is constant outside of a finite interval, then by §18, (6)

$$\frac{1}{2\omega} \int_{-\omega}^{\omega} h(x)\, dx = \int \frac{\sin \alpha\omega}{\alpha\omega}\, dS(\alpha) \quad .$$

If $a < 0$ and $b > 0$, then by §18, (8)

$$\left| \int^{a} + \int_{b} \frac{\sin \alpha\omega}{\alpha\omega}\, dS(\alpha) \right| \leq \frac{S(\infty) - S(b)}{b\omega} + \frac{S(a) - S(-\infty)}{-a\omega} \quad ,$$

and by §18, (4)

$$\left| \int_a^b \frac{\sin \alpha\omega}{\alpha\omega} \, dS(\alpha) \right| \leqq S(b) - S(a) \leqq \epsilon(b, a) \quad ,$$

where $\epsilon(b, a)$, because of the continuity of $S(\alpha)$, can be made arbitrarily small by choosing suitable values of b and a. Hence

$$\lim_{\omega \to \infty} \frac{1}{2\omega} \int_{-\omega}^{\omega} h(x) \, dx = 0 \quad ,$$

or in the notation of §9, 2

(9) $$\underline{\mathfrak{M}}\{h(x)\} = 0 \quad .$$

But if $S(\alpha)$ does not vanish outside of a finite interval, then we consider the functions

$$h_n(x) = \int_{-n}^{n} e(x\alpha) \, dS(\alpha) \quad .$$

By what has just been proved

$$\underline{\mathfrak{M}}\{h_n(x)\} = 0 \quad .$$

If a set of functions, each of which has a "mean value", converges uniformly in $[-\infty, \infty]$, then the limit function also has a "mean value" which can be calculated by taking the limit. Hence (9) is valid for the most general function (7). Since $h(x)e(-\lambda x)$ has the continuous distribution function $S(\alpha + \lambda)$, the following is also valid

(10) $$\underline{\mathfrak{M}}\{h(x)e(-\lambda x)\} = 0 \quad .$$

On the other hand

$$\mathfrak{M}\{e(0x)\} = \underline{\mathfrak{M}}\{1\} = 1$$

$$\mathfrak{M}\{e(\rho x)\} = 0, \qquad \rho \neq 0 \quad .$$

The following assertion is therefore valid for the function

(11) $$g_n(x) = \sum_0^n a_\nu e(\lambda_\nu x) \quad .$$

$\mathfrak{M}\{g_n(x)e(-\lambda x)\} = 0$ for $\lambda \neq \lambda_\nu$, $\nu = 0, 1, \ldots, n$ and $= a_\nu$ for

$\lambda = \lambda_\nu$. Since as $n \longrightarrow \infty$, the functions (11) converge uniformly to (8) in $[-\infty, \infty]$, we have

$$(12) \qquad \mathfrak{M}\{g(x)e(-\lambda x)\} = \begin{cases} 0 & \text{for } \lambda \neq \lambda_\nu, \ \nu = 0, 1, 2, \ldots \\ a_\nu & \text{for } \lambda = \lambda_\nu \end{cases} .$$

We therefore obtain by (10) and (12)

THEOREM 24. For each function

$$f(x) = \int e(x\alpha)\, dV(\alpha) \quad ,$$

the following relation holds for all real λ

$$\mathfrak{M}\{f(x)e(-\lambda x)\} = V(\lambda + 0) - V(\lambda - 0) .$$

4. If $f(x)$ is positive-definite, then $\overline{f(x)}$ is too. By Theorem 19, the product $h(x)\overline{h(x)}$ is a function of $\mathfrak{P}$, whose distribution function is again continuous. By (9), it therefore follows that

$$(13) \qquad \mathfrak{M}\{|h(x)|^2\} = 0 \quad .$$

5. Let $\varphi(t)$ be a given continuous function, bounded in $[-\infty, \infty]$, with the property that the limit

$$(14) \qquad f(x) = \lim_{T \to \infty} \frac{1}{2T} \int_{-T}^{T} \varphi(t)\overline{\varphi(t - x)}\, dt = \mathfrak{M}_t\{\varphi(t)\overline{\varphi(t - x)}\}$$

exists uniformly in each finite x-interval. One easily finds that $f(x)$ is bounded and continuous, and that for each c

$$(15) \qquad \mathfrak{M}_t\{\varphi(t)\overline{\varphi(t - x)}\} = \mathfrak{M}_t\{\varphi(t + c)\overline{\varphi(t + c - x)}\} \quad .$$

In the last we may put $c = x$. Hence

$$\overline{f(-x)} = \mathfrak{M}_t\{\overline{\varphi(t)}\varphi(t + x)\} = \mathfrak{M}_t\{\varphi(t)\overline{\varphi(t - x)}\} = f(x) \quad .$$

Moreover if in (15), x is replaced by $x_\mu - x_\nu$ and c by x_μ, and we then sum over μ and ν, we obtain

$$\sum_{\mu, \nu=1}^{m} f(x_\mu - x_\nu)\rho_\mu\overline{\rho_\nu} = \mathfrak{M}_t\left\{\left|\sum_{\mu=1}^{m} \varphi(t + x_\mu)\rho_\mu\right|^2\right\} \geq 0 \quad .$$

Hence $f(x)$ is positive-definite.

Setting as before

$$f(x) = g(x) + h(x) \quad , \tag{16}$$

we shall compute

$$a(\lambda) = \mathfrak{M}_u \{g(u)e(-\lambda u)\} = \mathfrak{M}_u \{f(u)e(-\lambda u)\} \quad .$$

Its value is

$$a(\lambda) = \lim_{\omega \to \infty} \frac{1}{2\omega} \int_{-\omega}^{\omega} \left[\lim_{T \to \infty} \frac{1}{2T} \int_{-T}^{T} \varphi(t)\overline{\varphi(t - u)}e(-\lambda u)\, dt \right] du \quad .$$

Since the inner limit in each interval $-\omega \leq u \leq \omega$ is uniform, it is permissible to interchange the limit with respect to t and the integration with respect to u. Therefore

$$a(\lambda) = \lim_{\omega \to \infty} \lim_{T \to \infty} \frac{1}{2T} \int_{-T}^{T} \varphi(t) \left[\frac{1}{2\omega} \int_{-\omega}^{\omega} \overline{\varphi(t - u)}e(-\lambda u)\, du \right] dt \quad . \tag{17}$$

We now make a further assumption in regard to $\varphi(t)$, namely that for each λ, the limit

$$c(\lambda) = \lim_{\omega \to \infty} \frac{1}{2\omega} \int_{-\omega+t}^{\omega+t} \varphi(\xi)e(-\lambda\xi)\, d\xi \equiv \mathfrak{M}\{\varphi(\xi)e(-\lambda\xi)\}$$

exists uniformly with respect to all t in $[-\infty, \infty]$. Passing over to the conjugate-complex, and replacing ξ by $-u$, there results

$$\lim_{\omega \to \infty} \frac{1}{2\omega} \int_{-\omega}^{\omega} \overline{\varphi(t - u)}e(-\lambda u)\, du = e(-\lambda t)\overline{c(\lambda)} \quad .$$

Hence

$$\frac{1}{2\omega} \int_{-\omega}^{\omega} \overline{\varphi(t - u)}e(-\lambda u)\, du = e(-\lambda t)\overline{c(\lambda)} + \epsilon(t, \omega) \quad ,$$

where the error term $\epsilon(t, \omega)$ converges uniformly to zero, as $\omega \to \infty$, for all t in $[-\infty, \infty]$. Substituting this in (17), we have

$$a(\lambda) = \overline{c(\lambda)} \cdot \lim_{T \to \infty} \frac{1}{2T} \int_{-T}^{T} \varphi(t) e(-\lambda t)\, dt = |c(\lambda)|^2 .$$

Those values of λ for which $a(\lambda)$ or $c(\lambda)$ vanishes, are the "particular oscillations" of the function $f(x)$ or $\varphi(t)$. It follows from what has just been proved that both functions have the same "eigenvalues" λ_ν, and that the amplitudes belonging to them $a_\nu = a(\lambda\nu)$ and $c_\nu = c(\lambda\nu)$ are connected by the relation

$$a_\nu = |c_\nu|^2 .$$

Substituting the special value $x = 0$ in (16) and going back to the definition of $f(x)$ and $g(x)$, we obtain

$$\mathfrak{M}\{|\varphi(t)|^2\} = \sum_\nu |c_\nu|^2 + h(0) .$$

Because $h(0) \geqq 0$, the Bessel inequality

$$\sum_\nu |c_\nu|^2 \leqq \mathfrak{M}\{|\varphi(t)|^2\} ,$$

therefore holds for the function $\varphi(t)$. There was actually no need for this lengthy discussion in order to deduce this inequality; it could have been obtained directly with greater simplicity. But our method of deduction enables us to decide for which functions $\varphi(t)$, the Parseval equation

$$\sum_\nu |c_\nu|^2 = \mathfrak{M}\{|\varphi(t)|^2\}$$

is valid. Our criterion states: $h(0) = 0$, and since $h(x)$ is positive-definite, it follows by §18, (12) that

$$h(x) \equiv 0 .$$

An application will be made of this presently.

As is well known, the Parseval equation is valid for periodic functions. More generally, it is valid for the almost periodic functions of H. Bohr, a statement which we shall prove. From the definition of almost periodic functions [73], the following properties, which we shall assume as known, are obtained in a relatively simple manner: 1) A function $\varphi(t)$ which is almost periodic, fulfills all the above conditions; and the function $f(x)$ as defined in (14) is likewise almost periodic.

2) Sums, products, and a uniform limit in $[-\infty, \infty]$ of almost periodic functions are again almost periodic functions. 3) If, for an almost periodic function $F(x)$, the mean value

$$\mathfrak{M}\{|F(x)|^2\}$$

vanishes, then $F(x) \equiv 0$.

We can now reason as follows [74]. Since the function $e(\lambda x)$ is almost periodic for real λ, therefore by 1) and 2), the function

$$h(x) = f(x) - g(x) = f(x) - \sum_{\nu} a_\nu e(\lambda_\nu x)$$

is almost periodic. By (13) the "mean value" of $|h(x)|^2$ is zero. Hence by 3)

$$h(x) \equiv 0 \quad ,$$

Q.E.D.

CHAPTER V

OPERATIONS WITH FUNCTIONS OF THE CLASS $\mathfrak{F}_0$

§22. The Question [75]

1. Let us consider, for any real number δ_σ and any (complex) constant $a_{\rho\sigma}$, the difference-differential equation[1]

$$\text{(A)} \qquad \sum_{\rho=0}^{r} \sum_{\sigma=0}^{s} a_{\rho\sigma} y^{(\rho)}(x + \delta_\sigma) = f(x) \quad .$$

If all δ_σ are equal to each other (i.e., $s = 0$), and say equal to zero, there results the pure differential equation:

$$\text{(B)} \qquad c_r y^{(r)}(x) + c_{r-1} y^{(r-1)}(x) + \dots + c_0 y(x) = f(x) \quad .$$

On the other hand, if $r = 0$, there results the pure difference equation

$$\text{(C)} \qquad a_0 y(x + \delta_0) + a_1 y(x + \delta_1) + \dots + a_s y(x + \delta_s) = f(x) \quad .$$

The easiest case of (C) is the one in which the spans δ_σ form an arithmetical sequence, the simplest perhaps when $\delta_\sigma = \sigma$. In this case, the equation becomes

$$\text{(D)} \qquad a_0 y(x) + a_1 y(x + 1) + \dots + a_s y(x + s) = f(x) \quad .$$

A special case of (A) which is worthy of notice, and which embraces (B) is

---

[1] We recall that the zero derivative of a function is understood to be the function itself, that is, $y^{(0)}(x) \equiv y(x)$.

Also compare with the observation in §13.

$$\text{(E)} \qquad y^{(r)}(x) + \sum_{\rho=0}^{r-1} \sum_{\sigma=0}^{s} a_{\rho\sigma} y^{(\rho)}(x + \delta_\sigma) = f(x), \qquad (r \geqq 1) \quad .$$

Equation (C) is not a special case of (E); and we shall see later that equation (E) even for arbitrary large r and arbitrary δ_σ is simpler than (D).

For brevity, we shall frequently denote any one of our equations by

$$(1) \qquad \Lambda_y = f(x) \quad ,$$

and the functional Λ_y will denote the left side of whichever equation we consider.

2. <u>We call a function</u> $y(x)$ <u>r-times differentiable in</u> $\mathfrak{F}_0$, <u>if it is defined in</u> $[-\infty, \infty]$, <u>has derivatives to (at least) the rth order, and together with its first</u> r <u>derivatives is absolutely integrable</u>. If the transform of $y(x)$ is denoted by $\varphi(\alpha)$, then by §3, 4, the transform of $y^{(\rho)}(x)$, $\rho = 0, 1, \ldots, r$, has the value $(i\alpha)^\rho \varphi(\alpha)$. In other words, if the representation

$$(2) \qquad y(x) \sim \int \varphi(\alpha) e(x\alpha)\, d\alpha$$

is formally differentiated ρ-times, then the result is the "correct" representation

$$y^{(\rho)}(x) \sim \int (i\alpha)^\rho \varphi(\alpha) e(x\alpha)\, d\alpha \quad .$$

Hence follows, cf. §13, 4,

$$(3) \qquad y^{(\rho)}(x + \delta) \sim \int (i\alpha)^\rho e(\delta\alpha) \varphi(\alpha) e(x\alpha)\, d\alpha \quad ,$$

and therefore for each function $y(x)$ which is r-times differentiable in $\mathfrak{F}_0$ we obtain

$$(4) \qquad \Lambda y \sim \int G(\alpha) \varphi(\alpha) e(x\alpha)\, d\alpha \quad ,$$

where $G(\alpha)$ stands for the function

$$(5) \qquad \sum_{\rho=0}^{r} \sum_{\sigma=0}^{s} a_{\rho\sigma} (i\alpha)^\rho e(\delta_\sigma \alpha) \quad .$$

If the contrary is not expressly emphasized, we shall consider the function $G(\alpha)$ only for real values $-\infty < \alpha < \infty$. We call it the characteristic function of equation (A). The function $G(\alpha)$ can also be interpreted as the "operator" belonging to the operation Λy, i.e., if within the function class $\mathfrak{F}_0$, a function $y(x)$ r-times differentiable in $\mathfrak{F}_0$ passes over to a function Λy, then a very simple operation in the function class $\mathfrak{T}_0$, namely the multiplication of $\varphi(\alpha)$ by $G(\alpha)$, corresponds to this change over. Hence we shall also say that Λy has resulted from $y(x)$ by "multiplication" with the "multiplier" $G(\alpha)$, cf. §23, 1.

There are still other operations Λy, to some extent more general ones, which correspond in the $\mathfrak{T}_0$-domain, to the multiplication by a suitable operator $G(\alpha)$. For example, let

$$\text{(F)} \qquad \Lambda y \equiv \lambda y(x) - \frac{1}{2\pi}\int y(\xi)K(x-\xi)\,d\xi \quad ,$$

where λ is a constant, and $K(\xi)$ is a fixed absolutely integrable function in $[-\infty, \infty]$. If the transform of $K(\xi)$ is denoted by $\gamma(\alpha)$, then a function (2) goes over into a function (4) with

$$\text{(6)} \qquad G(\alpha) = \lambda - \gamma(\alpha) \quad .$$

All types of operations considered thus far are contained in the general operation

$$\Lambda y \equiv \sum_{\rho=0}^{r}\sum_{\sigma=0}^{s} a_{\rho\sigma}y^{(\rho)}(x+\delta_\sigma) + \frac{1}{2\pi}\sum_{\mu=0}^{m}\sum_{\nu=0}^{n} b_{\mu\nu}\int y^{(\mu)}(\xi+\omega_\nu)K_{\mu\nu}(x-\xi)\,d\xi \ ,$$

where $a_{\rho\sigma}$ and $b_{\mu\nu}$ are arbitrary constants, δ_σ and ω_r real constants, and $K_{\mu\nu}(\xi)$ absolutely integrable functions. If the transform of the last functions are denoted by $\gamma_{\mu\nu}(\alpha)$, then the operator belonging to it reads

$$\sum_{\rho=0}^{r}\sum_{\sigma=0}^{s} a_{\rho\sigma}(i\alpha)^\rho e(\delta_\sigma\alpha) + \sum_{\mu=0}^{m}\sum_{\nu=0}^{n} b_{\mu\nu}(i\alpha)^\mu e(\omega_\nu\alpha)\gamma_{\mu\nu}(\alpha) \quad .$$

However we shall not consider our operator in versions of this or similar generality. Our purpose is to show how, for a given function $f(x)$ of $\mathfrak{F}_0$, one can solve (1) for $y(x)$ by means of (4); and the versions A to F will offer ample opportunity to describe the method of solution and its advantages.

2. In the present chapter we stipulate that a function $y(x)$ shall be a solution of (1) if it is r-times differentiable in $\mathfrak{F}_0$,

[r <u>is the order of the highest derivative of</u> $y(x)$ <u>actually occurring in</u> Λy], <u>and satisfies</u> (1) <u>in</u> $[-\infty, \infty]$. If

$$f(x) \sim \int E(\alpha) e(x\alpha)\, d\alpha \tag{7}$$

and if (2) is a solution of (1), then we have by (4)

$$G(\alpha)\varphi(\alpha) = E(\alpha) \quad . \tag{8}$$

The question as to whether at least one solution exists can be separated from the question as to how many solutions exist. Since the operation Λy is additive

$$\Lambda(c_1 y_1 + c_2 y_2) = c_1 \Lambda y_1 + c_2 \Lambda y_2 \quad ,$$

the general solution of (1) can therefore be obtained out of a particular solution by adding to the latter any solution of the homogeneous equation

$$\Lambda y = 0 \quad . \tag{9}$$

3. The homogeneous equation is settled immediately since it corresponds to the relation

$$G(\alpha)\varphi(\alpha) = 0 \quad . \tag{10}$$

If now $G(\alpha)$ is different from zero in $[-\infty, \infty]$, then it follows that

$$\varphi(\alpha) = 0 \quad . \tag{11}$$

This means that (9) has only the trivial solution

$$y(x) = 0 \quad .$$

If $G(\alpha)$ does vanish at points of $[-\infty, \infty]$, then the trivial solution remains the only one, if the zero points of $G(\alpha)$, - that is the values of α for which $G(\alpha) = 0$ -, are nowhere dense, as for instance if they are isolated. Because in this case, (11) is valid for an everywhere dense set of values of α, and since $\varphi(\alpha)$ as a function of ξ_0 is continuous, therefore (11) must be valid throughout. - The function (5) is an analytic function of the variable α. Hence its zero points are isolated, and therefore the general equation (A), in its homogeneous case, has no solution (in our sense). - If we admit "solutions" other than those stipulated here then the homogeneous equation may well have some. In the case of equation (B) say, if the (complex) zeros λ_ρ of the polynomial

$$c_r(i\tau)^r + c_{r-1}(i\tau)^{r-1} + \dots + c_0 = 0$$

are all simple, then the totality of all "solutions" are the expressions

$$C_1 e(\lambda_1 x) + C_2 e(\lambda_2 x) + \dots + C_r e(\lambda_r x)$$

for arbitrary constants C_ρ, but these solutions, by what we have just proved, could not possibly have the property of being r-times differentiable in $\mathfrak{F}_0$. There is a difference however in the case of (F). Here $G(\alpha)$ can be constant in intervals, and when this occurs there are indeed solutions of the homogeneous equation. We shall return to this later, cf. §26.

4. Let us consider the non-homogeneous case (A). At most one solution can exist. For a solution to exist, the following two conditions, by reason of (8), must be satisfied

1) the function

$$\varphi(\alpha) = G(\alpha)^{-1} E(\alpha)$$

must be a function of $\mathfrak{T}_0$, i.e., $f(x)$ can be "multiplied" by $G(\alpha)^{-1}$ (cf. the next paragraph).

2) If $r > 0$, then the function $y(x)$ of $\mathfrak{F}_0$ belonging to $\varphi(\alpha)$ must be r-times differentiable in $\mathfrak{F}_0$.

These conditions are not only necessary, but also sufficient. For if they are satisfied, then the function

$$y(x) \sim \int G(\alpha)^{-1} E(\alpha) e(x\alpha)\, d\alpha$$

is a solution of (A) as can be verified by substitution. We have the problem, therefore, of stating criteria for the fulfillment of these conditions.

§23. Multipliers

1. Let

$$f(x) \sim \int E(\alpha) e(x\alpha)\, d\alpha \tag{1}$$

be a given function. By a <u>multiplier</u> of $E(\alpha)$ or of $f(x)$, we mean a continuous function $\Gamma(\alpha)$ defined in $-\infty < \alpha < \infty$ which is so constituted that the function $\Gamma(\alpha) E(\alpha)$ again belongs to $\mathfrak{T}_0$. We shall also denote the function

$$\int \Gamma(\alpha) E(\alpha) e(x\alpha)\, d\alpha \tag{2}$$

by

$$\Gamma[f] \quad .$$

We shall say further that we have "multiplied" $f(x)$ by $\Gamma(\alpha)$, and if $\Gamma(\alpha)$ can be written in the form $G(\alpha)^{-1}$, we shall call the function $G(\alpha)$ a "divisor" of $f(x)$.

If $\Gamma(\alpha)$ is a multiplier for all functions of $\mathfrak{T}_0$, then we shall call $\Gamma(\alpha)$ a (general) multiplier (of the class $\mathfrak{T}_0$ or $\mathfrak{F}_0$).

2. If Γ_1 and Γ_2 are multipliers and c_1 and c_2 constants, then $\Gamma = c_1\Gamma_1 + c_2\Gamma_2$ is also a multiplier,

$$\Gamma[f] = c_1\Gamma_1[f] + c_2\Gamma_2[f] \quad .$$

Moreover the product $\Gamma = \Gamma_1\Gamma_2$ of two general multipliers is again a general multiplier, and we have namely

$$\Gamma[f] = \Gamma_1(\Gamma_2[f]) = \Gamma_2(\Gamma_1[f]) \quad .$$

3. In particular

$$\Gamma(\alpha) = c; \quad \Gamma[f] = cf(x)$$

belong to the general multipliers; for real λ also the function $\Gamma(\alpha) = e(\lambda\alpha) : \Gamma[f] = f(x + \lambda)$. If an infinite series

$$c_1e(\mu_1\alpha) + c_2e(\mu_2\alpha) + \ldots + c_ne(\mu_n\alpha) + \ldots \tag{3}$$

is given in which the μ_n are arbitrary real numbers, and the c_n arbitrary complex numbers, then for each n the partial sum

$$H_n(\alpha) = \sum_{\nu=1}^{n} c_\nu e(\mu_\nu\alpha) \tag{4}$$

is a (general) multiplier:

$$H_n[f] = \sum_{\nu=1}^{n} c_\nu f(x + \mu_\nu) \quad . \tag{5}$$

If the series

$$\sum_{\nu=0}^{\infty} |c_\nu|$$

converges, then on the one hand the series (3) is absolutely and uniformly convergent - we denote its sum by $H(\alpha)$ -, and on the other hand it follows from

$$\int |H_{n+p}[f] - H_n[f]|\, dx \leqq \sum_{\nu=n+1}^{n+p} |c_\nu| \int |f(x + \mu_\nu|\, dx \leqq$$

$$\leqq \sum_{\nu=n+1}^{n+p} |c_\nu| \cdot \int |f(x)|\, dx \quad ,$$

that the series (5) converges in the integrated mean (cf. §13, 2). Its limit function $F(x)$ again belongs to $\mathfrak{F}_0$. The transform of $F(x)$ is equal to the value of the limit of the transform of (5). The transform of (5) is $H_n(\alpha)E(\alpha)$, and since $H_n(\alpha)$ converges to $H(\alpha)$ it follows that $H(\alpha)E(\alpha)$ is the transform of $F(x)$. But $f(x)$ can now be any function of $\mathfrak{F}_0$. We have the result therefore, that the sum $H(\alpha)$ of an absolutely convergent series (3) is a general multiplier, and that

$$H[f] = F \quad .$$

If [for a function $f(x)$ of $\mathfrak{F}_0$], the series

$$\sum_{\nu=0}^{\infty} c_\nu f(x + \mu_\nu)$$

converges for almost all x, as is the case, for example, when $f(x)$ is a bounded function, then its sum is identical with $F(x)$ and hence with $H[f]$ (cf. Appendix 11).

4. Let

$$G(\alpha) = \sum_{\sigma=0}^{s} a_\sigma e(\delta_\sigma \alpha) \tag{6}$$

be a given exponential polynomial with arbitrary real δ_σ and complex a_σ. Under a specifiable restriction, the function $G(\alpha)^{-1}$ is expandable in an absolutely convergent series of the form (3). This restriction states that $G(\alpha)$ shall be essentially different from zero, i.e., that a constant $s > 0$ exist such that

$$|G(\alpha)| \geqq s > 0, \qquad (-\infty < \alpha < \infty) \quad . \tag{7}$$

That this condition is necessary follows immediately, because the relation

$$|H(\alpha)| \leqq \sum_0^\infty |c_\nu|$$

implies that the function $H(\alpha)$ is bounded. That the condition is also sufficient is not difficult to prove. However, in order to avoid going far afield, we shall here briefly discuss only the special case that $G(\alpha)$ can be written in the form

$$G(\alpha) = a_s \prod_{\chi=1}^{k} \left[e(\omega_\chi \alpha) - \lambda_\chi \right] \tag{8}$$

with real $\omega_\chi \neq 0$ and complex λ_χ [76]. In particular, this special case occurs if in $G(\alpha)$, the spans δ_σ form an arithmetical progression, say $\delta_\sigma = \sigma\delta$. One can then set: $k = s$, $\omega_\chi = \delta$, and the λ_χ (which need not be different from one another) are then the complex zero points of the polynomial

$$\sum_{\sigma=0}^{s} a_\sigma \lambda^\sigma \quad .$$

If a λ_χ has the absolute value 1, then for a suitable real β, $\lambda_\chi = e(\omega_\chi \beta)$, and the number $\alpha = \beta$ is a zero point of $G(\alpha)$. But if

$$|\lambda_\chi| \neq 1 \quad \text{for} \quad \chi = 1, 2, \ldots, k \quad , \tag{9}$$

then each individual factor of (8) is essentially different from zero; hence $G(\alpha)$ is also essentially different from zero. In the case of (8) therefore, condition (7) is equivalent to condition (9). If $|\lambda| \neq 1$, then the function

$$[e(\omega\alpha) - \lambda]^{-1}$$

has the absolutely convergent expansion

$$\sum_{n=0}^{\infty} \lambda^n e(-(1+n)\omega\alpha) \qquad \text{or} \qquad \sum_{n=0}^{\infty} \lambda^{-1-n} e(n\omega\alpha) \quad .$$

Such an expansion is an absolutely convergent series of the form (3). The product of two (and therefore of several) absolutely convergent series of the form (3) is again an absolutely convergent series of this form. The product series results if we multiply term by term and then collect such terms in which the same exponential $e(\mu\alpha)$ occurs. Hence under assumption (9), i.e., under assumption (7), the reciprocal of the function (8) is indeed expandable in an absolutely convergent series (3).

5. By Theorem 13, the product of two functions of $\mathfrak{T}_0$ is again a function of $\mathfrak{T}_0$. Therefore each function $\Gamma(\alpha)$ of $\mathfrak{T}_0$ is a multiplier. We recall the observations of §13, 9. If $K_\Gamma(x)$ is the function of $\mathfrak{F}_0$

belonging to $\Gamma(\alpha)$, then for the function defined in 1

$$\Gamma[f] \sim \int E(\alpha)\Gamma(\alpha)e(x\alpha)\,d\alpha$$

we have by the Faltung rule of §13, 4

$$\Gamma[f] = \frac{1}{2\pi}\int f(\xi)K_\Gamma(x - \xi)\,d\xi \quad .$$

6. The function

$$\gamma(\alpha) = \frac{1}{\alpha - i}$$

is the transform of the function

$$K_\gamma(x) = \begin{cases} 2\pi i e^{-x} & , \quad x > 0 \\ 0 & , \quad x < 0 \end{cases} \quad .$$

In particular

$$K_\gamma(x) = O(|x|^{-r}) \, , \qquad\qquad (r > 0) \quad .$$

This same estimate is valid for the kernel which belongs to $(\alpha + i)^{-1}$.

7. Let $\Gamma(\alpha)$ be a function r-times differentiable, $r \geq 2$, and for a suitable constant c, let

$$\Gamma(\alpha) = \frac{c}{\alpha} + H(\alpha)$$

outside of an interval $- A \leq \alpha \leq A$, where the function $H(\alpha)$ together with its first r-derivatives, is absolutely integrable. The function

$$\Gamma(\alpha) - \frac{c}{\alpha - i}$$

is r-times differentiable, and since it has the form

$$\frac{- ci}{\alpha(\alpha - i)} + H(\alpha)$$

outside of $[- A, A]$, it, together with the first r derivatives, is absolutely integrable. It is therefore a function of $\mathfrak{F}_0$,[1] and the

[1] For the purpose of this chapter, we depend principally on this fact. We could have, therefore, assumed from the outset that $\Gamma(\alpha)$ is two times differentiable, instead of making the seemingly more troublesome hypothesis that there is a number $r \geq 2$ such that $\Gamma(\alpha)$ is r-times differentiable. But by the introduction of the letter r, certain practical relations result, as for example the relation (10) soon to follow. While we shall not require this immediately, we shall need it in the next chapter though.

associated kernel has the order of magnitude $O(|x|^{-r})$. Hence the function $\Gamma(\alpha)$ itself is also a function of $\mathfrak{T}_0$; therefore it is a multiplier, and we have for the kernel belonging to it

$$K_\Gamma(x) = O(|x|^{-r}) \quad . \tag{10}$$

8. For what follows we need the following lemma. Let $\Gamma(\alpha)$ be a given r-times differentiable function in an interval (α_1, α_2). One can construct a function $\Gamma^*(\alpha)$ which is defined and is r-times differentiable in $[-\infty, \infty]$, agreeing with $\Gamma(\alpha)$ in (α_1, α_2), and vanishing outside of a certain finite interval (α_1', α_2') - naturally with $\alpha_1' < \alpha_1$ and $\alpha_2 < \alpha_2'$ (Appendix 14). This extension of $\Gamma(\alpha)$ to $\Gamma^*(\alpha)$ comes about by the adjoining of a bell shaped extension piece on the left end point α_1 and on the right end point α_2 of $\Gamma(\alpha)$. One can even prescribe that the numbers α_1' and α_2' be arbitrarily near to α_1 and α_2 respectively.

9. Let the function (1) and an interval (α_1, α_2) be given. That function in α which agrees with $E(\alpha)$ in (α_1, α_2) and otherwise vanishes, is in general not a function of $\mathfrak{T}_0$, because it is in general not a continuous function.[1] If one therefore cuts out a continuous piece from a function of $\mathfrak{T}_0$ (i.e., out of the group of oscillations of a function of $\mathfrak{F}_0$), then this piece need not again be a function of $\mathfrak{T}_0$ (i.e., it need not again form the oscillation group of a function of $\mathfrak{F}_0$). But for many problems, it is important to be able to "isolate" a finite piece of a function $E(\alpha)$. This "isolation" can be accomplished, provided the cut is not carried out too sharply on both interval ends α_1 and α_2. For as α_1' is laid close to α_1 on its left and α_2' close to α_2 on its right, there is a function $\Gamma(\alpha)$ which has the value 1 in (α_1, α_2), vanishes outside of (α_1', α_2'), and in $[-\infty, \infty]$ has derivatives up to a previously assigned order $r \geq 2$. The product $\Gamma(\alpha)E(\alpha)$ is a function of $\mathfrak{T}_0$ which agrees with $E(\alpha)$ in (α_1, α_2) and vanishes outside of the interval (α_1', α_2'). For the function of $\mathfrak{F}_0$ belonging to it, we have

$$\Gamma[f] = \frac{1}{2\pi}\int K_\Gamma(\xi)f(x - \xi)\, d\xi \quad , \tag{11}$$

where $K_\Gamma(\xi)$ has the order of magnitude $O(|x|^{-r})$. One can even construct the function $\Gamma(\alpha)$ in such a way that it has derivatives of arbitrary higher order (Appendix 15). Then

$$K_\Gamma(x) = O(|x|^{-r}), \qquad (r = 1, 2, 3, \ldots) \quad , \tag{12}$$

and by reason of (11), many "smoothness properties" of $f(x)$ will carry

[1] And for each function of $\mathfrak{T}_0$, continuity is a necessary condition.

over to the function $\Gamma[f]$ obtained from it, see for instance §24, 1.

10. Let

$$(13)\qquad \tau_1, \tau_2, \ldots, \tau_x, \ldots$$

be a finite set of real numbers which differ from one another. For each index x, we can determine a multiplier $\Gamma_x(\alpha)$ which is differentiable arbitrarily often, which has the value 1 in some (sufficiently small) neighborhood of τ_x and which vanishes outside of a (somewhat larger) closed interval, the latter interval not including any of the other points (13). The multiplier

$$\Gamma_0(\alpha) = 1 - \sum_x \Gamma_x(\alpha)$$

is differentiable arbitrarily often, has the property that it vanishes in a certain neighborhood of each point (13) and has the value 1 outside of a (sufficiently large) finite interval.

11. Let a function (1) be given. Two continuous functions $\Gamma_1(\alpha)$ and $\Gamma_2(\alpha)$ which differ from one another only at such points α for which $E(\alpha) = 0$, are either both multipliers of (1) or neither of them is. We call them "equivalent" in regard to (1).

12. If the transform of (1) vanishes outside of a finite interval (a, b), then by Theorem 12

$$(14)\qquad f(x) = \int_a^b E(\alpha)e(x\alpha)\,d\alpha\ .$$

If $\Gamma(\alpha)$ is a function defined and r-times $(r \geq 2)$ differentiable in (a, b), then

$$(15)\qquad \int_a^b \Gamma(\alpha)E(\alpha)e(x\alpha)\,d\alpha$$

is again a function of $\mathfrak{F}_0$. For if one extends $\Gamma(\alpha)$ to a function $\Gamma^*(\alpha)$ which is defined and r-times differentiable in $[-\infty, \infty]$, and which vanishes outside of a finite interval, then $\Gamma(\alpha)$ and $\Gamma^*(\alpha)$ are equivalent as regards (14), and (15) is then the function $\Gamma^*[f]$. $\Gamma^*[f]$ can also be written in the form (11) where $K_{\Gamma^*}(x) = O(|x|^{-r})$.

§24. Differentiation and Integration

1. If $K(x)$ is r-times differentiable in $\mathfrak{F}_0$ and $f(x)$ belongs

to $\mathfrak{F}_0$, then the function

$$g(x) = \frac{1}{2\pi}\int K(\xi)f(x-\xi)\,d\xi$$

is also r-times differentiable in $\mathfrak{F}_0$, and indeed

$$g^{(\rho)}(x) = \frac{1}{2\pi}\int K^{(\rho)}(\xi)f(x-\xi)\,d\xi, \qquad 0 \leqq \rho \leqq r .$$

As is easily seen, it is sufficient to prove this assertion for the case $r = 1$. The function

$$h(x) = \frac{1}{2\pi}\int K'(\xi)f(x-\xi)\,d\xi = \frac{1}{2\pi}\int f(\eta)K'(x-\eta)\,d\eta$$

is by §13, 3 a function of $\mathfrak{F}_0$. As has already been shown in the proof of Theorem 15, 2

$$\int_{x_0}^{x} h(x)\,dx = g(x) - g(x_0) ,$$

and therefore $h(x) = g'(x)$, Q.E.D.

2. To each function $f(x)$ there is, for each integer $r > 0$, an rth integral, i.e., a function $F_r(x)$ such that $F_r^{(r)}(x) = f(x)$. The function $F_r(x)$ is unique excepting for a random additive polynomial of (r-1)th degree in x. We call a function $f(x)$ r-times integrable in $\mathfrak{F}_0$ if $f(x)$ belongs to $\mathfrak{F}_0$, and the additive polynomial in $F_r(x)$ can be selected in such a way that $F_r(x)$ together with derivatives $F_r^{(\rho)}(x)$, $0 \leqq \rho \leqq r$, belong to $\mathfrak{F}_0$, i.e., $F_r(x)$ is r-times differentiable in $\mathfrak{F}_0$. In §3, 4 we showed that if $f(x)$ is 1-time integrable in $\mathfrak{F}_0$, then the integral $F_1(x)$ is unique and has the value

$$\int^{x} f(x)\,dx = -\int_{x} f(x)\,dx \quad . \tag{1}$$

More generally the following is valid. If $f(x)$ is r-times integrable in $\mathfrak{F}_0$, then $F_r(x)$ is unique and has the value

$$\int^{x} dx_1 \int^{x_1} dx_2 \cdots \int^{x_{r-2}} dx_{r-1} \int^{x_{r-1}} f(x_r)\,dx_r$$

$$= (-1)^r \int_{x} dx_1 \int_{x_1} dx_2 \cdots \int_{x_{r-2}} dx_{r-1} \int_{x_{r-1}} f(x_r)\,dx_r \quad .$$

The proof is very simple. We prove it say for $r = 2$. By hypothesis $F_2'(x)$ belongs to $\mathfrak{F}_0$ and is the integral of the function $F_2''(x) = f(x)$, hence $F_2'(x)$ is unique and has the value (1). Now $F_2(x)$ as the integral of $F_2'(x)$ is again unique and results from $F_2'(x)$ in the same way as $F_2'(x)$ results from $f(x)$. Therefore

$$F_2(x) = \int^{x} dx_1 \int^{x_1} f(x_2)\, dx_2 = (-1)^2 \int_{x} dx_1 \int_{x_1} f(x_2)\, dx_2 \quad ,$$

THEOREM 25. If the function

$$f(x) \sim \int E(\alpha)e(x\alpha)\, d\alpha \tag{2}$$

is r-times differentiable in $\mathfrak{F}_0$, then

$$f^{(\rho)}(x) \sim \int (i\alpha)^\rho E(\alpha)e(x\alpha)\, d\alpha, \qquad \rho = 0, 1, \ldots, r. \tag{3}$$

Conversely if a function

$$\varphi(x) \sim \int (i\alpha)^r E(\alpha)e(x\alpha)\, d\alpha \tag{4}$$

exists (i.e., if $(i\alpha)^r$ is a multiplier of $E(\alpha)$), then $f(x)$ is r-times differentiable in $\mathfrak{F}_0$ (and hence $f^{(r)}(x) = \varphi(x)$).

PROOF. The first part of the theorem has already been proved in §3, 4, and used many times previously. The second part is more difficult. We shall pre-require the following two theorems which are also useful of themselves.

THEOREM 26. If $E(\alpha)$ vanishes outside of a finite interval (A, B), then $f(x)$ is differentiable arbitrarily often in $\mathfrak{F}_0$.

PROOF. That $f(x)$ is differentiable arbitrarily often follows from §4, 2, c). Indeed

$$f^{(\rho)}(x) = \int_A^B (i\alpha)^\rho E(\alpha)e(x\alpha)\, d\alpha, \qquad \rho = 0, 1, 2, \ldots \quad .$$

And that these derivatives belong to $\mathfrak{F}_0$ follows from §23, 12.

THEOREM 27. If $E(\alpha)$ vanishes in an interval $[A, B]$ which contains the point $\alpha = 0$ $(A < 0 < B)$, then $f(x)$ is integrable arbitrarily often in $\mathfrak{F}_0$, and the ρth integral has the form

$$F_\rho(x) \sim \int \frac{E(\alpha)}{(i\alpha)^\rho} e(x\alpha)\, d\alpha \quad . \tag{5}$$

PROOF. Since together with $E(\alpha)$, also $(i\alpha)^{-1}E(\alpha)$ vanishes in $[A, B]$, it is sufficient to prove that $f(x)$ is 1-times integrable in $\mathfrak{F}_0$ and that this integral has the value

$$g(x) \sim \int \frac{E(\alpha)}{i\alpha} e(x\alpha)\, d\alpha \quad . \tag{6}$$

The function $(i\alpha)^{-1}$ is regular oustide of $[A, B]$. Therefore in $[-\infty, \infty]$ a function $\Gamma(\alpha)$ can be found (cf. Appendix 16) which is two times differentiable, and outside of $[A, B]$ agrees with $(i\alpha)^{-1}$. This function is a multiplier by §23, 7, and since in regard to $E(\alpha)$, it is "equivalent" with $(i\alpha)^{-1}$, cf. §23, 11, the function (6) exists in any case, and indeed $g(x) = \Gamma(f)$. We still have to prove that $g(x)$ is differentiable and that $g'(x) = f(x)$, and it suffices to prove only that it is differentiable in $\mathfrak{F}_0$. For if this is so, then the derivative, by the first part of Theorem 25, has the transform $(i\alpha)\Gamma(\alpha)E(\alpha) = E(\alpha)$ and hence $g'(x) = f(x)$. Set

$$\Gamma_1(\alpha) = \frac{1}{2i(\alpha-i)}, \qquad \Gamma_2(\alpha) = \frac{1}{2i(\alpha+i)}$$

and write

$$\Gamma(\alpha) = \Gamma_1(\alpha) + \Gamma_2(\alpha) + \Gamma_3(\alpha) \quad .$$

Then correspondingly

$$g(x) = \Gamma_1[f] + \Gamma_2[f] + \Gamma_3[f] = g_1 + g_2 + g_3 \quad .$$

Outside of $[A, B]$, we have

$$\Gamma_3(\alpha) = \Gamma(\alpha) - \Gamma_1(\alpha) - \Gamma_2(\alpha) = \frac{1}{i\alpha(\alpha^2+1)} \quad ,$$

and therefore by §13, 9 the associated kernel $K_3(x)$ is 1-times differentiable in $\mathfrak{F}_0$. Hence by 1. $g_3(x)$ is also 1-times differentiable in $\mathfrak{F}_0$. Furthermore by §23, 6

$$2g_1(x) = \int_0 e^{-\xi} f(x - \xi)\, d\xi = e^{-x} \int_{-x} e^{-\eta} f(-\eta)\, d\eta \quad ,$$

from which we obtain by differentiation (Appendix 8)

$$2g_1'(x) = -e^{-x} \int_{-x} e^{-\eta} f(-\eta)\, d\eta + e^{-x} e^{x} f(x) = -2g_1(x) + f(x) \quad .$$

Hence $g_1'(x)$ is a function of $\mathfrak{F}_0$. An analogous reasoning is valid also for $g_2'(x)$. Therefore Theorem 27 is proved.

We can now prove finally Theorem 25. In accordance with §23, 10, we attach to the single point $\tau_1 = 0$, the multipliers $\Gamma_1(\alpha)$ and $\Gamma_0(\alpha) = 1 - \Gamma_1(\alpha)$, and set

$$f_\chi = \Gamma_\chi[f], \quad \varphi_\chi = \Gamma_\chi[\varphi], \qquad (\chi = 0, 1) \quad .$$

Then $f = f_0 + f_1$, $\varphi = \varphi_0 + \varphi_1$. By Theorem 26, $f_1(x)$ is differentiable arbitrarily often in $\mathfrak{F}_0$, and we have

$$f_1^{(r)}(x) = \varphi_1(x) \quad . \tag{7}$$

By Theorem 27, there is a function $F_0(x)$ of $\mathfrak{F}_0$ such that

$$F_0^{(r)}(x) = \varphi_0(x) \quad . \tag{8}$$

By comparing (7) and (8), we infer that $\varphi(x)$ is the rth derivative of a certain function $g(x) = f_1(x)$ which is r-times differentiable in $\mathfrak{F}_0$. We must still show that $g(x) = f(x)$. If the transform of $g(x)$ is denoted by $F(\alpha)$, then as has already been proved, $(i\alpha)^r F(\alpha)$ is the transform of $\varphi(x)$. By comparing it with (4), we have actually $F(\alpha) = E(\alpha)$.

3. The true content of Theorem 25 can be expressed as follows. In order that the function (2) be r-times differentiable in $\mathfrak{F}_0$, it is necessary and sufficient that $(i\alpha)^r E(\alpha)$ belong to $\mathfrak{T}_0$. If the roles of $f(x)$ and $\varphi(x)$ are interchanged, one can also make the following statement. In order that (2) be r-times integrable in $\mathfrak{F}_0$, it is necessary and sufficient that $(i\alpha)^{-r} E(\alpha)$ belong to $\mathfrak{T}_0$, i.e., that a function $\Phi(\alpha)$ of $\mathfrak{T}_0$ exist for which

$$E(\alpha) = (i\alpha)^r \Phi(\alpha) \quad . \tag{9}$$

The functions $E(\alpha)$ and $\Phi(\alpha)$, as functions of $\mathfrak{T}_0$, are continuous. Therefore, if for given $E(\alpha)$ there is to exist such a function $\Phi(\alpha)$, we must have $E(0) = 0$. But this requirement is not sufficient. By Theorem 27, it is sufficient for example, that $E(\alpha)$ vanish in an entire neighborhood of $\alpha = 0$.

4. The function

$$e(-\lambda x)f(x) \quad , \qquad (\lambda \text{ real}) \quad , \tag{10}$$

has the transform $E(\alpha + \lambda)$. If (10) is r-times differentiable in $\mathfrak{F}_0$, then the rth derivative has the representation

$$\int (i\alpha)^r E(\alpha + \lambda) e(x\alpha)\, d\alpha \quad .$$

The product of the rth derivative with $e(\lambda x)$ has then the representation

$$\int \{i(\alpha - \lambda)\}^r E(\alpha) e(x\alpha)\, d\alpha \quad . \tag{11}$$

Hence there exists a function with the representation (11). Conversely one finds: If a function (11) exists, then (10) is r-times differentiable in $\mathfrak{F}_0$. A similar statement holds for integration. We have therefore the following theorem.

THEOREM 28. Let

$$f(x) \sim \int E(\alpha) e(x\alpha)\, d\alpha$$

be a given function. In order that a function with the representation

$$\int \{i(\alpha - \lambda)\}^r E(\alpha) e(x\alpha)\, d\alpha$$

or

$$\int \{i(\alpha - \lambda)\}^{-r} E(\alpha) e(x\alpha)\, d\alpha$$

exist, it is necessary and sufficient that the function $e(-\lambda x)f(x)$ be r-times differentiable or integrable in $\mathfrak{F}_0$.

5. Let it be known of two functions $f_1(x)$ and $f_2(x)$ of $\mathfrak{F}_0$, that their transforms agree in the neighborhood of a point $\alpha = \lambda$. For the function

$$e(-\lambda x)f_1(x) - e(-\lambda x)f_2(x) \tag{12}$$

the transform vanishes in the neighborhood of $\alpha = 0$. Hence by Theorem

27, (12) is integrable arbitrarily often in $\mathfrak{F}_0$. This easily implies as follows.

If the transforms $E_1(\alpha)$ and $E_2(\alpha)$ of two functions $f_1(x)$ and $f_2(x)$ agree in the neighborhood of a point $\alpha = \lambda$,

$$E_1(\alpha) = E_2(\alpha), \qquad [(A < \alpha \leqq \lambda), \ (\lambda \leqq \alpha < B)]$$

and if a function with the representation

$$\int \{i(\alpha - \lambda)\}^{-r} E_1(\alpha) e(x\alpha)\, d\alpha$$

exists, then there also exists a function with the representation

$$\int \{i(\alpha - \lambda)\}^{-r} E_2(\alpha) e(x\alpha)\, d\alpha \quad .$$

§25. The Difference-Differential Equation

1. The characteristic function $G(\alpha)$ of the general equation (A), cf. §22, 1, can have a finite or a countably infinite number of zeros. If zeros exist, we denote them arbitrarily by

$$(1) \qquad \tau_1, \tau_2, \ldots, \tau_\chi, \ldots$$

and their multiplicities by

$$(2) \qquad \ell_1, \ell_2, \ldots, \ell_\chi, \ldots \quad .$$

The multiplicity of the zeros will enter into our observations in the following manner. In each neighborhood (A_χ, B_χ) of τ_χ which contains no other zeros

$$(3) \qquad G(\alpha) = \{i(\alpha - \tau_\chi)\}^{\ell_\chi} G_*(\alpha) \quad ,$$

where in (A_χ, B_χ), $G_*(\alpha)$ differs from zero and is differentiable arbitrarily often. Actually we will require the existence of derivatives only up to a specifiable order $r \geqq 2$.

THEOREM 29. In order that equation (A) have a solution, it is necessary that the function

$$(4) \qquad e(-\tau_\chi x) f(x)$$

be ℓ_χ-times integrable in $\mathfrak{F}_0$ for al χ, i.e., that all functions

(5) $$[i(\alpha - \tau_\chi)]^{-\ell_\chi} E(\alpha)$$

belong to $\mathfrak{T}_0$.

REMARK. For example, the given necessary condition is fulfilled if $E(\alpha)$ vanishes in a neighborhood of each zero τ_χ, cf. Theorem 27 and §24, 4.

PROOF. If (A) is solvable, then there exists a function $\varphi(\alpha)$ of $\mathfrak{T}_0$ for which

(6) $$G(\alpha)\varphi(\alpha) = E(\alpha) \quad .$$

We select a zero point τ_χ and consider a multiplier $\Gamma_\chi(\alpha)$ which vanishes outside of (A_χ, B_χ) and has the value 1 in a smaller neighborhood of τ_χ. Then, cf. (3)

(7) $$[i(\alpha - \tau_\chi)]^{\ell_\chi} G_*(\alpha)\Gamma_*(\alpha)\varphi(\alpha) = \Gamma_*(\alpha)E(\alpha) \quad .$$

The functions

(8) $$\Gamma_*(\alpha)E(\alpha)$$

and

$$G_*(\alpha) \cdot [\Gamma_*(\alpha)\varphi(\alpha)]$$

both belong to $\mathfrak{T}_0$, the last by §23, 12. The relation (7) says therefore, that (8) has the divisor $[i(\alpha - \tau_\chi)]^{\ell_\chi}$, and since by construction of $\Gamma_*(\alpha)$, (8) agrees with $E(\alpha)$ in a certain neighborhood of τ_χ, hence by §24, 5 $E(\alpha)$ must also of necessity have this divisor, Q.E.D.

2. To obtain also sufficient conditions for the solvability of (A), we observe that by §22, 4 in conjunction with Theorem 25, it is necessary and sufficient for the solvability of (A), that the functions

(9) $$(i\alpha)^\rho G(\alpha)^{-1}E(\alpha)$$

belong to $\mathfrak{T}_0$ for $\rho = 0, 1, \ldots, r$.

3. We now insert a lemma.

LEMMA. Let $p \geqq 0$, $q \geqq p + 2$, m an integer $\geqq 0$ and $\alpha_0 > 0$. Let $g(\alpha)$ and $h(\alpha)$ be two given functions in $\alpha_0 \leqq \alpha < \infty$ which are m-times continuously differentiable, and together with the first m derivatives bounded,

$$|g^{(\mu)}(\alpha)| \leq M, \qquad |h^{(\mu)}(\alpha)| \leq M, \qquad 0 \leq \mu \leq m .$$

Moreover let the function

$$A\alpha^q + \alpha^{q-1}g(\alpha)$$

differ from zero in $(\alpha_0, \infty]$, where $A \neq 0$ is a constant.

Then the function

$$r(\alpha) = \frac{\alpha^p h(\alpha)}{A\alpha^q + \alpha^{q-1}g(\alpha)} , \tag{10}$$

together with its first m derivatives, is absolutely integrable in $[\alpha_0, \infty]$.

PROOF. We write

$$r(\alpha) = \frac{1}{\alpha^{q-p}} \frac{h(\alpha)}{A + \alpha^{-1}q(\alpha)} = \frac{1}{\alpha^{q-p}} k(\alpha) .$$

By hypothesis, $k(\alpha)$ is a bounded, continuous function. Because $q - p \geq 2$, $r(\alpha)$ is therefore absolutely integrable in $[\alpha_0, \infty]$. For $m = 0$, therefore, the proof is finished. For $m \geq 1$, one can differentiate $r(\alpha)$, and set for the derivative

$$r^1(\alpha) = \frac{\alpha^{p_1} h_1(\alpha)}{A_1\alpha^{q_1} + \alpha^{q_1-1} g_1(\alpha)} \tag{11}$$

where $p_1 = p + q$, $q_1 = 2q$, $A_1 = A^2$ and $h_1(\alpha)$ and $g_1(\alpha)$ are certain detailed expressions from which one can easily see that these functions have $m - 1$ continuous and bounded derivatives, and that the denominator of (11) differs from zero in $(\alpha_0, \infty]$. Since $p_1 \geq 0$ and $q_1 \geq p_1 + 2$, the quantities p_1, q_1, $m_1 = m - 1$, α_0, $g_1(\alpha)$, $h_1(\alpha)$, A_1 again satisfy the hypothesis of the theorem. It is evident therefore, that the proof of the conclusion can be arrived at by induction from $m - 1$ to m, Q.E.D.

We will now treat equations (E), (B), (C) and (A) in this order. The first three are special cases of (A), to be sure, but we will also obtain special statements in return.

4. Equation (E).

Since the exponential function $e(\delta\alpha)$ is bounded for real δ, the order of magnitude of the characteristic function

$$G(\alpha) = (i\alpha)^r + \sum_{\rho=0}^{r-1} \sum_{\sigma=0}^{s} a_{\rho\sigma}(i\alpha)^r e(\delta_\sigma \alpha)$$

is determined, as $\alpha \longrightarrow \pm\infty$, by the highest term α^r. There is, therefore an α_0, such that for $|\alpha| \geqq \alpha_0$

$$|G(\alpha)| \geqq M|\alpha|^r \quad .$$

In particular, $G(\alpha) \neq 0$ for $|\alpha| \geqq \alpha_0$, and therefore $G(\alpha)$ as an analytic function, can have only a finite number of zeros.[1]

We assume for the time being that no real zeros at all exist. We shall prove in this case, that the functions

$$H_\rho(\alpha) = (i\alpha)^\rho G(\alpha)^{-1} \tag{12}$$

are general multipliers for $0 \leqq \rho \leqq r$. For its proof we distinguish three cases: 1. $0 \leqq \rho \leqq r - 2$. 2. $\rho = r - 1$ and 3. $\rho = r$. The above lemma can be applied directly in case 1. In it we set $p = \rho$, $q = r$, $A = i^r$, $h(\alpha) = i^\rho$,

$$g(\alpha) = \frac{1}{\alpha^{r-1}} \sum_{\rho=0}^{r-1} \sum_{\sigma=0}^{s} a_{\rho\sigma}(i\alpha)^\rho e(\delta_\sigma \alpha) \quad .$$

The derivative of each order of $g(\alpha)$ is bounded in $(0 <)\ \alpha_0 \leqq \alpha < \infty$ because the derivative of each order of $\alpha^{-\lambda}e(\delta\alpha)$, $\lambda \geqq 0$ is itself bounded. Hence by the corollary $H_\rho(\alpha)$, together with all derivatives, especially with the first two derivatives, is absolutely integrable in $[\alpha_0, \infty]$. The same assertion also holds as $\alpha \longrightarrow -\infty$. Therefore $H_\rho(\alpha)$ is a multiplier. It is not possible to employ the lemma directly in case 2. But one very easily finds that it can be applied to the difference

$$H_{r-1}(\alpha) - \frac{1}{i\alpha}$$

and then only §23, 7 need be considered. In case 3 we put

$$H_r(\alpha) = \frac{(i\alpha)^r}{G(\alpha)} = 1 - \frac{G(\alpha) - (i\alpha)^r}{G(\alpha)}$$

and we can write

$$H_r(\alpha) = 1 - \sum_{\rho=0}^{r-1} \sum_{\sigma=0}^{s} a_{\rho\sigma} H_\rho(\alpha) e(\delta_\sigma \alpha) \quad .$$

Since the $H_\rho(\alpha)$ have already been recognized as multipliers for $0 \leqq \rho \leqq r - 1$, it therefore follows that $H_r(\alpha)$ is also a multiplier, §23, 2.

---

[1] We recall that in general $G(\alpha)$ is considered only for real values of α.

THEOREM 30. For the solvability of (E), the given necessary condition in Theorem 29 is also sufficient.

In particular, if the characteristic function has no zeros at all, then a solution always exists, and one can write for the solution

$$y(x) = \frac{1}{2\pi}\int K(\xi)f(x - \xi)\, d\xi \quad , \tag{13}$$

where

$$K(\xi) = \int G(\alpha)^{-1}e(x\alpha)\, d\alpha \quad .$$

5. PROOF. If there are no zeros at all, the assertion follows fro what has just been proved, and it may even be supplemented as follows. Denote by $K_\rho(x)$, the function of $\mathfrak{F}_0$ belonging to $H_\rho(\alpha)$, $0 \leqq \rho \leqq r - 1$. Hence by our observations, it follows that

$$y^{(\rho)}(x) = \frac{1}{2\pi}\int K_\rho(\xi)f(x - \xi)\, d\xi \quad . \tag{14}$$

But because $H_\rho(x) = (i\alpha)^\rho H_0(\alpha)$, therefore $K_\rho(\xi) = K^{(\rho)}(\xi)$. Hence not only (13), but also

$$y^{(\rho)}(x) = \frac{1}{2\pi}\int K^{(\rho)}(\xi)f(x - \xi)\, d\xi, \qquad (0 \leqq \rho \leqq r - 1)$$

is valid.

6. If now $G(\alpha)$ has points (1) as zeros, then by the hypothesis of the theorem, the functions

$$\{i(\alpha - \tau_\chi)\}^{-\ell_\chi}E(\alpha) \tag{15}$$

are in $\mathfrak{T}_0$ and we have to prove, cf. 2, that the functions

$$(i\alpha)^\rho G(\alpha)^{-1}E(\alpha) \ , \qquad 0 \leqq \rho \leqq r \ , \tag{16}$$

also belong to $\mathfrak{T}_0$. We attach multipliers $\Gamma_1(\alpha)$, $\Gamma_2(\alpha)$, ... and $\Gamma_0(\alpha)$ to the zero points (1) in accordance with §23, 10, and set

$$E_\chi(\alpha) = \Gamma_\chi(\alpha)E(\alpha) \ , \qquad \chi = 0, 1, 2, \ldots \ . \tag{17}$$

Since $E(\alpha) = E_0(\alpha) + E_1(\alpha) + \ldots$, it will be sufficient to prove that the functions

$$(i\alpha)^\rho G(\alpha)^{-1}E_\chi(\alpha) \ , \qquad 0 \leqq \rho \leqq r \ , \tag{18}$$

belong to $\mathfrak{T}_0$ for $\chi = 0, 1, 2, \ldots$. For $\chi \geqq 1$, the proof follows at once by §23, 12, if for (18) one writes

$$(i\alpha)^{\rho} G_*(\alpha)^{-1} [\Gamma_\chi(\alpha)\{i(\alpha - \tau_\chi)\}^{-\ell_\chi} E(\alpha)] .$$

If $\chi = 0$, we make use of the fact that $E_0(\alpha)$ vanishes in the neighborhood of each zero of $G(\alpha)$. We alter $G(\alpha)$ in these neighborhoods in such a way that the new function which we call $G_0(\alpha)$ vanishes nowhere, and is two times differentiable. Because

$$(i\alpha)^{\rho} G_0(\alpha)^{-1} E_0(\alpha) = (i\alpha)^{\rho} G(\alpha)^{-1} E_0(\alpha), \qquad \rho = 0, 1, \ldots, r ,$$

it is sufficient to prove that the functions

$$H_\rho(\alpha) = (i\alpha)^{\rho} G_0(\alpha)^{-1} , \qquad \rho = 0, 1, \ldots, r ,$$

are multipliers. If $0 \leqq \rho \leqq r - 1$, the proof proceeds exactly as it did for the functions (12), that is again by the use of the lemma under 3, and by considering that

$$G(\alpha) = G_0(\alpha)$$

outside of a certain finite interval. In regard to $H_r(\alpha)$, we note that the function

$$H_r(\alpha) + \sum_{\rho=0}^{r-1} \sum_{\sigma=0}^{s} a_{\rho\sigma} H_\rho(\alpha) e(\delta_\sigma \alpha) - 1$$

is zero outside of a finite interval; therefore by §23, 5 it is a multiplier, and consequently $H_r(\alpha)$ is also a multiplier, Q.E.D.

7. For example, consider the equation

$$y(x) - \frac{c}{2} [y(x + 1) + y(x - 1)] - y''(x) = f(x) .$$

Here

$$G(\alpha) = 1 + \alpha^2 - c \cos \alpha .$$

If, for example $|c| < 1$, then $G(\alpha)$ has no zeros. There is therefore a solution for each $f(x)$ of $\mathfrak{F}_0$, and indeed

$$y(x) = \frac{1}{2\pi} \int K(\xi) f(x - \xi)\, d\xi ,$$

where

$$K(x) = \int \frac{e(x\alpha)\, d\alpha}{1 + \alpha^2 - c \cos \alpha} \quad .$$

On the other hand, if $c = 1$, $G(\alpha)$ has a zero of multiplicity $\ell = 2$ at the origin. Hence there exists a solution then and only then, if $f(x)$ is (at least) two times integrable in $\mathfrak{F}_0$.

8. Equation (B).

In this special case of (E)

$$G(\alpha) = \sum_{\rho=0}^{r} c_\rho (i\alpha)^\rho \quad .$$

We assume that the complex zeros of this polynomial are all single and denote them by

$$\lambda_1, \ldots, \lambda_r \quad .$$

For the existence of a solution of (B), it is necessary and sufficient by Theorem 30, that for each real zero λ_ρ, so far as such zeros occur at all, the function

$$e(\lambda_\rho x) \int^x e(-\lambda_\rho \xi) f(\xi)\, d\xi \tag{19}$$

belong to $\mathfrak{F}_0$. If this happens, one can also write for (19), cf. §24, 2,

$$- e(\lambda_\rho x) \int_x e(-\lambda_\rho \xi) f(\xi)\, d\xi \quad . \tag{20}$$

For displaying the solution effectively, consider the partial fraction decomposition of $G(\alpha)^{-1}$

$$\frac{1}{G(\alpha)} = \sum_{\rho=1}^{r} \frac{A_\rho}{i(\alpha - \lambda_\rho)} \quad . \tag{21}$$

It corresponds to splitting

$$y(x) = \sum_{\rho=0}^{r} A_\rho y_\rho(x) \quad ,$$

where

$$y_\rho(x) \sim \int \frac{E(\alpha)}{i(\alpha - \lambda_\rho)} e(x\alpha)\, d\alpha \quad .$$

By §14, 3, $y_\rho(x)$ has the value (19) if $\Im(\lambda_\rho) > 0$, the value (20) if $\Im(\lambda_\rho) < 0$; the values (19) and (20) in common if $\Im(\lambda_\rho) = 0$.

If among the complex zeros of $G(\alpha)$, multiple ones occur, then the more general expression

$$\sum_{\rho, \ell} \frac{A_{\rho \ell}}{\{i(\alpha-\lambda_\rho)\}^\ell}$$

replaces the partial fraction decomposition (21). Correspondingly, if a solution exists, it has the form

$$y(x) = \sum_{\rho, \ell} A_{\rho \ell} y_{\rho \ell}(x) \quad ,$$

where by §14, 3 and §24, 2 and 4, $y_{\rho \ell}(x)$ has the value (19) for $\Im(\lambda_\rho) \geq 0$ and the value (20) for $\Im(\lambda_\rho) \leq 0$, provided the integral with respect to ξ is taken as an ℓ-fold in (19) and (20).

9. Equation (C).

Here infinitely many zeros of $G(\alpha)$ can occur. In order that a solution exist, the hypothesis of Theorem 29 must be satisfied for each individual zero point.

In the simplest case for example

$$\frac{y(x+\delta) - y(x)}{\delta} = f(x) \tag{22}$$

we have

$$G(\alpha) = \frac{e(\delta\alpha) - 1}{\delta} \quad .$$

This function has zeros at $\frac{2\pi}{\delta}\nu$, $\nu = 0, \pm 1, \pm 2, \ldots$. Therefore, if a solution of (22) is to exist, all the functions

$$e\left(-\frac{2\pi}{\delta}\nu x\right) f(x), \qquad \nu = 0, \pm 1, \pm 2, \ldots$$

must first of all be integrable in $\mathfrak{F}_0$.

But in contrast to equation (E), in the case of equation (C) the requirement made in Theorem 29 is not sufficient for solvability. The assertion cannot even be made that the equation is always solvable if $G(\alpha)$ is everywhere different from zero. For example, if

$$G(\alpha) = \sum_{\sigma=0}^{s} a_\sigma e(\delta_\sigma \alpha)$$

and the δ_σ are irrationally related to each other, then it is possible for $G(\alpha)$ to be never literally zero in $-\infty < \alpha < \infty$ and yet to become "arbitrarily small" at "almost periodic" intervals. If (C) is to be solvable for arbitrary $f(x)$, then by §22, 4, the function $G(\alpha)^{-1}$ must be a general multiplier. But if $G(\alpha)$ becomes arbitrarily small, then $G(\alpha)^{-1}$ becomes arbitrarily large, meaning that it does not remain bounded in $[-\infty, \infty]$. It can be shown however [which we will not undertake here] that each general multiplier must be bounded of necessity.

It is not so however, if $G(\alpha)$ is essentially different from zero. By §23, 4, the following theorem is namely valid.

THEOREM 31. For the solvability of (C), it is sufficient that $G(\alpha)$ be essentially different from zero, i.e., it satisfy a relation

$$|G(\alpha)| \geqq s > 0, \qquad (-\infty < \alpha < \infty) \ .$$

10. Equation (A).

Let it be known of the function

$$G(\alpha) = \sum_{\rho=0}^{r} \sum_{\sigma=0}^{s} a_{\rho\sigma}(i\alpha)^{\rho} e(\delta_\sigma \alpha) \ , \tag{23}$$

that the "principal part"

$$\gamma(\alpha) = \sum_{\sigma=0}^{s} a_{r\sigma} e(\delta_\sigma \alpha) \tag{24}$$

satisfies the relation

$$|\gamma(\alpha)| \geqq s > 0, \qquad (-\infty < \alpha < \infty) \tag{25}$$

One can then set

$$G(\alpha) = \gamma(\alpha) \left\{ (i\alpha)^r + \sum_{\rho=0}^{r-1} \sum_{\sigma=0}^{s} a_{\rho\sigma}(i\alpha)^{\rho} \frac{e(\delta_\sigma \alpha)}{\gamma(\alpha)} \right\} = \gamma(\alpha) G_1(\alpha) \ .$$

Each function

$$\frac{e(\delta_\sigma \alpha)}{\gamma(\alpha)}$$

together with all its derivatives is, by (25), bounded. Therefore $G_1(\alpha)$ has at most a finite number of zeros with well determined finite multiplicities. If no zeros at all exist, then in a manner entirely similar to that of equation (E), one concludes that the functions

$$(i\alpha)^\rho G_1(\alpha)^{-1}, \qquad 0 \leqq \rho \leqq r$$

are general multipliers. Since $\gamma(\alpha)^{-1}$ is also a general multiplier, so also are the functions

$$(i\alpha)^\rho G(\alpha)^{-1} = \gamma(\alpha)^{-1}(i\alpha)^\rho G_1(\alpha)^{-1}, \qquad 0 \leqq \rho \leqq r .$$

Furthermore, if the function $G_1(\alpha)$ does have zeros, then the procedure used for equation (E) can still be applied, and if one considers that $G_1(\alpha)$ has the same zeros with the same multiplicities as $G(\alpha)$, the following theorem ensues.

THEOREM 32. If the "principal part" (24) of the characteristic function (23) is essentially different from zero, then the hypothesis of Theorem 29 is also sufficient for the solvability of (A).

11. If the assumption (25) is dropped altogether, the following remark can still be made.

THEOREM 33. If the given function $f(x)$ is so constituted that its transform $E(\alpha)$ vanishes outside of an interval (a, b)

$$f(x) = \int_a^b E(\alpha)e(x\alpha)\, d\alpha , \tag{26}$$

then it is sufficient for the solvability of (A), that the hypothesis of Theorem 29 be satisfied in regard to the zeros of $G(\alpha)$ which fall in the interval (a, b).

REMARK. In regard to the zeros of $G(\alpha)$ which fall outside of (a, b), the given condition of Theorem 29, by the remark to Theorem 29, is satisfied by the function (26) of itself.

PROOF. If no zeros at all are contained in the interval (a, b), then the functions

$$(i\alpha)^\rho G(\alpha)^{-1}E(\alpha), \qquad 0 \leqq \rho \leqq r ,$$

by §23, 12, belong to $\mathfrak{T}_0$, and our equation is therefore solvable. The case in which zeros are present is settled by the same artifice as was used in 6.

§26. The Integral Equation

1. Let us consider the integral equation (F) previously introduced in §22

$$\lambda y(x) - \frac{1}{2\pi}\int K(\xi)y(x - \xi)\, d\xi = f(x) \quad . \tag{1}$$

In this case

$$G(\alpha) = \lambda - \gamma(\alpha) \tag{2}$$

where $\gamma(\alpha)$ denotes the transform of $K(\xi)$.

In the homogeneous case, $f(x) \equiv 0$, a non-trivial solution exists only if there is an entire α-interval in which $G(\alpha)$ vanishes. Indeed each function

$$y(x) \sim \int \varphi(\alpha)e(x\alpha)\, d\alpha$$

satisfies (1) whose transform vanishes everywhere where $G(\alpha)$ is different from zero. If therefore $G(\alpha)$ vanishes among other things, in an interval $A \leq \alpha \leq B$, and if $\psi(\alpha)$ is two times differentiable in this interval, and vanishes on the interval ends along with the first derivative, then the function

$$\int_A^B \psi(\alpha)e(x\alpha)\, d\alpha$$

is a solution of (1).

If λ is looked at as a parameter, and those values of λ, for which a non-trivial solution of the homogeneous equation exists, interpreted as eigenvalues of the equation, then the totality of eigenvalues consists of those numbers λ for which the equation

$$\gamma(\alpha) = \lambda$$

holds in an entire α-interval. Since each group of disjoint intervals on the straight line is countable, there is at most a countably infinite number of eigenvalues. However there is no need at all to have an eigenvalue. Rather if $\gamma(\alpha)$ in the most literal sense is "regular", namely analytic, then there is generally no eigenvalue at all because in that case $\gamma(\alpha)$ cannot be constant in an interval.

2. We remark that $\gamma(\alpha)$ is certainly analytic if there is a constant $a > 0$ such that

$$K(x) = O(e^{-a|x|})$$

as $|x| \longrightarrow \infty$. It is not difficult to show, although we shall not attempt it, that the integral

$$\gamma(\alpha) = \frac{1}{2\pi} \int K(\xi) e(-\alpha\xi)\, d\xi$$

then converges for all complex values of the strips

$$|\Im(\alpha)| < a \quad ,$$

and forms an analytic function there, which together with all its derivatives, is bounded in each partial strip $|\Im(\alpha)| \leqq a_0 < a$.

3. If $K(\xi)$ is "hermitian", i.e., $\overline{K(-\xi)} = K(\xi)$, then $\gamma(\alpha)$ is real, and only real eigenvalues enter the picture.

4. We shall now consider the non-homogeneous equation (1). For a given $\lambda \neq 0$, $G(\alpha)$ is either nowhere zero, as for example in the case where $|\lambda| > \operatorname{Max}|\gamma(\alpha)|$, or the zeros all lie in a finite interval since, because $\gamma(\alpha) \longrightarrow 0$ as $|\alpha| \longrightarrow \infty$,

$$\lim_{|\alpha| \to \infty} [\lambda - \gamma(\alpha)] = \lambda \neq 0 \quad .$$

THEOREM 34. For a given $\lambda \neq 0$, it is sufficient for the solvability of (1), that $\gamma(\alpha)$ be two times differentiable, be absolutely integrable together with the first two derivatives, and that one of the following three conditions be satisfied.

1) $G(\alpha)$ is nowhere zero.

2) $G(\alpha)$ has (what is always the case if $\gamma(\alpha)$ is analytic) only a finite number of zeros τ_χ with well determined finite multiplicities ℓ_χ, and the function

$$e(-\tau_\chi x) f(x)$$

is ℓ_χ-times integrable in $\mathfrak{F}_0$ for each zero τ_χ.

3) It is possible to specify a finite number of finite intervals $A_\chi \leqq \alpha \leqq B_\chi$ with the following properties. Each zero of $G(\alpha)$ is contained in the interior of one of the intervals, and the transform

$E(\alpha)$ of $f(x)$ vanishes in this interval.

REMARK IN REGARD TO ASSUMPTION 2. By the assumption that at each of the finitely many points τ_χ at which $G(\alpha)$ vanishes, a zero with multiplicity ℓ_χ be displayed, we mean that in a certain neighborhood of τ_χ it is possible to set

$$G(\alpha) = \{i(\alpha - \tau_\chi)\}^{\ell_\chi} G_*(\alpha) \quad ,$$

where $G_*(\alpha)$ is different from zero and is two times differentiable.

PROOF OF 1). Because $\gamma(\alpha)$ is continuous, $G(\alpha)$ is essentially different from zero. Setting

$$G^{-1} = \frac{1}{\lambda - \gamma(\alpha)} = \frac{1}{\lambda} + \frac{\gamma(\alpha)}{\lambda[\lambda - \gamma(\alpha)]} = \frac{1}{\lambda} + \gamma_\lambda(\alpha) \quad ,$$

we have the result that $\gamma_\lambda(\alpha)$, together with the first two derivatives with respect to α, is absolutely integrable, and is therefore a multiplier. Setting

$$K_\lambda(\xi) = \int \frac{\gamma(\alpha)}{\lambda[\lambda - \gamma(\alpha)]} e(\xi\alpha)\, d\alpha \quad ,$$

we obtain

$$y(x) = \frac{1}{\lambda} f(x) + \frac{1}{2\pi} \int K_\lambda(\xi) f(x - \xi)\, d\xi \quad .$$

For small values of λ^{-1} the solving kernel $K_\lambda(\xi)$ can be developed in a power series in the usual manner.

PROOF OF 2). The proof of the case where a finite number of zeros with finite multiplicity is considered follows exactly as it did in §25, 5.

PROOF OF 3). This case is even simpler than 2. Changing the function $G(\alpha)$ in the intervals (A_χ, B_χ) to a twice differentiable function $G_0(\alpha)$ which vanishes nowhere (cf. Appendix 16), we see that $G_0(\alpha)^{-1}$ is a multiplier. And because $G(\alpha)$ and $G_0(\alpha)$ are "equivalent" in regard to $f(x)$, cf. §23, 11, it follows that

$$y(x) \sim \int G_0(\alpha)^{-1} E(\alpha) e(x\alpha)\, d\alpha$$

is a solution of our equation, Q. E.D.

5. We shall now consider a very special integral equation occurring in the literature which does not come under (1), the equation

namely [77]

$$y'(x) = \frac{1}{2}\int_0 \frac{J_1(t)}{t}\left\{y(x + t) - y(x - t)\right\} dt \quad , \tag{3}$$

where $J_1(t)$ denotes the Bessel function of the first order. Writing for the right side

$$\frac{1}{2}\int^0 \frac{J_1(-t)}{-t}\, y(x - t)\, dt - \frac{1}{2}\int_0 \frac{J_1(t)}{t}\, y(x - t)\, dt \quad ,$$

it is evident that it consists of two combinations of $y(x)$ with functions of $\mathfrak{F}_0$. If we introduce a would-be representation

$$y(x) \sim \int \varphi(\alpha) e(x\alpha)\, d\alpha$$

then we obtain a "representation" of the equation (3) itself, if on the right side we insert

$$y(x \pm t) = \int \varphi(\alpha) e(\pm t\alpha) e(x\alpha)\, d\alpha$$

and interchange the integrations with respect to t and α. The permissibility of this interchange was established in the proof of Theorem 13. In fact the content of Theorem 13 is essentially identical with the statement that this interchange is admissible. In this way the relation

$$\int i\alpha\varphi(\alpha) e(x\alpha)\, d\alpha \sim \int \gamma(\alpha)\varphi(\alpha) e(x\alpha)\, d\alpha \tag{4}$$

is obtained, where

$$\gamma(\alpha) = \frac{1}{2}\int_0 \frac{J_1(t)}{t}\, [e(\alpha t) - e(-\alpha t)]\, dt = i\int_0 \frac{J_1(t)}{t} \sin \alpha t\, dt \quad .$$

Setting $\nu = 1$ in §16, (5), we obtain

$$\gamma(\alpha) = \begin{cases} i\alpha \text{ for } 0 \leq \alpha \leq 1 \quad , \\[2ex] \dfrac{i}{\alpha + \sqrt{\alpha^2 - 1}} \text{ for } 1 \leq \alpha < \infty \quad , \end{cases}$$

and in addition

$$\gamma(-\alpha) = -\gamma(\alpha) \quad .$$

From (4), it follows that

$$[i\alpha - \gamma(\alpha)]\varphi(\alpha) = 0 \quad .$$

The factor in front of $\varphi(\alpha)$ vanishes for $-1 \leq \alpha \leq 1$, and is different from zero for other values of α. From this fact, it is very easily seen that the totality of solutions of (3) consists exactly of those functions of $\mathfrak{F}_0$ whose transforms vanish for $|\alpha| > 1$.

§27. Systems of Equations

1. We consider the system of differential equations

$$(1) \qquad y'_\mu(x) = \sum_{\nu=1}^{m} a_{\mu\nu} y_\nu(x) + f_\mu(x), \qquad \mu = 1, 2, \ldots, m \quad ,$$

where $f_\mu(x)$ are functions of $\mathfrak{F}_0$. By a "solution", we mean a system of functions $y_1(x), \ldots, y_m(x)$ which are 1-time differentiable in $\mathfrak{F}_0$, and satisfy the system of equations in $[-\infty, \infty]$.[1] A characteristic function $G(\alpha)$ again plays a role in the question of its solvability, and it is the polynomial of the mth degree

$$G(\alpha) = \begin{vmatrix} i\alpha - a_{11} & -a_{12} & \cdots & -a_{1m} \\ -a_{21} & i\alpha - a_{22} & \cdots & -a_{2m} \\ \vdots & \vdots & & \\ -a_{m1} & -a_{m2} & \cdots & i_\alpha - a_{mm} \end{vmatrix}$$

THEOREM 35. In order that the system (1) have a solution, one of the following two conditions is sufficient.

1. The characteristic function $G(\alpha)$ have no zeros.

2. At each zero τ_χ of $G(\alpha)$ with multiplicity ℓ_χ, the functions

$$e(-\tau_\chi x)f_\mu(x), \qquad \mu = 1, 2, \ldots, m$$

be ℓ_χ-times integrable in $\mathfrak{F}_0$.

There is always one solution at most.

---

[1] It is sufficient to assume that the $y_\mu(x)$ belong to $\mathfrak{F}_0$, are differentiable, and satisfy the system of equations. The derivatives y'_μ are then of themselves functions of $\mathfrak{F}_0$ by reason of equation (1).

REMARK. In case 2., as contrasted with equation (A), cf. Theorem 29, the given condition is not necessary. This can be seen by an example. Consider the equations

$$y_1'(x) = i[y_1(x) + y_2(x)] + f_1(x)$$

$$y_2'(x) = -i[y_1(x) + y_2(x)] + f_2(x) .$$

Here

$$G(\alpha) = \begin{vmatrix} i\alpha - i & -i \\ i & i\alpha + i \end{vmatrix} = -\alpha^2 .$$

Hence $G(\alpha)$ has a two-fold zero at the point $\alpha = 0$. But if $p(x)$ is twice and $q(x)$ is once differentiable in $\mathfrak{F}_0$, then the equation system is solvable for the functions

$$f_1(x) = p''(x) + q'(x), \quad f_2(x) = p''(x) - q'(x) .$$

Its solution is

$$y_1(x) = 2ip(x) + p'(x) + q(x)$$

$$y_2(x) = -2ip(x) + p'(x) - q(x) .$$

But the functions $f_1(x)$ and $f_2(x)$ are two times integrable in $\mathfrak{F}_0$ only if $q(x)$ is one time integrable in $\mathfrak{F}_0$; however this does not hold if for example

$$q(x) = \left(\frac{\sin x}{x}\right)^2$$

since

$$\int q(x)\, dx \neq 0 ,$$

cf. §3, 4.

PROOF. Set, for $\mu = 1, 2, \ldots, m$

$$f_\mu(x) \sim \int E_\mu(\alpha) e(x\alpha)\, d\alpha \tag{3}$$

$$y_\mu(x) \sim \int \varphi_\mu(\alpha) e(x\alpha)\, d\alpha . \tag{4}$$

Substituting in (1), we obtain the equations

$$(5) \qquad i\alpha\varphi_\mu(\alpha) - \sum_{\nu=1}^{m} a_{\mu\nu}\varphi_\nu(\alpha) = E_\mu(\alpha), \qquad \mu = 1, 2, \ldots, m .$$

Denoting the algebraic complement of the determinant (2) by $G_{\mu\nu}(\alpha)$, it follows by (5) that

$$(6) \qquad G(\alpha)\varphi_\nu(\alpha) = \sum_{\mu=1}^{m} G_{\mu\nu}(\alpha)E_\mu(\alpha), \qquad \nu = 1, 2, \ldots, m .$$

Hence for all α, apart from the (isolated) zeros of $G(\alpha)$,

$$(7) \qquad \varphi_\nu(\alpha) = \sum_{\mu=1}^{m} \frac{G_{\mu\nu}(\alpha)}{G(\alpha)} E_\mu(\alpha) , \qquad \nu = 1, 2, \ldots, m .$$

Therefore there can be only one solution at most.

Assume now that the hypotheses of the theorem are fulfilled. Then by §25, 4-6, the functions

$$\frac{(i\alpha)^\rho}{G(\alpha)} E_\mu(\alpha)$$

belong to $\mathfrak{T}_0$ for $0 \leqq \rho \leqq m$. But since the polynomials $G_{\mu\nu}(\alpha)$ are at most of degree $(m - 1)$, it follows that the right sum in (7) is a function of $\mathfrak{T}_0$, and that the function $y_\nu(x)$ of $\mathfrak{F}_0$ belonging to it, is differentiable in $\mathfrak{F}_0$. It is verifiable at once by substitution that the $y_\nu(x)$ thus obtained, actually form a solution. Q.E.D.

2. The proceedings just described can be employed in the solution of very general systems of functional equations. We shall treat one more in extenso, namely the system of difference-equations

$$(8) \qquad y_\mu(x + \delta_\mu) = \sum_{\nu=1}^{m} a_{\mu\nu}y_\nu(x) + f_\mu(x), \qquad \mu = 1, 2, \ldots, m ,$$

with arbitrary real differences δ_μ. The characteristic function reads

$$(9) \qquad G(\alpha) = \begin{vmatrix} e(\delta_1\alpha) - a_{11} & - a_{12} & \cdots & - a_{1m} \\ - a_{21} & e(\delta_2\alpha) - a_{22} & \cdots & - a_{2m} \\ \vdots & \vdots & & \vdots \\ - a_{m1} & - a_{m2} & & e(\delta_m\alpha) - a_{mm} \end{vmatrix}$$

It is an expression of the form

$$\sum_{\sigma=0}^{m} a_\sigma e(\omega_\sigma \alpha) \tag{10}$$

with real exponents a_σ. We now make the assumption that the expression does not vanish identically. This always occurs for example if all δ_μ have one and the same sign, say $\delta_\mu > 0$.

THEOREM 36. If the characteristic function $G(\alpha)$ is essentially different from zero

$$|G(\alpha)| \geq S > 0, \qquad (-\infty < \alpha < \infty) \quad , \tag{11}$$

then the system (8) has a solution.

PROOF. Introducing representations (3) and (4) we obtain

$$e(\delta_\mu \alpha)\varphi_\mu(\alpha) - \sum_{\nu=1}^{m} a_{\mu\nu}\varphi_\nu(\alpha) = E_\mu(\alpha) \quad .$$

From this, equations (6) and (7) again follow, provided $G_{\mu\nu}$ is understood to be the algebraic complement of (9). Since further $G(\alpha)$ has only isolated zeros even if (11) is not satisfied, there is again only one solution. If now assumption (11) is satisfied, then $G(\alpha)^{-1}$ is a multiplier. Since the $G_{\mu\nu}(\alpha)$, as exponential polynomials of the form (10), are multipliers, it follows that functions (7) in the present meaning of the letters occurring there, are contained in $\mathfrak{T}_0$, and the functions of $\mathfrak{F}_0$ belonging to them form a solution of (8).

3. If assumption (11) is dropped, then as in §25, 11, one can state a sufficient condition for its solvability, whenever all the transforms $E_\mu(\alpha)$ vanish outside of an interval $A \leq \alpha \leq B$. It reads that for each zero τ_χ of $G(\alpha)$ lying in (A, B) with multiplicity ℓ_χ, the functions $e(-\tau_\chi x)f_\mu(x)$, $\mu = 1, 2, \ldots, m$, shall be ℓ_χ-times integrable in $\mathfrak{F}_0$.

CHAPTER VI

GENERALIZED TRIGONOMETRIC INTEGRALS

§28. Definition of the Generalized Trigonometric Integrals

1. In the previous chapter, we had to limit ourselves to such "solutions" of the functional equations under consideration which along with the required derivatives were absolutely integrable as $x \longrightarrow \pm \infty$. We shall now take into consideration functions of a more general behavior in the infinite domain.

Let $k = 0, 1, 2, \ldots$. We denote as functions class $\mathfrak{F}_k$ the totality of the functions $f(x)$ which are integrable in the finite region, and after division by x^k are absolutely integrable as $x \longrightarrow \pm \infty$, i.e., for which

$$(1) \qquad \int_{-1}^{1} |f(x)|\, dx + \int^{-1} \left| \frac{f(x)}{x^k} \right| dx + \int_{1} \left| \frac{f(x)}{x^k} \right| dx$$

is finite. It is evident that $\mathfrak{F}_k$ is contained in $\mathfrak{F}_{k+1}$, and that each function which as $x \longrightarrow \pm \infty$ does not grow stronger than a power $|x|^\ell$ $(\ell = 0, 1, 2, \ldots)$, is contained in $\mathfrak{F}_{\ell+2}$. Since

$$\frac{1}{2} \leqq \frac{1}{1 + |x|^k} \leqq 1, \quad \text{for} \quad |x| \leqq 1$$

$$\frac{1}{2|x|^k} \leqq \frac{1}{1 + |x|^k} \leqq \frac{1}{|x|^k}, \quad \text{for} \quad |x| \geqq 1 ,$$

the requirement that (1) be finite is equivalent to the requirement that

$$\int |f(x)| p_k(x)\, dx$$

be finite, where for brevity we have set

(2)
$$p_k(x) = \frac{1}{1 + |x|^k} \quad .$$

From the above, it is at once recognized that the functions class $\mathfrak{F}_k$ is additive. We shall frequently make use of the fact that along with $f(x)$, also $f(x + \lambda)$ belongs to $\mathfrak{F}_k$; also $f(x)g(x)$ if $g(x)$ is bounded (and integrable in the finite region).

2. In what follows, the letter k will be used only to designate the class index. If in certain explicit observations, the value of k will be limited to $k = 1, 2, 3, \ldots$ or $k = 2, 3, \ldots$, it will be understood that the case $k = 0$ or $k = 0, 1$ can be completed easily, and this completion will then be left to the reader.

3. Let $\varphi(\alpha)$ and $\psi(\alpha)$ be two given functions continuous in a finite or infinite interval $[A, B]$. We call $\varphi(\alpha)$ and $\psi(\alpha)$ "k-equivalent" in the considered interval, or for short "equivalent", and write

$$\varphi(\alpha) \overset{k}{\asymp} \psi(\alpha)$$

or shorter

$$\varphi(\alpha) \asymp \psi(\alpha) \quad ,$$

if the difference $\varphi(\alpha) - \psi(\alpha)$ is a polynomial of the (k-1)th degree in α.[1] If $k = 0$, equivalent is no different from identical. The k-equivalence has the three properties of an equality:

1) $\varphi(\alpha) \asymp \varphi(\alpha)$
2) $\varphi(\alpha) \asymp \psi(\alpha)$ implies $\psi(\alpha) \asymp \varphi(\alpha)$
3) $\varphi(\alpha) \asymp \psi(\alpha)$ and $\psi(\alpha) \asymp \chi(\alpha)$ implies $\varphi(\alpha) \asymp \chi(\alpha)$.

The following observation is very important. If $\varphi(\alpha)$ and $\psi(\alpha)$ are equivalent in the adjacent intervals $[A, C]$, $[C, B]$, they need not be equivalent in $[A, B]$ for $k \geq 2$, even if the functions themselves are continuous. For the polynomials $P(\alpha)$ and $Q(\alpha)$ by which $\varphi(\alpha)$ and $\psi(\alpha)$ differ in $[A, C]$ and $[C, B]$ need not be identical in spite of the continuous union at $\alpha = C$. But if the two intervals $[A, C_1]$ and $[C_2, B]$, $C_2 < C_1$, overlap, then $P(\alpha)$, $Q(\alpha)$ are identical and $\varphi(\alpha)$, $\psi(\alpha)$ are equivalent in $[A, B]$.

4. Let us consider a function $f(x)$ of $\mathfrak{F}_0$ whose transform $E(\alpha)$ of $\mathfrak{T}_0$ we shall hereafter denote by $E(\alpha, 0)$. By $E(\alpha, k)$, $k \geq 1$, we shall mean the k-th integral of $E(\alpha, 0)$ to within the unique k-equivalence, i.e.,

[1] By a polynomial of the (k-1)th degree, we mean a polynomial of degree $(k - 1)$ at most.

$$E^{(k)}(\alpha, k) = E(\alpha, 0) \quad . \tag{3}$$

If in particular we set

$$E(\alpha, k) = \int_0^\alpha E(\beta, k - 1)\, d\beta \quad ,$$

then by the substitution

$$E(\alpha) = \frac{1}{2\pi}\int f(x) e(-\alpha x)\, dx$$

and by the interchange of the order of integration, successively for $k = 1, 2, 3, \ldots,$ we obtain

$$E(\alpha, k) = \frac{1}{2\pi}\int f(x)\, \frac{e(-\alpha x) - h_k(\alpha x)}{(-ix)^k}\, dx \quad ,$$

where

$$h_k(t) = \sum_{x=0}^{k-1} \frac{(-it)^x}{x!} \quad .$$

Since however $h_k(\alpha x)$ is a polynomial of degree $(k - 1)$ in α, we have

$$2\pi E(\alpha, k) \overset{k}{\asymp} \int_{-1} f(x)\, \frac{e(-\alpha x) - h_k(\alpha x)}{(-ix)^k}\, dx + \left(\int^{-1} + \int_1\right) \frac{f(x)}{(-ix)^k}\, e(-\alpha x)\, dx \quad . \tag{4}$$

Setting for brevity

$$L_k \equiv L_k(\alpha, x) = \begin{cases} \displaystyle\sum_{x=0}^{k-1} \frac{(-i\alpha x)^x}{x!} & \text{for } |x| \leq 1 \quad , \\ 0 & \text{for } |x| > 1 \end{cases} \tag{4'}$$

and $L_0 = 0$, we can write

$$E(\alpha, k) \overset{k}{\asymp} \frac{1}{2\pi}\int f(x)\, \frac{e(-\alpha x) - L_k}{(-ix)^k}\, dx \quad . \tag{5}$$

As is recognizable by (4), the integral on the right is now also absolutely convergent if $f(x)$ belongs to $\mathfrak{F}_k$. This motivates the following definition.

Let $f(x)$ be a function of $\mathfrak{F}_k$. By the k-transform of $f(x)$ we mean the function of α determined by (5) to within the polynomial of degree $(k - 1)$. We shall denote it by $E(\alpha, k)$, $\Phi(\alpha, k)$, etc., and if the value of the index k is fixed by the content, then we shall also write $E(\alpha)$, $\Phi(\alpha)$, etc.

The totality of all k-transforms will be denoted by $\mathfrak{T}_k$.

If a function of $\mathfrak{F}_k$ is considered as a function of $\mathfrak{F}_{k+\ell}$, $\ell \geq 1$, then for the transforms $E(\alpha, k)$ and $E(\alpha, k + \ell)$ belonging to them, there is, as one may easily verify, the important relation

$$E(\alpha, k) \overset{k}{\asymp} E^{(\ell)}(\alpha, k + \ell) \quad . \tag{6}$$

The class $\mathfrak{T}_k$ is additive: $f = c_1 f_1 + c_2 f_2$ implies

$$E(\alpha, k) \overset{k}{\asymp} c_1 E_1(\alpha, k) + c_2 E_2(\alpha, k) \quad .$$

If the first term on the right of (4) is denoted by $2\pi\Phi(\alpha)$ and the second by $2\pi\Psi(\alpha)$, then the function

$$\Phi^{(k)}(\alpha) = \frac{1}{2\pi} \int_{-1}^{1} f(x) e(-\alpha x)\, dx$$

is a 0-transform, and the function $\Psi(\alpha)$ is likewise contained in $\mathfrak{T}_0$. Hence both are continuous and converge to 0 as $\alpha \longrightarrow \pm\infty$. By the well known formula (for an arbitrary k-times continuously differentiable function $\Phi(\alpha)$)

$$\Phi(\alpha) \overset{k}{\asymp} \frac{1}{(k-1)!} \int_{0}^{\alpha} (\alpha - \beta)^{k-1} \Phi^{(k)}(\beta)\, d\beta \quad ,$$

one concludes without difficulty that $\Phi(\alpha) = o(|\alpha|^k)$ as $\alpha \longrightarrow \pm\infty$. It therefore follows that each function $E(\alpha, k)$ is continuous, and

$$E(\alpha, k) = o(|\alpha|^k) \quad \text{as} \quad \alpha \longrightarrow \pm\infty \quad . \tag{7}$$

5. A first justification for the introduction of the k-transforms is offered by the following theorem.

THEOREM 37. If the k-transforms of two functions of $\mathfrak{F}_k$ agree (i.e., are k-equivalent), then the functions are identical [79].

PROOF. Consider the difference function

$$\Delta^k\varphi(\alpha) = \varphi(\alpha + k) - \binom{k}{1}\varphi(\alpha + k - 1) + \dots + (-1)^k\binom{k}{k}\varphi(\alpha) \quad .$$

It vanishes for a polynomial of degree $(k - 1)$; hence

$$\Delta^k L_k(\alpha, x) = 0$$

where $L_k(\alpha, x)$ is defined by (4'). On the other hand

$$\Delta^k e(-\alpha x) = e(-\alpha x)[e(-x) - 1]^k \quad .$$

It therefore follows from (5) that

$$\Delta^k E(\alpha, k) = \frac{1}{2\pi}\int g(x)e(-\alpha x)\,dx \quad ,$$

where

$$g(x) = f(x)\left(\frac{e(-x) - 1}{-ix}\right)^k \quad . \tag{8}$$

The function $g(x)$ belongs to $\mathfrak{F}_0$ and $\Delta^k E(\alpha, k)$ is its 0-transform. If now two functions $f(x)$ have equivalent k-transforms, then the corresponding functions $g(x)$ have the same 0-transform and are identical by Theorem 14. But then the functions $f(x)$ are also identical, Q.E.D.

If we are given a function $f(x)$ and its k-transform $E(\alpha)$ then the fact of their belonging together will be denoted by the symbolic relation[1]

$$f(x) \sim \int e(x\alpha)d^k E(\alpha) \quad .$$

We also call this relation or its right side a "representation" of $f(x)$. Cf. §31 in regard to the possible "convergence" of this representation to $f(x)$.

---

[1] If it were not for typographical considerations, we should have written

$$\int e(x\alpha)\,\frac{d^k E(\alpha)}{d\alpha^{k-1}}$$

instead of

$$\int e(x\alpha)d^k E(\alpha) \quad .$$

6. One easily finds

$$(9)\quad \int \frac{e(-\alpha x) - L_2(\alpha,x)}{(ix)^2}\,dx = \left(\int_0^1 + \int_1\right) \frac{e(-\alpha x) + e(\alpha x) - 2}{(-ix)^2}\,dx + 2\int_1 \frac{dx}{(-ix)^2}$$

$$= 4\int_0 \left(\frac{\sin\frac{\alpha}{2}x}{x}\right)^2 dx - 2 = 2|\alpha| \cdot \frac{\pi}{2} - 2 \quad .$$

Hence for $f(x) = 1$

$$E(\alpha,\ 2) \asymp \frac{1}{2}\ |\alpha| \quad .$$

From (6) there follows more generally for $k \geq 2$

$$E(\alpha,\ k) \asymp \frac{1}{2}\,\frac{1}{(k-1)!}\ |\bar{\alpha}|^{k-1} \quad ,$$

where by

$$|\bar{\gamma}|^k$$

we mean, both here and in what follows, that function which has the value γ^k for $\gamma > 0$ and the value $-\gamma^k$ for $\gamma < 0$.

<u>For</u> $f(x) = x^\mu$, $\mu = 0, 1, 2, \ldots,$ <u>and</u> $k \geq \mu + 2$, <u>we have</u>

$$2\pi E(\alpha,\ k) \overset{k}{\asymp} i^\mu \int \frac{e(-\alpha x) - L_k(\alpha x)}{(-ix)^{k-\mu}}\,dx \overset{k}{\asymp} i^\mu \int \frac{e(-\alpha x) - L_{k-\mu}}{(-ix)^{k-\mu}}\,dx \quad .$$

<u>Therefore</u>

$$E(\alpha,\ k) \asymp \frac{1}{2}\,\frac{i^\mu}{(k-\mu-1)!}\ |\bar{\alpha}|^{k-\mu-1} \quad .$$

Let $f(x)$ be a function of $\mathfrak{F}_k$, τ real and $\Phi(\alpha)$ the k-transform of $f(x) \cdot e(\tau x)$. Then

$$2\pi\Phi(\alpha) \asymp \int f(x)\,\frac{e(-\ (\alpha-\tau)x) - e(\tau x)L_k(\alpha,x)}{(-ix)^k}\,dx$$

$$\asymp \int f(x)\,\frac{e(-\ (\alpha-\tau)x) - L_k(\alpha-\tau,x)}{(-ix)^k}\,dx + \int f(x)\,\frac{L_k(\alpha-\tau,x) - e(\tau x)L_k(\alpha,x)}{(-ix)^k}\,dx$$

The second integral is a polynomial of degree $(k - 1)$ in α. Hence

$$(10)\qquad \Phi(\alpha) \overset{k}{\asymp} E(\alpha - \tau,\ k) \quad ,$$

where $E(\alpha, k)$ denotes the k-transform of $f(x)$. In particular, for $k \geq \mu + 2$, $x^\mu e(\tau x)$ has the k-transform

$$\frac{1}{2} \frac{i^\mu}{(k-\mu-1)!} \overline{|\alpha - \tau|}^{k-\mu-1} . \tag{11}$$

7. Let real numbers $\tau_1, \tau_2, \ldots, \tau_n$ be given which differ from one another, and also complex numbers $c_{\mu\nu}$. The function

$$f(x) = \sum_{\mu=0}^{k-2} \sum_{\nu=0}^{n} c_{\mu\nu} x^\mu e(\tau_\nu x), \qquad (k \geq 2) \tag{12}$$

belongs to $\mathfrak{F}_k$. <u>We call each such function a trivial function of</u> $\mathfrak{F}_k$ (with exponents τ_ν). Its k-transform amounts to

$$\frac{1}{2} \sum_{\mu=0}^{k-2} \sum_{\nu=1}^{n} c_{\mu\nu} \frac{i^\mu}{(k-\mu-1)!} \overline{|\alpha - \tau_\nu|}^{k-\mu-1} + P(\alpha) , \tag{13}$$

where $P(\alpha)$ is any polynomial of degree $(k - 1)$. <u>We call each function</u> (13) <u>a trivial function of</u> $\mathfrak{T}_k$ (with exponents τ_ν). The function (13) has the following property. In each of the $n + 1$ intervals

$$[-\infty, \tau_1], [\tau_1, \tau_2], \ldots, [\tau_{n-1}, \tau_n], [\tau_n, \infty] ,$$

it is equal to a polynomial of degree $k - 1$. The polynomials are continuously joined at the points τ_ν, yet their derivatives are not. The λ-th derivative, $\lambda = 1, 2, \ldots, k - 1$, has the jump

$$S_{\lambda\nu} = c_{k-\lambda-1,\nu} \cdot i^{k-\lambda-1}$$

at the point τ_ν.

By the inverse formula

$$c_{\mu\nu} = (-i)^\mu S_{k-\mu-1,\nu}$$

one can obtain any prescribed jumps of the derivatives at the points τ_ν through suitable values of the constants $c_{\mu\nu}$. Therefore the totality of the trivial functions of $\mathfrak{T}_k$ consists of all those functions in α which are composed in the stated manner, of finitely many continuously joined polynomial pieces of degree $(k - 1)$. - A trivial function of $\mathfrak{T}_k$ can also be characterized as follows. The function is continuous, and if finitely many suitable α-points are removed, the function is k-times continuously differentiable and the k-th derivative vanishes.

§29. Further Particulars About the Functions of $\mathfrak{F}_k$

1. We call a sequence of functions in $\mathfrak{F}_k$

$$f_1(x),\ f_2(x),\ f_3(x),\ \dots \tag{1}$$

k-convergent or k-convergent to $f(x)$ (also in $\mathfrak{F}_k$) if, cf. §28, (2),

$$\lim_{\substack{m\to\infty\\ n\to\infty}} \int |f_m(x) - f_n(x)|\ p_k(x)\ dx = 0 \tag{2}$$

or

$$\lim_{n\to\infty} \int |f_n(x) - f(x)|\ p_k(x)\ dx = 0\ . \tag{3}$$

A function $\varphi(x)$ then and only then belongs to $\mathfrak{F}_k$ if $\varphi(x)p_k(x)$ belongs to $\mathfrak{F}_o$. Relation (2) states that the functions $f_n(x)p_k(x)$ are o-convergent, and relation (3) states that they are o-convergent to $f(x)p_k(x)$. By §13, 2, the following is therefore valid. If a sequence (1) is k-convergent, then there is a certain function $f(x)$ of $\mathfrak{F}_k$ - we call it the k-limit of (1) - towards which it k-converges. Each function $f(x)$ of $\mathfrak{F}_k$ is the k-limit of functions of $\mathfrak{F}_o$, perhaps of the "severed" functions

$$f_n(x) = \begin{cases} f(x) & \text{for } |x| \leq n \\ 0 & \text{for } |x| > n\ . \end{cases} \tag{4}$$

2. For the function

$$P(\eta) = \int \frac{dx}{(1+|x|^k)(1+|x-\eta|^{k+2})}\ , \tag{5}$$

we have

$$P(\eta) \leq \frac{c_k}{1 + |\eta|^k}\ , \tag{6}$$

where c_k is a constant dependent only on k. Writing the integral (5) in the form

$$\int \frac{dx}{(1+|x-\eta|^k)(1+|x|^{k+2})} , \tag{7}$$

it is immediately seen that it is bounded in η, and for say $\eta > 1$, the valuation (6) results if the limits of integration of (7) are decomposed into the three intervals

$$-\infty < x \leqq \tfrac{1}{2}\eta, \quad \tfrac{1}{2}\eta \leqq x \leqq \tfrac{3}{2}\eta, \quad \tfrac{3}{2}\eta \leqq x < \infty .$$

3. Let $f(x)$ be a function of $\mathfrak{F}_k$. Because of (6)

$$\int |f(\eta)|P(\eta)\, d\eta \leqq C_k \int |f(\eta)p_k(\eta)|\, d\eta ,$$

and if, after substitution of the integral (5), the order of integration is interchanged on the left with respect to x and η (Appendix 7, 10), there results

$$\int \frac{dx}{1+|x|^k} \int \frac{|f(\eta)|\, d\eta}{1+|x-\eta|^{k+2}} \leqq C_k \int |f(\eta)|p_k(\eta)\, d\eta .$$

Hence the function

$$\varphi(x) \doteq \int \frac{f(x+\xi)\, d\xi}{1+|\xi|^{k+2}} = \int \frac{f(\eta)\, d\eta}{1+|x-\eta|^{k+2}} \tag{8}$$

is a function of $\mathfrak{F}_k$, and what is more

$$\int |\varphi(x)|p_k(x)\, dx \leqq C_k \int |f(x)|p_k(x)\, dx . \tag{9}$$

4. We deduce from this the following theorem

THEOREM 38. Let the sequence $f_1(x)$, $f_2(x)$, $f_3(x)$, ... be k-convergent to $f_0(x)$. With a function $K(\xi)$ for which

$$|K(\xi)| \leqq A(1+|\xi|^{k+2})^{-1} , \tag{10}$$

we form the functions

$$g_\nu(\xi) = \frac{1}{2\pi}\int f_\nu(x-\xi)K(\xi)\, d\xi = \frac{1}{2\pi}\int f_\nu(x+\xi)K(-\xi)\, d\xi , \tag{11}$$

$$\nu = 0, 1, 2, 3, \ldots .$$

Then the $g_\nu(x)$ are likewise functions of $\mathfrak{F}_k$ and the sequence $g_1(x), g_2(x), g_3(x), \ldots$ is k-convergent to $g_0(x)$.

PROOF. Because

$$|g_\nu(x)| \leqq \frac{A}{2\pi} \int \frac{|f_\nu(x+\xi)| \, d\xi}{1 + |\xi|^{k+2}} ,$$

it follows by 3 that the $g_\nu(\lambda)$ belong to $\mathfrak{F}_k$.

In consequence of

$$|g_n(x) - g_0(x)| \leqq \frac{A}{2\pi} \int \frac{|f_n(x+\xi) - f_0(x+\xi)|}{1 + |\xi|^{k+2}} \, d\xi ,$$

the inequality

$$(12) \qquad \int |g_n(x) - g_0(x)| p_k(x) \, dx \leqq \frac{AC_k}{2\pi} \int |f_n(x) - f_0(x)| p_k(x) \, dx$$

holds because of 3. Hence the asserted convergence behavior results.

THEOREM 39. Let $f(x)$ be a function of $\mathfrak{F}_k$ and $K(\xi)$ a function which satisfies (10). We consider, say for $n = 1, 2, 3, \ldots$, the functions belonging likewise to $\mathfrak{F}_k$ (by Theorem 38)

$$(13) \qquad f_n(x) = \int f\left(x + \frac{\xi}{n}\right) K(\xi) \, d\xi .$$

These functions are k-convergent to the function

$$(14) \qquad f_*(x) = f(x) \cdot \int K(\xi) \, d\xi$$

i.e.,

$$(15) \qquad \lim_{n \to \infty} \int |f_n(x) - f_*(x)| p_k(x) \, dx = 0 .$$

In particular

$$(16) \qquad \lim_{n \to \infty} \int_{x_0}^{x_1} |f_n(x) - f_*(x)| \, dx = 0$$

is valid for finite numbers x_0, and x_1.

PROOF. We shall make use (without proof) of the following theorem of Lebesgue [80]. If $f(x)$ is integrable in (A, B), then

$$\lim_{\xi \to 0} \int_a^b |f(x) - f(x + \xi)| \, dx = 0 ,$$

for $A < a < b < B$. It expresses the fact that each integrable function possesses a certain "continuity in the mean".

Let $f(x)$ be a function of $\mathfrak{F}_k$. By the theorem just stated,

$$\lim_{\xi \to 0} \int_a^b |f(x) - f(x + \xi)| p_k(x) \, dx = 0$$

for every two finite numbers a, b. On the other hand

$$\lim_{b \to \infty} \int_b |f(x + \xi)| p_k(x) \, dx = \lim_{a \to -\infty} \int^a |f(x + \xi)| p_k(x) \, dx = 0 ,$$

and it is found without difficulty that this limit relation holds uniformly for all ξ in each interval $|\xi| \leqq \xi_0$. From this, we obtain the result that the function

$$\delta(\xi) = \int |f(x) - f(x + \xi)| p_k(x) \, dx \tag{17}$$

converges to zero as $\xi \longrightarrow 0$.

Let ξ_0 be a fixed number > 0. For variable n, set

$$K_n(\xi) = \begin{cases} nK(n\xi) & \text{for } |\xi| \leqq \xi_0 \\ 0 & \text{for } |\xi| > \xi_0 , \end{cases} \tag{18}$$

and

$$H_n(\xi) = nK(n\xi) - K_n(\xi) . \tag{19}$$

We set correspondingly[1]

$$g_n(x) = \int f(x + \xi) K_n(\xi) \, d\xi, \quad h_n(x) = \int f(x + \xi) H_n(\xi) \, d\xi$$

$$g_*(x) = f(x) \int K_n(\xi) \, d\xi, \qquad h_*(x) = f(x) \int H_n(\xi) \, d\xi ,$$

[1] The functions $g_*(x)$ and $h_*(x)$ also depend on n.

so that

$$f_n(x) - f_*(x) = g_n(x) - g_*(x) + h_n(x) - h_*(x) \quad .$$

Since

$$\int |f_n - f_*| p_k \, dx \leqq \int |g_n - g_*| p_k \, dx + \int |h_n - h_*| p_k \, dx \quad ,$$

it is sufficient for the proof of (15) to show that

$$(20) \qquad \lim_{n \to \infty} \int |h_n - h_*| p_k \, dx = 0$$

and

$$(21) \qquad \int |g_n - g_*| p_k \, dx \leqq \eta(\xi_0) \quad ,$$

where $\eta(\xi_0)$ is independent of n and approaches zero as $\xi_0 \longrightarrow 0$. By (10), it follows that

$$(22) \qquad |H_n(\xi)| \leqq \frac{A_n}{1 + |\xi|^{k+2}}$$

where

$$(23) \qquad \lim_{n \to \infty} A_n = 0 \quad .$$

By 3.

$$\int |h_n(x)| p_k(x) \, dx \leqq A_n C_k \int |f(x)| p_k(x) \, dx$$

$$\int |h_*(x)| p_k(x) \, dx \leqq A_n \int |f(x)| p_k(x) \, dx \cdot \int \frac{d\xi}{1 + |\xi|^{k+2}} \quad ,$$

and because of (23), (20) follows. Further, denoting the upper limit of the function $\delta(\xi)$ defined by (17) in the interval $-\xi_0 \leqq \xi \leqq \xi_0$, by $\epsilon(\xi_0)$, we have

$$\int |g_n - g_*| p_k \, dx \leqq \iint |f(x + \xi) - f(x)| p_k(x) |K_n(\xi)| \, dx \, d\xi$$

$$\leqq \int_{-\xi_0}^{\xi_0} \delta(\xi) |K_n(\xi)| \, d\xi \leqq \epsilon(\xi_0) \cdot n \int |K(n\xi)| \, d\xi = \epsilon(\xi_0) \cdot \int |K(\xi)| \, d\xi = \eta(\xi_0)$$

Q.E.D.

5. If $f(x)$ and $f'(x)$ both belong to $\mathfrak{F}_k$, then

$$f(x) = o(|x|^k), \qquad (x \longrightarrow \pm\infty) . \tag{24}$$

For if this relation did not hold say as $x \longrightarrow +\infty$, then there would be a number $A > 0$, and a set of points

$$1 < x_1 < x_2 < x_3 < \cdots \longrightarrow \infty , \tag{25}$$

such that

$$|f(x_\nu)| \geqq Ax_\nu^k . \tag{26}$$

We can assume that

$$x_{\nu+1} - x_\nu \geqq 1 , \qquad \nu = 1, 2, 3, \ldots , \tag{27}$$

[otherwise one takes a suitable subset of (25)]. By the finiteness of

$$\int_1 \frac{|f'(x)|}{x^k}\, dx \tag{28}$$

it follows easily that

$$\int_{x_\nu}^{x_\nu+1} |f'(x)|\, dx = o(x_\nu^k) .$$

Now if $x_\nu \leqq x \leqq x_\nu + 1$, then

$$|f(x) - f(x_\nu)| \leqq \int_{x_\nu}^{x_\nu+1} |f'(x)|\, dx = o(x_\nu^k) ;$$

therefore by (26) there is a ν_0 such that for $\nu > \nu_0$,

$$|f(x)| \geqq \tfrac{1}{2} Ax_\nu^k \geqq \tfrac{1}{4} Ax^k$$

in the interval

$$x_\nu \leqq x \leqq x_\nu + 1 .$$

Because of (27), we would now have

$$\int_1^\infty \frac{|f(x)|}{x^k}\,dx \geq \sum_{\nu=1}^{\infty} \int_{x_\nu}^{x_\nu+1} \frac{|f(x)|}{x^k}\,dx = \infty$$

in contradiction to the assumption that $f(x)$ belongs to $\mathfrak{F}_k$.

6. We call $f(x)$ r-times differentiable in $\mathfrak{F}_k$ if the derivatives $f'(x)$, $f''(x)$, ..., $f^{(r)}(x)$ exist and together with $f(x)$ belong to $\mathfrak{F}_k$. We call $f(x)$ r-times integrable in $\mathfrak{F}_k$ if the k-th integral of $f(x)$ can be so normalized that it is r-times differentiable in $\mathfrak{F}_k$.

7. Let $K(\xi)$ be r-times differentiable, and together with the first r-derivatives satisfy (10). The function

$$g(x) = \frac{1}{2\pi}\int K(\xi)f(x - \xi)\,d\xi = \frac{1}{2\pi}\int f(\xi)K(x - \xi)\,d\xi \quad ,$$

formed with a function $f(x)$ of $\mathfrak{F}_k$, is r-times differentiable in $\mathfrak{F}_k$, and what is more

$$g^{(\rho)}(x) = \frac{1}{2\pi}\int K^{(\rho)}(\xi)f(x - \xi)\,d\xi \ , \qquad \rho = 0, 1, 2, \ldots, r \ .$$

The proof can be limited to the case $r = 1$; indeed it is sufficient to show that the function (belonging to $\mathfrak{F}_k$)

$$h(x) = \frac{1}{2\pi}\int K'(\xi)f(x - \xi)\,d\xi = \frac{1}{2\pi}\int f(\xi)K'(x - \xi)\,d\xi$$

is the derivative of $g(x)$. We form the function

$$\varphi(x) = \int_0^x h(\eta)\,d\eta = \frac{1}{2\pi}\int_0^x d\eta \int f(\xi)K'(\eta - \xi)\,d\xi \ .$$

Interchanging the integrations with respect to ξ and η (cf. the observations in 3. for the admissibility of this interchange), we obtain

$$\varphi(x) = \frac{1}{2\pi}\int f(\xi)K(x - \xi)\,d\xi - \frac{1}{2\pi}\int f(\xi)K(0 - \xi)\,d\xi = g(x) - g(0) \ ,$$

Q.E.D.

Moreover, if $f(x)$ is r-times differentiable in $\mathfrak{F}_k$, then because of

$$f^{(\rho)}(x) = o(|x|^k) \ , \qquad \rho = 0, 1, 2, \ldots, r - 1 \ ,$$

we obtain by ρ-times partial integration, cf. 5,

$$g^{(\rho)}(x) = \frac{1}{2\pi}\int K(\xi) f^{(\rho)}(x - \xi)\, d\xi , \qquad \rho = 0, 1, \ldots, r .$$

8. If $f(x)$ belongs to $\mathfrak{F}_0$, then of the function

$$F(x) = \int_{x_0}^{x} f(\xi)\, d\xi$$

one can in general only assert that it is bounded; therefore it belongs to $\mathfrak{F}_2$. Over against this we shall show the following. <u>If</u> $f(x)$ <u>belongs to</u> $\mathfrak{F}_k$, $k \geqq 1$, <u>then</u> $F(x)$ <u>belongs to</u> $\mathfrak{F}_{k+1}$. Since a constant belongs to $\mathfrak{F}_{k+1}$, $k \geqq 1$, it does not matter which fixed value is selected for x_0. We therefore take say $x_0 = 0$. Further it is sufficient to show that the numbers

$$A_n = \int_1^n \frac{|F(x)|}{x^{k+1}}\, dx , \qquad n = 1, 2, 3, \ldots$$

and correspondingly the numbers

$$\int_{-n}^{-1} \frac{|F(x)|}{|x|^{k+1}}\, dx$$

are bounded. If one sets

$$a_\nu = \int_\nu^{\nu+1} |f(x)|\, dx ,$$

then because of the membership of $f(x)$ in $\mathfrak{F}_k$, the series

$$a_0 + \frac{a_1}{1^k} + \frac{a_2}{2^k} + \frac{a_3}{3^k} + \cdots \tag{29}$$

is convergent. But for $\mu \leqq x \leqq \mu + 1$, $\mu \geqq 0$,

$$|F(x)| \leqq \sum_{\nu=0}^{\mu} a_\nu = F_\mu .$$

Therefore

$$A_n \leq \sum_{\mu=1}^{n} \frac{F_\mu}{\mu^{k+1}} .$$

This yields

$$A_n \leq \sum_{\nu=0}^{n} c_\nu a_\nu ,$$

where for $\nu \geq 1$

$$c_\nu = \sum_{\mu=\nu}^{\infty} \frac{1}{\mu^{k+1}} \leq \frac{c}{\nu^k}$$

and $c_0 = c_1$. Because of the convergence of (29), A_n is therefore bounded.

§30. Further Particulars About the Functions of $\mathfrak{T}_k$

1. Let us recall §28, 3. If the function sets $\varphi_n(\alpha)$, $\psi_n(\alpha)$ converge in $A < \alpha < B$, to the functions $\varphi(\alpha)$, $\psi(\alpha)$, and if (for fixed k)

$$\varphi_n(\alpha) \asymp \psi_n(\alpha) , \qquad n = 1, 2, 3, \ldots,$$

then

$$\varphi(\alpha) \asymp \psi(\alpha) .$$

If namely the polynomials of degree $(k - 1)$

$$\varphi_n(\alpha) - \psi_n(\alpha)$$

converge in $[A, B]$ as $n \longrightarrow \infty$, then the limit function is also a polynomial of this type. This follows if k fixed points $\alpha_1, \alpha_2, \ldots, \alpha_k$ are selected, and in the Lagrange interpolation formula

$$\varphi_n(\alpha) - \psi_n(\alpha) = \sum_{\chi=1}^{k} \frac{(\varphi_n(\alpha_\chi) - \psi_n(\alpha_\chi))\Omega(\alpha)}{(\alpha - \alpha_\chi)\Omega'(\alpha_\chi)}, \qquad \Omega(\alpha) = \prod_{\chi=1}^{k} (\alpha - \alpha_\chi)$$

one allows $n \longrightarrow \infty$ on both sides.

We hold an arbitrary point α_0 of the interval $[A, B]$ fixed. For any function $\Phi(\alpha)$ in $[A, B]$, we shall understand by

$$\int_{(r)}^{\alpha} \Phi(\alpha)\, d\alpha$$

the definite r-th integral of $\Phi(\alpha)$ in the normalization

$$\int_{\alpha_0}^{\alpha} d\alpha_1 \int_{\alpha_0}^{\alpha_1} \cdots \int_{\alpha_0}^{\alpha_{r-2}} d\alpha_{r-1} \int_{\alpha_0}^{\alpha_{r-1}} \Phi(\alpha_r)\, d\alpha_r \quad .$$

In particular therefore

$$\int_{(1)}^{\alpha} \Phi(\alpha)\, d\alpha = \int_{\alpha_0}^{\alpha} \Phi(\alpha)\, d\alpha \quad .$$

For $k = 1$, the formula of partial integration

$$\int_{(1)}^{\alpha} \chi(\alpha)\psi'(\alpha)\, d\alpha = \chi(\alpha)\psi(\alpha) - \int_{(1)}^{\alpha} \chi'(\alpha)\psi(\alpha)\, d\alpha - \chi(\alpha_0)\psi(\alpha_0)$$

is valid.

We can also write for this

$$\int_{(1)}^{\alpha} \chi(\alpha)\psi'(\alpha)\, d\alpha \overset{1}{\asymp} \chi(\alpha)\psi(\alpha) - \int_{(1)}^{\alpha} \chi'(\alpha)\psi(\alpha)\, d\alpha \quad .$$

For general k, the analagous formula reads

$$(1) \qquad \int_{(k)}^{\alpha} \chi(\alpha)\psi^{(k)}(\alpha)\, d\alpha \overset{k}{\asymp} \chi(\alpha)\psi(\alpha) + \sum_{r=1}^{k} (-1)^r \binom{k}{r} \int_{(r)}^{\alpha} \chi^{(r)}(\alpha)\psi(\alpha)\, d\alpha \quad .$$

We leave its verification, by differentiating both sides k-times, to the reader.

Now let $\psi(\alpha)$ be a continuous and $\chi(\alpha)$ a k-times continuously differentiable function. We then consider the function

$$(2) \qquad \varphi(\alpha) \overset{k}{\asymp} \chi(\alpha)\psi(\alpha) + \sum_{r=1}^{k} (-1)^r \binom{k}{r} \int_{(r)}^{\alpha} \chi^{(r)}(\alpha)\psi(\alpha)\, d\alpha$$

determined to within a polynomial of degree $(k - 1)$. If $\psi(\alpha)$ is also k-times continuously differentiable, then by (1)

$$(3) \qquad \varphi(\alpha) \overset{k}{\asymp} \int_{(k)}^{\alpha} \chi(\alpha)\psi^{(k)}(\alpha)\, d\alpha \quad ,$$

for which we can write more incisively

$$(4) \qquad \varphi^{(k)}(\alpha) = \chi(\alpha)\psi^{(k)}(\alpha) \quad .$$

In analogy to (3) and (4), we write, if only continuity of $\psi(\alpha)$ is assumed,[1]

$$(5) \qquad \varphi(\alpha) \overset{k}{\asymp} \int_{(k)}^{\alpha} \chi(\alpha) d^k\psi(\alpha)$$

or

$$(6) \qquad \varphi^k(\alpha) = \chi(\alpha) d^k\psi(\alpha) \quad ,$$

and agree that each of the symbolic relations (5) and (6) is only another way of writing (2).

If $\chi(\alpha) \equiv 0$, then it follows by (2), for any $\psi(\alpha)$, whatsoever,

$$(7) \qquad \varphi(\alpha) \overset{k}{\asymp} 0 \quad .$$

We also write for this

$$(8) \qquad d^k\varphi(\alpha) = 0 \quad ,$$

so that this relation is equivalent to the relation

$$d^k\varphi(\alpha) = 0 d^k\psi(\alpha) \quad ,$$

where $\psi(\alpha)$ is an arbitrary continuous function. The following observation is very important. If an everywhere continuous function $\varphi(\alpha)$ satisfies (8) in two adjoining intervals [A, C], [C, B], it does not follow that it satisfies (8) in the whole interval [A, B]. It is

[1] If it were not for typographical considerations, we should have written instead of (5)

$$\varphi(\alpha) \overset{k}{\asymp} \int_{(k)}^{\alpha} \chi(\alpha) \frac{d^k\psi(\alpha)}{d\alpha^{k-1}} \quad .$$

Cf. also the footnote on page 142.

otherwise however if the intervals overlap, cf. §28, 3. A trivial function $E(\alpha)$ of $\mathfrak{T}_k$ with exponents $\tau_1, \tau_2, \ldots, \tau_n$ is characterized by the fact that it is continuous, and satisfies the relation

$$d^k E(\alpha) = 0$$

in each of the intervals

$$[-\infty, \tau_1], [\tau_1, \tau_2], \ldots, [\tau_{n-1}, \tau_n], [\tau_n, \infty] .$$

3. Let $\varphi(\alpha)$, $\psi(\alpha)$, $\chi(\alpha)$ and $\varphi_n(\alpha)$, $\psi_n(\alpha)$, $\chi_n(\alpha)$ be given functions in $[A, B]$ for which

1) $d^k \varphi_n(\alpha) = \chi_n(\alpha) d^k \psi_n(\alpha), \qquad n = 1, 2, 3, \ldots$

and

2) $\varphi_n(\alpha) \longrightarrow \varphi(\alpha), \; \psi_n(\alpha) \longrightarrow \psi(\alpha)$
$\chi_n^{(r)}(\alpha) \longrightarrow \chi^{(r)}(\alpha) \qquad r = 0, 1, 2, \ldots, k$

uniformly in each subinterval (A', B') of $[A, B]$ as $n \longrightarrow \infty$.

Then the relation

$$d^k \varphi(\alpha) = \chi(\alpha) d^k \psi(\alpha) \tag{9}$$

also holds. For by (1)

$$\varphi_n(\alpha) \overset{k}{\asymp} \int_{(k)}^{\alpha} \chi_n(\alpha) d^k \psi_n(\alpha) , \tag{10}$$

and by (2), the left side or the right side is convergent to

$$\varphi(\alpha) \quad \text{or} \quad \int_{(k)}^{\alpha} \chi(\alpha) d^k \psi(\alpha) .$$

Hence by (1)

$$\varphi(\alpha) \overset{k}{\asymp} \int_{(k)}^{\alpha} \chi(\alpha) d^k \psi(\alpha) , \tag{11}$$

Q.E.D.

4. If the functions $\psi_n(\alpha)$ converge uniformly to $\psi(\alpha)$ in each subinterval (A', B') of $[A, B]$, and if the functions $\varphi_n(\alpha)$ are defined by the relation

$$d^k \varphi_n(\alpha) = \chi(\alpha) d^k \psi_n(\alpha) ,$$

then the $\varphi_n(\alpha)$ can be normalized in such a way that they converge uniformly to a limit function $\varphi(\alpha)$ in each interval (A', B') for which (9) holds, say by the stipulation

$$\varphi_n(\alpha) = \int_{(k)}^{\alpha} \chi(\alpha) d^k \psi_n(\alpha)$$

5. The following <u>associative law</u> holds

$$\chi_1(\alpha)\left\{\chi_2(\alpha) d^k \psi(\alpha)\right\} = \left\{\chi_1(\alpha)\chi_2(\alpha)\right\} d^k \psi(\alpha) \quad ,$$

and it is to be understood as follows. Let $\psi(\alpha)$ be continuous and $\chi_1(\alpha)$ and $\chi_2(\alpha)$ k-times continuously differentiable. Set $\chi(\alpha) = \chi_1(\alpha)\chi_2(\alpha)$. If the functions $\Psi(\alpha)$, $\Phi(\alpha)$, $\varphi(\alpha)$ are determined by the relations

$$\begin{aligned} d^k \Psi(\alpha) &= \chi_2(\alpha) d^k \psi(\alpha) \\ d^k \Phi(\alpha) &= \chi_1(\alpha) d^k \psi(\alpha) \\ d^k \varphi(\alpha) &= \chi(\alpha) d^k \psi(\alpha), \end{aligned}$$

then

$$d^k \Phi(\alpha) = d^k \varphi(\alpha) \quad , \tag{12}$$

i.e.,

$$\Phi(\alpha) \overset{k}{\asymp} \varphi(\alpha) \quad .$$

If $\psi(\alpha)$ is k-times continuously differentiable the assertion reduces to

$$\chi_1\left\{\chi_2 \psi^{(k)}\right\} = \left\{\chi_1 \chi_2\right\} \psi^{(k)}$$

and is trivially correct. In the general case, there is in each subinterval (A', B') of $[A, B]$, by the Weierstrass approximation theorem, a sequence of k-times continuously differentiable functions $\psi_n(\alpha)$ which converge uniformly to $\psi(\alpha)$ as $n \longrightarrow \infty$. In the relations

$$\begin{aligned} d^k \Psi_n(\alpha) &= \chi_2(\alpha) d^k \psi_n(\alpha) \\ d^k \Phi_n(\alpha) &= \chi_1(\alpha) d^k \Psi_n(\alpha) \\ d^k \varphi_n(\alpha) &= \chi(\alpha) d^k \psi_n(\alpha) \end{aligned} \tag{13}$$

one can by (4) normalize the functions $\Psi_n(\alpha)$, $\Phi_n(\alpha)$ and $\varphi_n(\alpha)$ in such a way that in each subinterval (A'', B'') of $[A', B']$, they converge uniformly to $\Psi(\alpha)$, $\Phi(\alpha)$ and $\varphi(\alpha)$. Since the $\psi_n(\alpha)$ are k-times continuously differentiable, we have

$$\Phi_n^{(k)}(\alpha) = \varphi_n^{(k)}(\alpha) \quad ,$$

and by (1) the validity of (12) in (A", B") follows. But this last interval can be an arbitrarily large subinterval of (A, B).

6. If in a subinterval [a, b] of [A, B], the function $\chi(\alpha)$ is different from zero, then $\chi(\alpha)^{-1}$ is continuously differentiable just as often as $\chi(\alpha)$ itself. Mutiplying both sides of

$$d^k\varphi(\alpha) = \chi(\alpha) d^k\psi(\alpha) \tag{14}$$

by $\chi(\alpha)^{-1}$ and applying the associative law to the right side, we obtain

$$\chi(\alpha)^{-1} d^k\varphi(\alpha) = d^k\psi(\alpha) \quad .$$

In particular, from the relation

$$G(\alpha) d^k E(\alpha) = 0 \tag{14'}$$

we obtain the relation

$$E(\alpha) \overset{k}{\asymp} 0$$

in each interval in which $G(\alpha) \neq 0$. If therefore $E(\alpha)$ is the k-transform of $f(x)$, and if $G(\alpha)$, apart from a finite number of points $\tau_1, \ldots, \tau_n$, is different from zero in $[-\infty, \infty]$, then as a consequence of (14'), $f(x)$ is a trivial function of $\mathfrak{F}_k$ with exponents τ_ν, cf. §28, 7.

7. Let $A < A' < A'' < B'' < B' < B$. Let $\psi(\alpha)$ be continuous in [A, B] and $d^k\psi(\alpha) = 0$ in (A', B'). Let $\chi_1(\alpha)$ and $\chi_2(\alpha)$ be k-times continuously differentiable in (A, B) and $\chi_1(\alpha) = \chi_2(\alpha)$ in [A, A"] and [B", B]. For the functions $\varphi_1(\alpha)$ and $\varphi_2(\alpha)$ defined by

$$d^k\varphi_n(\alpha) = \chi_n(\alpha) d^k\psi(\alpha) , \qquad n = 1, 2,$$

we have

$$\varphi_1(\alpha) \overset{k}{\asymp} \varphi_2(\alpha), \qquad A < \alpha < B .$$

For the function $\Phi(\alpha) = \varphi_1(\alpha) - \varphi_2(\alpha)$ satisfies the relation

$$d^k\Phi(\alpha) = \chi_1(\alpha) - \chi_2(\alpha)\, d^k\psi(\alpha) \quad ,$$

and therefore

$$d^k\Phi(\alpha) = 0 \tag{15}$$

in $[A, A'']$, $[A', B']$, $[B'', B]$. Since these intervals overlap, and together yield the interval $[A, B]$, it follows that (15) is also valid in $[A, B]$.

8. Setting for $\ell = 1, 2, 3, \ldots$

$$\Psi(\alpha) = \int_{(\ell)}^{\alpha} \psi(\alpha)\, d\alpha \;, \qquad \Phi(\alpha) = \int_{(\ell)}^{\alpha} \varphi(\alpha)\, d\alpha \;,$$

it follows by (14), if $\chi(\alpha)$ is $(k+\ell)$-times continuously differentiable, that

$$d^{k+\ell}\Phi(\alpha) = \chi(\alpha) d^{k+\ell}\Psi(\alpha) \quad .$$

This relation also can be proved most quickly if $\psi(\alpha)$ is approximated by k-times continuously differentiable functions $\psi_n(\alpha)$, if $\varphi_n(\alpha)$ is defined by (13) and then n is allowed to become infinite.

9. If the functions $f_n(x)$ of $\mathfrak{F}_k$ k-converge to $f(x)$, then their k-transforms $E_n(\alpha)$ are convergent in suitable normalization to the k-transform $E(\alpha)$ of $f(x)$, and indeed uniformly in each finite interval. On the other hand, each function $f(x)$ of $\mathfrak{F}_k$ can be represented as a k-limit of functions $f_n(x)$, each of which belongs to $\mathfrak{F}_0$, cf. §29, 1.

We shall make two applications from the above.

10. Let us denote the k-transforms of $f(x)$ and $f(x + \lambda)$ by $E(\alpha, k)$ and $\Phi(\alpha, k)$. Then

$$d^k\Phi(\alpha, k) = e(\lambda\alpha) d^k E(\alpha, k) \quad . \tag{16}$$

In fact, if $f(x)$ belongs to $\mathfrak{F}_0$, then

$$\Phi(\alpha, 0) = e(\lambda\alpha) E(\alpha, 0) \quad ,$$

and by 8., (16) follows. In the general case, we represent $f(x)$ as the k-limit of functions of $\mathfrak{F}_0$, and make $n \longrightarrow \infty$ in

$$d^k\Phi_n(\alpha, k) = e(\lambda\alpha) d^k E_n(\alpha, k) \quad .$$

11. THEOREM 40. With a function $f(x)$ of $\mathfrak{F}_k$ and a function $K(\xi)$ for which

(17) $$|K(\xi)| \leq A(1 + |\xi|^{k+2})^{-1} ,$$

we form the function

(18) $$g(x) = \frac{1}{2\pi}\int f(\xi)K(x - \xi)\, d\xi ,$$

which also belongs to $\mathfrak{F}_k$. Between the k-transforms $E(\alpha)$ and $\Phi(\alpha)$ of $f(x)$ and $g(x)$, there is the relation

(19) $$d^k\Phi(\alpha) = \gamma(\alpha)d^kE(\alpha) ,$$

where $\gamma(\alpha)$ denotes the 0-transform of $K(\xi)$. (That $\gamma(\alpha)$ is k-times continuously differentiable follows from §13, 9.)

PROOF. If $f(x)$ even belongs to $\mathfrak{F}_0$, (19) is the Faltung rule of §13. For general $f(x)$, we consider functions $f_n(x)$ of $\mathfrak{F}_0$ which are k-convergent to $f(x)$. The corresponding functions $g_n(x)$, which likewise belong to $\mathfrak{F}_0$, are by Theorem 38, k-convergent to $g(x)$. Hence (19) results from

$$d^k\Phi_n(\alpha) = \gamma(\alpha)d^kE_n(\alpha)$$

by letting $n \longrightarrow \infty$.

§31. Convergence Theorems

1. THEOREM 41. Let $\Phi(\alpha)$ be absolutely integrable in $[-\infty, \infty]$. Then the function

(1) $$g(x) = \int e(x\alpha)\Phi(\alpha)\, d\alpha ,$$

since it is bounded, is contained in $\mathfrak{F}_2$; its 2-transform $E(\alpha)$ is two times differentiable, and satisfies the relation

$$E''(\alpha) = \Phi(\alpha)$$

PROOF. We shall need only the special case, and therefore shall only prove this part, in which $\Phi(\alpha)$ vanishes outside of a finite interval. However the general case can be deduced from the special one without difficulty, if one makes use of the fact that the functions

$$g_n(x) = \int_{-n}^{n} e(x\alpha)\Phi(\alpha)\, d\alpha$$

are 2-convergent to $g(x)$.

The function

$$\psi(\beta, \alpha) = \frac{1}{2\pi}\int e(x\beta)\, \frac{e(-x\alpha) - L_2(\alpha, x)}{-x^2}\, dx \quad ,$$

for fixed β, is the 2-transform of $e(x\beta)$. Therefore by §28, 6

$$\Psi(\beta, \alpha) = \frac{1}{2}\, |\alpha - \beta| + A(\beta) + \alpha B(\beta) \quad .$$

The coefficients $A(\beta)$ and $B(\beta)$ are continuous since $\Psi(\beta, \alpha)$ is continuous in β for $\alpha = 0$ and say for $\alpha = 1$. By introducing the function

$$\Delta(\alpha, \beta) = \begin{cases} \alpha - \beta & \text{for } \alpha > \beta \\ 0 & \text{for } \alpha < \beta \end{cases} ,$$

we obtain

$$\Psi(\beta, \alpha) = \Delta(\alpha, \beta) + A(\beta) + \frac{\beta}{2} + \alpha\left(B(\beta) - \frac{1}{2}\right) \quad .$$

If one now substitutes (1) in

$$E(\alpha) \asymp \frac{1}{2\pi}\int g(x)\, \frac{e(-\alpha x) - L_2}{-x^2}\, dx$$

and (as is permissible) interchanges the order of integration, one obtains

$$E(\alpha) \asymp \int \Phi(\beta)\Psi(\beta, \alpha)\, d\beta \asymp \int \Delta(\alpha, \beta)\Phi(\beta)\, d\beta = \int^{\alpha} (\alpha - \beta)\Phi(\beta)\, d\beta \quad .$$

Hence $E(\alpha)$ is actually the second integral of $\Phi(\alpha)$.

2. THEOREM 42. Let the function $f(x)$ of $\mathfrak{F}_k$ be such that its k-transform $E(\alpha)$ is k-times continuously differentiable not only on a left half line $[-\infty, a_0]$, but also on a right half line $[b_0, \infty]$. For $a < a_0$, $b_0 < b$, we define the generalized integral

$$J(x, a, b) = \int_a^b e(x\alpha)\, d^k E(\alpha)$$

by the expression

$$e(xb)\sum_{r=0}^{k-1}(-ix)^r E^{(k-r-1)}(b) - e(xa)\sum_{r=0}^{k-1}(-ix)^r E^{(k-r-1)}(a)$$

(2)

$$+ (-ix)^k \int_a^b e(x\alpha)E(\alpha)\, d\alpha \quad .$$

Then in regard to the limit value

(3) $$\lim_{A \to \infty} J(x, -A, A)$$

the following assertion holds.

If this limit exists for almost all x in an interval (x_0, x_1), then its value agrees with $f(x)$ for almost all x in (x_0, x_1) [81].

REMARK. Our definition of the generalized integral $J(x, a, b)$ is such that it has the following properties.

1) If $E(\alpha)$ is k-times differentiable throughout (not necessarily continuously), then

$$J(x, a, b) = \int_a^b e(x\alpha)E^{(k)}(\alpha)\, d\alpha \quad .$$

2) For $a' < a_0$, $b_0 < b'$

(4) $$J(x, a', b') = J(x, a, b) + J(x, a', a) + J(x, b, b') \quad .$$

3) If, in particular, $E(\alpha)$ has its k-th derivative equal to zero in $[-\infty, a_0]$ and $[b_0, \infty]$, (and thus is $\asymp 0$ on each of the half lines), then

$$J(x, a', a) = J(x, b, b') = 0 \quad .$$

Hence the function

(5) $$F(x) = J(x, a, b)$$

is independent of the special values of a and b. In particular, we have also

(6) $$F(x) = \lim_{A \to \infty} J(x, -A, A) \quad .$$

PROOF. For the actual proof, we first introduce the following theorem.

3. THEOREM 43. If the k-transform $E(\alpha)$ of $f(x)$ has its k-th derivative equal to zero in $[-\infty, a_0]$ and $[b_0, \infty]$, then

$$f(x) = \lim_{A \to \infty} J(x, -A, A) \quad , \tag{7}$$

(and therefore, among other things, $f(x)$ is differentiable arbitrarily often).

PROOF. Set

$$J(x, a, b) = t(x) + (-ix)^k g(x)$$

where $t(x)$ means the "trivial" part of the function (2), and

$$g(x) = \int_a^b e(x\alpha)E(\alpha)\, d\alpha \quad .$$

Since $g(x)$ is bounded, the function $F(x)$ defined by (5) or (6) is contained in $\mathfrak{F}_{k+2}$. We now examine the (k+2)-transform $\Phi(\alpha)$ of $F(x)$. In the interval $a < \alpha < b$, the part of $\Phi(\alpha)$ resulting from $t(x)$, is equivalent to zero, cf. §27, 7. Hence

$$\Phi(\alpha) \overset{k+2}{\asymp} \int (-ix)^k g(x) \frac{e(-\alpha x) - L_{k+2}}{(-ix)^{k+2}}\, dx = \int g(x) \frac{e(-\alpha x) - L_{k+2}}{(-ix)^2}\, dx$$

$$\overset{k+2}{\asymp} \int g(x) \frac{e(-\alpha x) - L_2}{(-ix)^2}\, dx \overset{k+2}{\asymp} \Psi(\alpha) \quad ,$$

where $\Psi(\alpha)$ means the 2-transform of $g(x)$. But by Theorem 41

$$\Psi''(\alpha) = E(\alpha) \quad .$$

Consequently

$$\Phi(\alpha) \overset{k+2}{\asymp} \int_{(2)}^{\alpha} E(\alpha)\, d\alpha \quad , \qquad\qquad a < \alpha < b \quad .$$

Now since (a, b) can be an arbitrarily large interval, and since $E(\alpha)$ is the k-transform of $f(x)$, it follows by Theorem 37 that $f(x)$ and $F(x)$, considered as functions of $\mathfrak{F}_{k+2}$, agree. But this says that the functions $f(x)$ and $F(x)$ are identical.

4. We can now prove Theorem 42. Consider in $[-\infty, \infty]$, a function $\gamma(\alpha)$, $(k+2)$-times continuously differentiable and vanishing outside of a finite interval $(-\alpha_0, \alpha_0)$, which is even $[\gamma(-\alpha) = \gamma(\alpha)]$, monotonically decreasing in $0 \leq \alpha < \infty$ and has the value 1 when $\alpha = 0$. The function

$$K(\xi) = \int e(\xi\alpha)\gamma(\alpha)\,d\alpha$$

is such that

$$|K(\xi) \leq C(1 + |\xi|^{k+2})^{-1}$$

and

$$\frac{1}{2\pi}\int K(\xi)\,d\xi = \gamma(0) = 1 \quad .$$

The function

$$f_n(x) = \frac{n}{2\pi}\int f(\xi)K[n(x - \xi)]\,d\xi$$

has a k-transform $E_n(\alpha)$ which by Theorem 40 satisfies the relation

$$d^kE_n(\alpha) = \gamma\left(\frac{\alpha}{n}\right)d^kE(\alpha)$$

and therefore has the following three properties.

1) $E_n(\alpha)$ is k-times continuously differentiable outside of (a_0, b_0), and

$$E_n^{(k)}(\alpha) = \gamma\left(\frac{\alpha}{n}\right)E^{(k)}(\alpha) \quad .$$

2) In a suitable normalization, $E_n(\alpha)$ is uniformly convergent to $E(\alpha)$, as $n \longrightarrow \infty$, in each finite α-interval [because $\gamma(\alpha/n) \longrightarrow \gamma(0)$]. The derivatives of $E_n(\alpha)$ are likewise convergent to the corresponding derivatives of $E(\alpha)$ at each point $\alpha = a < a_0$ and $\alpha = b > b_0$.

3) Outside of the intervals $(-\alpha_0 n, \alpha_0 n)$

$$E_n^{(k)}(\alpha) = 0 \quad .$$

Because of 3), we have by Theorem 42

$$f_n(x) = \lim_{A \to \infty} J_n(x, -A, A) \quad , \tag{8}$$

where $J_n(x, a, b)$ denotes the expression (2) formed with $E_n(\alpha)$, i.e.,

$$(9) \qquad f_n(x) = J_n(x, -A, A) + \left(\int^{-A} + \int_A\right) e(x\alpha)\gamma\left(\frac{\alpha}{n}\right)E^{(k)}(\alpha)\, d\alpha$$

if $A > -a_o$ and $A > b_o$. Now let x be a fixed point in (x_o, x_1) at which the limit (3) exists, i.e., at which the integral

$$(10) \qquad \int_A [e(x\alpha)E^k(\alpha) + E(-x\alpha)E^k(-\alpha)]\, d\alpha$$

is convergent. Denote (for fixed x) the integrand by $\varphi(\alpha)$. Therefore (10) has the value

$$H = \int_A \varphi(\alpha)\, d\alpha \quad .$$

Next consider the quantity

$$H_n = \int_A \varphi(\alpha)\gamma\left(\frac{\alpha}{n}\right) d\alpha = \int_A^{\alpha_o n} \varphi(\alpha)\gamma\left(\frac{\alpha}{n}\right) d\alpha \quad .$$

We assert that

$$(11) \qquad \lim_{n \to \infty} H_n = H \quad .$$

For its proof, we introduce the function

$$H(\alpha) = \int_\alpha \varphi(\alpha)\, d\alpha \quad , \qquad \alpha \geq A \quad ,$$

and denote the maximum value of $|H(\alpha)|$ by M. Then

$$H_n = -\int_A H'(\alpha)\gamma\left(\frac{\alpha}{n}\right) d\alpha = \gamma\left(\frac{A}{n}\right)H(A) + \frac{1}{n}\int_A H(\alpha)\gamma'\left(\frac{\alpha}{n}\right) d\alpha \quad .$$

From this we obtain (11) because

$$\gamma\left(\frac{A}{n}\right) H(A) \longrightarrow \gamma(0)H(A) = 1 \cdot H(A) = H$$

and

$$\left|\int_A H(\alpha)\gamma'\left(\frac{\alpha}{n}\right) d\alpha\right| \leq M\int_A \left[-\gamma'\left(\frac{\alpha}{n}\right)\right] d\alpha \leq M \cdot \gamma(0) \quad .$$

Recalling that $\gamma(-\alpha) = \gamma(\alpha)$, (11) states that the second term in

$$J(x, -A, A) + \left(\int^{-A} + \int_{A} \right) e(x\alpha) E^{(k)}(\alpha)\, d\alpha$$

is the limit of the corresponding right term in (9). Moreover because of 2), the first term is the limit of the corresponding first term. We have shown therefore that for almost all x in (x_0, x_1)

$$\lim_{n \to \infty} f_n(x) = \lim_{A \to \infty} J(x, -A, A) \quad .$$

But since by Theorem 39, the sequence $f_n(x)$ converges in the mean to $f(x)$, the limit

$$\lim_{n \to \infty} f_n(x)$$

so far as it exists for almost all x in (x_0, x_1), is identical with $f(x)$, [Appendix 11], Q.E.D.

§32. Multipliers

The additional observations of this chapter will be similar in many respects to those of the previous chapter. Hence we shall often content ourselves with short hints and suggestions, and shall entirely suppress trivial generalizations from $k = 0$ to arbitrary k.

For the following paragraphs, cf. §23.

1. Let

$$f(x) \sim \int e(x\alpha) d^k E(\alpha) \tag{1}$$

be a given function of $\mathfrak{F}_k$. By a <u>multiplier</u> of $E(\alpha)$ or $f(x)$, we mean a k-times continuously differentiable function $\Gamma(\alpha)$ which is such that the function

$$\Phi(\alpha) \asymp \int_{(k)}^{\alpha} \Gamma(\alpha) d^k E(\alpha) \tag{2}$$

again belongs to $\mathfrak{T}_k$. We shall again denote the function of $\mathfrak{F}_k$ belonging to $\Phi(\alpha)$ by

$$\Gamma[f] \quad .$$

If $\Gamma(\alpha)$ is a general multiplier of the class $\mathfrak{F}_k$ (or $\mathfrak{T}_k$), then for brevity we speak of it also as a <u>k-multiplier</u>. If Γ_1 and Γ_2 are multipliers, then $\Gamma = c_1\Gamma_1 + c_2\Gamma_2$ is also a multiplier, and

$$\Gamma[f] = c_1\Gamma_1[f] + c_2\Gamma_2[f] \quad .$$

By the associative law (§30, 5), the product Γ of two k-multipliers Γ_1 and Γ_2 is again a k-multiplier, and

$$\Gamma[f] = \Gamma_1[\Gamma_2[f]] = \Gamma_2[\Gamma_1[f]] \quad .$$

2. By §30, 10, $e(\lambda\alpha)$ is a k-multiplier: $\Gamma[f(x)] = f(x + \lambda)$. Hence a finite series of the form

$$c_1 e(\mu_1\alpha) + c_2 e(\mu_2\alpha) + \dots + c_n e(\mu_n\alpha) + \dots \tag{3}$$

is a k-multiplier. We shall show that an infinite series of this form is then certainly a k-multiplier, if both series

$$\sum_{\nu=1}^{\infty} |c_\nu|, \quad \sum_{\nu=1}^{\infty} |\mu_\nu|^k |c_\nu| \tag{4}$$

converge. It is easy to verify the inequality

$$\frac{1}{1 + |x-\mu|^k} \leq 2^k \frac{1 + |\mu|^k}{1 + |x|^k} \quad ,$$

and if we also take into account the finiteness of the series (4) then, when putting

$$p_k(t) = (1 + |t|^k)^{-1} \quad ,$$

we obtain

$$\sum_{\nu=n+1}^{n+p} |c_\nu| p_k(x - \mu_\nu) \leq C_n p_k(x)$$

and where C_n is a number dependent only on n, for which

$$\lim_{n \to \infty} C_n = 0 \quad .$$

In the notation of §23, 3

$$\int |H_{n+p}[f] - H_n[f]| p_k \, dx \leqq \sum_{\nu=n+1}^{n+p} |c_\nu| \int |f(x + \mu_\nu)| p_k(x) \, dx$$

$$\leqq \int |f(x)| \sum_{n+1}^{n+p} |c_\nu| p_k(x - \mu_\nu) \, dx \leqq c_n \int |f(x)| p_k(x) \, dx \quad .$$

Hence the sequence of functions

$$H_n[f] \sim \int e(x\alpha) H_n(\alpha) d^k E(\alpha)$$

is k-convergent. We denote its k-limit by

$$F(x) \sim \int e(x\alpha) d^k \Phi(\alpha) \quad .$$

By §30, 9

$$\Phi(\alpha) \overset{k}{\asymp} \lim_{n \to \infty} \int_{(k)}^{\alpha} H_n(\alpha) d^k E(\alpha)$$

and therefore by §30, 4

$$d^k \Phi(\alpha) = H(\alpha) d^k E(\alpha) \quad .$$

Hence $H(\alpha)$ is a multiplier of the arbitrary function $f(x)$ of $\mathfrak{F}_k$.

If the function

$$G(\alpha) = \sum_{\sigma=0}^{s} a_\sigma e(\delta_\sigma \alpha) \tag{5}$$

is essentially different from zero,

$$|G(\alpha)| \geqq S > 0, \qquad (-\infty < \alpha < \infty) \quad , \tag{6}$$

then it can be expanded in an absolutely convergent series of the form (3). The function

$$\frac{d^k (G(\alpha)^{-1})}{d\alpha^k} \tag{7}$$

can be written as the quotient of two functions of the type (5). The denominator has the value $G(\alpha)^k$, and is also essentially different from zero. Therefore the function (7) can also be expanded in an absolutely convergent exponential series. As is known from the theory of almost periodic functions (and as can also be shown directly without difficulty), the development of (7) results from the development of $G(\alpha)^{-1}$ by k-times formal differentiation [82], and it therefore reads

$$\sum_{\nu=1}^{\infty} (i\mu_\nu)^k c_\nu e(\mu_\nu \alpha) \quad .$$

This series is absolutely convergent, but that means that the second series (4) also converges. We have therefore shown that for arbitrarily large k, the function $G(\alpha)^{-1}$ is a k-multiplier.

3. The O-transform of a function $K(\xi)$ which satisfies the relation

$$(8) \qquad |K(\xi)| \leqq A(1 + |\xi|^{k+2})^{-1}$$

is by Theorem 40, a k-multiplier. Such O-transforms are on the one hand, the functions

$$(9) \qquad \frac{1}{\alpha - i}, \quad \frac{1}{\alpha + i} \quad ;$$

on the other hand all such functions which with the first $k + 2$ derivatives are absolutely integrable, and generally $(k+2)$-times continuously differentiable functions $\Gamma(\alpha)$ which outside of an interval $-A \leqq \alpha \leqq A$, can be written as

$$\Gamma(\alpha) = \frac{c}{\alpha} + H(\alpha) \quad ,$$

where c is a constant and $H(\alpha)$ along with the first $k + 2$ derivatives is absolutely integrable.

The observations of §23, 10 carry over verbatim if "multiplier" is replaced by "k-multiplier". On the other hand the observations of §23, 11 are to be modified as follows. Let (1) be a given function. Two k-times continuously differentiable functions Γ_1, Γ_2 which differ from each other only for such points α in whose neighborhood

$$d^k E(\alpha) = 0 \quad ,$$

are either both multipliers of (1) or neither of them are. We call them "equivalent" in regard to (1).

For $k \geqq 1$, the result of §23, 12 can be maintained in the following way. Let a function $E(\alpha)$ of $\mathfrak{T}_k$ have the property that outside of a finite interval (a_0, b_0) the relation

$$d^k E(\alpha) = 0$$

holds, i.e., both in $[-\infty, a_0]$, and in $[b_0, \infty]$. Let $\Gamma^*(\alpha)$ be a given $(k+2)$-times differentiable function in a closed interval (a, b), $a < a_0$, $b_0 < b$. Then there is exactly one function $\Phi(\alpha)$ of $\mathfrak{T}_k$ which satisfies the relation $d^k\Phi(\alpha) = 0$ outside of (a_0, b_0), and the relation

$$d^k\Phi(\alpha) = \Gamma^*(\alpha) d^k E(\alpha)$$

in $[a, b]$.

4. THEOREM 44. Let $f(x)$ be a function of $\mathfrak{F}_k$ and $E(\alpha)$ its k-transform.

1) If $f(x)$ is r-times differentiable in $\mathfrak{F}_k$, then

$$f^{(\rho)}(x) \sim \int e(x\alpha)(i\alpha)^\rho d^k E(\alpha), \qquad \rho = 0, 1, 2, \ldots, r. \tag{10}$$

2) Conversely, if a function

$$\varphi(x) \sim \int e(x\alpha)(i\alpha)^r d^k E(\alpha) \tag{11}$$

exists (i.e., if $(i\alpha)^r$ is a multiplier of $E(\alpha)$), then $f(x)$ is r-times differentiable in $\mathfrak{F}_k$ (and hence $f^{(r)}(x) = \varphi(x)$).

3) If $d^k E(\alpha) = 0$ outside of a finite interval (a_0, b_0), then $f(x)$ is differentiable arbitrarily often in $\mathfrak{F}_k$, and hence by 1.,

$$f^{(\rho)}(x) = \int e(x\alpha)(i\alpha)^\rho d^k E(\alpha), \qquad \rho = 1, 2, 3, \ldots . \tag{12}$$

4) If $d^k E(\alpha) = 0$ inside an interval $[a_0, b_0]$ which contains the point $\alpha = 0$, then $f(x)$ is integrable arbitrarily often in $\mathfrak{F}_k$, and the rth integral $F_\rho(x)$ has the representation

$$F_\rho(x) \sim \int e(x\alpha)(i\alpha)^{-\rho} d^k E(\alpha) . \tag{13}$$

REMARK. Formula (10) or (13) is to be understood in such a manner that the k-transform $\Phi(\alpha)$ of the function standing on the left satisfies the relation

(14) $$d^k\Phi(\alpha) = (i\alpha)^\rho d^k E(\alpha)$$

or

(15) $$(i\alpha)^\rho d^k\Phi(\alpha) = d^k E(\alpha) \quad .$$

PROOF OF 1). It is sufficient to limit onself to the case $r = 1$, and indeed to prove for the k-transform $E_1(\alpha)$ of $f'(x)$ that

(16) $$d^k E_1(\alpha) = i\alpha d^k E(\alpha) \quad .$$

Hence

$$2\pi\, E_1(\alpha) \asymp \int f'(x)\, \frac{e(-\alpha x) - L_k}{(-ix)^k}\, dx \quad .$$

Considering that $if(x) = o(|x|^k)$, cf. §29, 5, we obtain by a judicious partial integration and a slight calculation following it

$$2\pi\, E_1(\alpha) \asymp - \int_{-1}^{1} f(x)\, \frac{d}{dx} \left(\frac{e(-\alpha x) - L_k(\alpha, x)}{(-ix)^k} \right) dx$$

$$- \left(\int^{-1} + \int_1^{\infty} \right) f(x)\, \frac{d}{dx} \left(\frac{e(-\alpha x)}{(-ix)^k} \right) dx$$

$$\asymp i\alpha \int f(x)\, \frac{e(-\alpha x) - L_k}{(-ix)^k}\, dx - i\alpha \int f(x)\, \frac{e(-\alpha x) - L_{k+1}}{(-ix)^{k+1}}\, dx \quad .$$

Therefore

$$E_1(\alpha) \asymp i\alpha\, E(\alpha) - i\alpha \int_0^{\alpha} E(\beta)\, d\beta$$

which is merely another way of writing (16).

PROOF OF 3). Let $\Gamma(\alpha)$ be (k+2)-times differentiable, equal to one in $(a_0 - \epsilon, b_0 + \epsilon)$, and equal to zero outside of a finite interval. We denote by $K(\xi)$ the function of $\mathfrak{F}_0$ belonging to it. By Theorem 40, the function

$$(17) \qquad r[f] = \frac{1}{2\pi}\int K(\xi)f(x-\xi)\,d\xi$$

is on the one hand identical with $f(x)$; on the other hand, by §29, 7, differentiable arbitrarily often in $\mathfrak{F}_k$.

PROOF OF 4). Let $a_0 < a < 0 < b < b_0$. One can find a function $r(\alpha)$ which is $(k+2)$-times differentiable, and which agrees with $(i\alpha)^{-1}$ outside of $[a, b]$. The conclusion now follows as in the proof of Theorem 27, making use of 3. and §29, 7.

PROOF OF 2). The conclusion is reached in a manner analogous to Theorem 25. Indeed we introduce successively the functions $r_\kappa(\alpha)$, $r_0(\alpha)$, f_κ, φ_κ, $F_0(x)$, $g(x)$, and the k-transform $F(\alpha)$ of $g(x)$. $g(x)$ is the r-th integral of $\varphi(x)$, and it must be proved that $f(x)$ is also an r-th integral of $\varphi(x)$. For this it must be shown that

$$h(x) = g(x) - f(x)$$

is a polynomial in x of degree $(r - 1)$. The k-transform of $h(x)$ amounts to $H(\alpha) = F(\alpha) - E(\alpha)$, and there is also the relation

$$(18) \qquad (i\alpha)^r d^k H(\alpha) = 0 \quad .$$

By §30, 6, $h(x)$ is a trivial function of $\mathfrak{F}_k$, and therefore differentiable arbitrarily often in $\mathfrak{F}_k$. Because of the already proved assertion 1), it follows by (18) that $h^{(r)}(x)$ vanishes, Q.E.D.

5. In a manner analogous to Theorem 28 and §24, 5, we arrive at the following. In order that a function

$$F(x) \sim \int e(x\alpha)\,\{i(\alpha - \lambda)\}^{\pm r} d^k E(\alpha)$$

exist, it is necessary and sufficient that $e(-\lambda x)f(x)$ be r-times differentiable or integrable in $\mathfrak{F}_k$ (+ is valid for differentiation, - for integration). What is more

$$F(x) = e(\lambda x)[e(-\lambda x)f(x)]^{(r)}$$

or

$$F(x) = e(\lambda x)\int_{(r)}^{\gamma} e(-\lambda x)f(x)\,dx \quad .$$

If the k-transforms $E_1(\alpha)$ and $E_2(\alpha)$ are k-equivalent in the neighborhood of a point $\alpha = \lambda$, and if a function

$$\int e(x\alpha)\ \{i(\alpha - \lambda)\}^{-r} d^k E_1(\alpha)$$

exists, then there also exists a function

$$\int e(x\alpha)\ \{i(\alpha - \lambda)\}^{-r} d^k E_2(\alpha) \quad .$$

§33. Operator Equations

1. Let $G(\alpha)$ be k-times continuously differentiable in $[-\infty, \infty]$. By a solution of the equation

$$G(\alpha) d^k \varphi(\alpha) = 0 \tag{1}$$

we mean a function $\varphi(\alpha)$ of $\mathfrak{T}_k$ which satisfies it, or the function of $\mathfrak{F}_k$ belonging to it, that is the function

$$y(x) \sim \int e(x\alpha) d^k \varphi(\alpha) \quad . \tag{2}$$

2. If $G(\alpha)$ has no (real) zeros, then by §30, 6, there is no solution (i.e., there is only the solution $y(x) \equiv 0$). If it has only a finite number of zeros

$$\tau_1, \ \ldots, \ \tau_\nu, \ \ldots, \ \tau_n \quad , \tag{3}$$

then only functions of the form

$$y(x) = \sum_{\nu=1}^{n} y_\nu(x) \tag{4}$$

with

$$y_\nu(x) = e(\tau_\nu x) \sum_{\mu=0}^{k-2} c_{\nu\mu} x^\mu, \qquad \nu = 1, 2, \ldots, n \tag{5}$$

can occur.

The following is a more precise statement. If the zeros (3) have multiplicity

$$\ell_1, \ \ldots, \ \ell_\nu, \ \ldots, \ \ell_n \quad , \tag{6}$$

then the general solution of (1) has the form

$$(7) \qquad y(x) = \sum_{\nu=1}^{n} \left(e(\tau_\nu x) \sum_{\mu=0}^{m_\nu} c_{\nu\mu} x^\mu \right) ,$$

where

$$m_\nu = \min (\ell_\nu - 1, k - 2), \qquad \nu = 1, 2, \ldots, n ,$$

and the $c_{\nu\mu}$ are arbitrary constants. For example, if Γ_ν is an arbitrary k-multiplier, then together with $\varphi(\alpha)$,

$$d_\nu(\alpha) \asymp \int_{(k)}^{\alpha} \Gamma_\nu(\alpha) d^k \varphi(\alpha)$$

is also a solution of (1). Now if $\Gamma_\nu(\alpha)$ "isolates" the zero τ_ν from the other zeros, and if $\varphi(\alpha)$ is the k-transform of (4), then according to the structure of the trivial functions discussed in §28, 7, $\varphi_\nu(\alpha)$ is the k-transform of $y_\nu(x)$. Hence each component $y_\nu(x)$ is by itself a solution. Conversely, together with the $y_\nu(x)$, $y(x)$ is also a solution. There remains, therefore, only to discuss for which $c_{\nu\mu}$ the function (5) is a solution. By hypothesis

$$(9) \qquad G(\alpha) = \{i(\alpha - \tau_\nu)\}^{\ell_\nu} G_*(\alpha) , \qquad G_*(\alpha) \neq 0$$

in the neighborhood of $\alpha = \tau_\nu$. From this it follows easily that the relation $G(\alpha) d^k \varphi_\nu(\alpha) = 0$ is synonomous with the relation

$$(10) \qquad \{i(\alpha - \tau_\nu)\}^{\ell_\nu} d^k \varphi_\nu(\alpha) = 0 .$$

By §32, 5, $y_\nu(x)$ is, then and only then, a solution of (10) if the ℓ_ν-th derivative of $e(- \tau_\nu x) y_\nu(x)$ vanishes. From this (7) follows.

3. If $\varphi_\chi(\alpha)$, $\chi = 1, 2, 3, \ldots$, are solutions of (1) which converge uniformly in each finite interval to a function $\varphi(\alpha)$, then the latter is also a solution. If therefore $G(\alpha)$ has infinitely many non-repetitive zeros

$$(11) \qquad \tau_1, \tau_2, \tau_3, \ldots , \qquad \lim_{\nu \to \infty} |\tau_\nu| = \infty$$

with well determined multiplicities

$$(12) \qquad \ell_1, \ell_2, \ell_3, \ldots ,$$

then the functions

$$(13)\qquad y_\chi(x) = \sum_{\nu=1}^{\chi}\left(e(\tau_\nu x)\sum_{\mu=0}^{m_\nu} c_{\nu\mu}^{(\chi)}x^\mu\right), \qquad \chi = 1, 2, 3, \ldots,$$

and each k-limit of such functions belong to the solution. Conversely, each solution $y(x)$ can then be k-approximated by functions (13). To prove the converse, let $\gamma(\alpha)$ and $K(\xi)$ be functions as defined in §31, 4, and consider the functions

$$y_p(x) = \frac{p}{2\pi}\int y(\xi)K(p(x - \xi))\, d\xi \quad .$$

By Theorem 39, they k-converge to $y(x)$ as $p \longrightarrow \infty$. Since

$$d^k\varphi_p(\alpha) = \gamma\left(\frac{\alpha}{p}\right)d^k\varphi(\alpha) \quad ,$$

we have

$$(14)\qquad G(\alpha)d^k\varphi_p(\alpha) = 0 \quad .$$

But now

$$(15)\qquad d^k\varphi_p(\alpha) = 0$$

outside of an interval (a_p, b_p). Moreover for each solution of (14) which satisfies (15), it follows just as in the case of finitely many τ_ν, that it is a trivial function of the form (13) — with exponents τ_ν in the interval (a_p, b_p) —. Q.E.D. If only such solutions are desired whose transforms outside of a finite interval (a, b) are equivalent to zero, and if no τ_ν has the value a or b, then the totality of the functions (13) result, formed with those τ_ν which lie in the interval (a, b).

4. If the zeros of $G(\alpha)$ are of a more complicated character, then the totality of solutions of (1) cannot be so easily described.

If, for instance, $G(\alpha)$ vanishes in an interval $[a_o, b_o]$, then the following belong to the solutions: each function (13) with exponents in the interval $[a_o, b_o]$ and each k-limit of such functions, but for example also each function

$$(16)\qquad (-ix)^k\int_a^b \gamma(\alpha)e(x\alpha)\, d\alpha$$

with $a_o < a < b < b_o$, provided $\gamma(\alpha)$ is two times differentiable, and together with the first derivative vanishes for $\alpha = a$ and $\alpha = b$

[because (16) is then a function of $\mathfrak{F}_k$ whose k-transform agrees with $\gamma(\alpha)$, and is therefore equivalent to zero for $\alpha < a$ and $b < \alpha$].

5. For a given function

$$f(x) \sim \int e(x\alpha)d^kE(\alpha) \tag{17}$$

we now consider more generally the non-homogeneous equation

$$G(\alpha)d^k\varphi(\alpha) = d^kE(\alpha) \quad . \tag{18}$$

If $\varphi_0(\alpha)$ is a particular solution of (18), then the general solution of (18) is obtained by adding to $\varphi_0(\alpha)$ any solution of (1). We shall therefore, only inquire further whether some one solution of (18) exists.

6. We assume that $G(\alpha)$ has a finite or a countably infinite number of zeros (11) with multiplicity (12). As in §25, 1, one shows the following. <u>In order that</u> (18) <u>have a solution, it is necessary that the function</u>

$$e(-\tau_\chi x)f(x) \tag{19}$$

<u>be ℓ-times integrable in</u> $\mathfrak{F}_k$ <u>for all</u> x, i.e., <u>that all functions</u>

$$\int_{(k)}^{\alpha} \left\{i(\alpha-\tau_\chi)\right\}^{-\ell_\chi} d^kE(\alpha)$$

<u>belong to</u> $\mathfrak{T}_k$.

In particular, let

$$G(\alpha) = (i\alpha)^r + \sum_{\rho=0}^{r-1}\sum_{\sigma=0}^{s} a_{\rho\sigma}(i\alpha)^\rho e(\delta_\sigma\alpha) \quad , \tag{20}$$

for $r \geqq 1$, or more generally let

$$G(\alpha) = \sum_{\rho=0}^{r}\sum_{\sigma=0}^{s} a_{\rho\sigma}(i\alpha)^\rho e(\delta_\sigma\alpha) \tag{21}$$

for $r \geqq 0$, where the "principal part"

$$\gamma(\alpha) = \sum_{\sigma=0}^{s} a_{r\sigma}e(\delta_\sigma\alpha)$$

is essentially different from zero,

$$(22) \qquad |\gamma(\alpha)| \geqq S > 0 , \qquad (-\infty < \alpha < \infty) .$$

By the same lemma and analogous observations as in §25, one shows the following. If $G(\alpha)$ has no zeros whatsoever, then the functions

$$(i\alpha)^{\rho}G(\alpha)^{-1} , \qquad \rho = 0, 1, \ldots, r,$$

are k-multipliers, and there exists a solution $y(x)$ <u>which is actually differentiable r-times in</u> $\mathfrak{F}_k$. In the case of (20), the following is again valid as in §25, 5,

$$y^{(\rho)}(x) = \frac{1}{2\pi}\int K^{(\rho)}(\xi)f(x - \xi)\, d\xi, \qquad \rho = 0, 1, \ldots, r.$$

More generally if $G(\alpha)$ has a finite number of zeros, then the above necessary condition is also sufficient for the existence of a solution $y(x)$ which is r-times differentiable in $\mathfrak{F}_k$. But it is determined only to within an arbitrary additive function of the form (7). For each function (7) is differentiable arbitrarily often (therefore especially r-times) in $\mathfrak{F}_k$.

If assumption (22) is not made in (21), then analogous to Theorem 33, the following is valid. If $E(\alpha)$ is equivalent to zero outside of a finite interval (a, b), then for the existence of a solution r-times differentiable in $\mathfrak{F}_k$, it is sufficient that the above necessary condition be satisfied in regard to the zeros of $G(\alpha)$ falling in the interval (a, b). But by (3), the arbitrary solution of (1) joined to a particular solution of (18) need not now be a trivial function. But if only such solutions of (18) are desired whose only trivial k-transform is equivalent to zero outside of (a, b), then only trivial functions with exponents in (a, b) are added.

7. In accordance with the integral equation examined in §26, we now consider the case

$$G(\alpha) = \lambda - \gamma(\alpha) ,$$

where λ is a parameter and $\gamma(\alpha)$ is the 0-transform of a function $K(\xi)$ which satisfies the valuation

$$|K(\xi)| \leq A(1 + |\xi|^{k+2})^{-1} .$$

If $k \geq 2$, there are always "eigenvalues", i.e., numbers λ, for which (1) is solvable. What is more, each value of $\gamma(\alpha)$ is an eigenvalue. For example, if $\alpha = \tau$ is a zero of $G(\alpha)$, then $y(x) = e(\tau x)$

is a solution of (1). This can be recognized by the relation

$$\lambda e(\tau x) - \frac{1}{2\pi}\int e(\tau\xi)K(x - \xi)\, d\xi = G(\tau) \cdot e(\tau x) \quad .$$

If $\gamma(\alpha)$ is analytic, as for example if

$$K(\xi) = O(e^{-a|\xi|}), \qquad a > 0 \quad ,$$

then for each eigenvalue λ, $G(\alpha)$ has finitely many zeros (3) with multiplicity (6), and moreover the eigen functions are given by (7).

Concerning the non-homogeneous equation, the following is valid, similarly as in §26, 4. For the solvability of

$$(\lambda - \gamma(\alpha))d^k\varphi(\alpha) = d^kE(\alpha)$$

for a given $\lambda \neq 0$, it is sufficient that $\gamma(\alpha)$ be $(k+2)$-times differentiable and be absolutely integrable together with the $K + 2$ derivatives, and that one of the following conditions be satisfied.

1) $G(\alpha)$ is nowhere zero.

2) $G(\alpha)$ has (what is always true of analytic $G(\alpha)$) only finitely many zeros τ_χ with multiplicity ℓ_χ, and for each χ, the function

$$e(-\tau_\chi x)f(x)$$

is ℓ_χ-times integrable in $\mathfrak{F}_k$.

3) Finitely many intervals $A_\chi \leqq \alpha \leqq B_\chi$ can be specified of the following character. Each zero of $G(\alpha)$ is contained in the interior of one of the intervals, and in each of these intervals $E(\alpha) \asymp 0$.

In the case of 1), we have again as in §26, the representation by means of the solving kernel.

§34. Functional Equations

1. If the given function $f(x)$ vanishes in one of the equations (A) - (F) of §22 (homogeneous case), or more generally belongs to $\mathfrak{F}_k$, and if the given function $K(\xi)$ satisfies a valuation

$$(1) \qquad |K(\xi)| \leqq A(1 + |\xi|^{k+2})^{-1}$$

in case (F), then one can inquire about those solutions which are r-times differentiable in $\mathfrak{F}_k$. Here it can be assumed that the index k is fixed, or it can be permitted to be arbitrarily large. If k is fixed, the equation

(2) $$G(\alpha)d^k\varphi(\alpha) = d^kE(\alpha)$$

holds for the k-transform, and the results of the previous paragraph are, basically, results concerning the functional equation

(3) $$\Lambda y = f(x)$$

belonging to (2). The results become considerably more significant if the index k is not held fixed.

By the function class $\mathfrak{F}$ we shall mean the union of all function classes $\mathfrak{F}_k$, so that each function of $\mathfrak{F}$ belongs to a certain class $\mathfrak{F}_k$. All functions therefore belong to $\mathfrak{F}$ which increase more weakly than a power $|x|^m$ for a sufficiently large m.

Let $f(x)$ now be a function of $\mathfrak{F}$, and in the case of equation (F), let

(4) $$|K(\xi)| \leqq A_n(1 + |\xi|^n)^{-1}, \qquad n = 1, 2, 3, \ldots .$$

By a solution of (3), we mean a function $y(x)$ which satisfies it and which is r-times differentiable in $\mathfrak{F}$, i.e., it, together with the first r-derivatives, belongs to $\mathfrak{F}$.

2. THEOREM 45. The equation

(5) $$y^{(r)}(x) + \sum_{\rho=0}^{r-1}\sum_{\sigma=0}^{s} a_{\rho\sigma}y^{(\rho)}(x + \delta_\sigma) = f(x), \qquad r \geqq 1 ,$$

always has a solution. This solution is unique to within an arbitrary additive function of the form

(6) $$\sum_{\nu=1}^{n}\left(e(\tau_\nu x)\sum_{\mu=0}^{\ell_\nu - 1} c_{\nu\mu}x^\mu \right) ,$$

where the numbers $\tau_1, \ldots, \tau_n$ are the zeros of $G(\alpha)$ and whose multiplicities are the $\ell_1, \ldots, \ell_n$ [83].

The same statement is valid for the general equation

(7) $$\sum_{\rho=0}^{r}\sum_{\sigma=0}^{s} a_{\rho\sigma}y^{(\rho)}(x + \delta_\sigma) = f(x) , \qquad r \geqq 0$$

whenever

$$(8) \qquad \left| \sum_{\sigma=0}^{s} a_{r\sigma} e(\delta_\sigma \alpha) \right| \geq S > 0, \qquad (-\infty < \alpha < \infty) .$$

PROOF. Let $f(x)$ be contained in $\mathfrak{F}_{k_0}$. Set

$$k_1 = k_0 + \ell_1 + \ell_2 + \dots + \ell_n + 2 ,$$

and consider a fixed $k \geq k_1$. By §29, 8, the function

$$e(-\tau_\chi x) f(x)$$

is ℓ_χ-times integrable in $\mathfrak{F}_k$ for each χ. Hence by §33, 6, our equation has a solution in $\mathfrak{F}_k$. Since $k \geq k_1$, $\min. (\ell_\nu - 1, k - 2) = \ell_\nu - 1$. Therefore the solution is unique to within an arbitrary additive function of the form (6). And since the special value of k does not enter the structure of (6), all solutions have thus been obtained, Q.E.D. — If no zeros at all of $G(\alpha)$ exist in (5), then one can again write

$$y^{(\rho)}(x) = \frac{1}{2\pi} \int K^{(\rho)}(\xi) f(x - \xi)\, d\xi , \qquad \rho = 0, 1, 2, \dots, r - 1.$$

One can gather from this representation, that not only the membership of $f(x)$ to a class $\mathfrak{F}_k$ is transmitted to $y(x)$, but also that other more intimate properties of $f(x)$ are transmitted to $y^{(\rho)}(x)$, $0 \leq \rho \leq r - 1$, and due to

$$y^{(r)}(x) = f(x) - \sum_{\rho=0}^{r-1} \sum_{\sigma=0}^{s} a_{\rho\sigma} y^{(\rho)}(x + \delta_\sigma)$$

also to $y^{(r)}(x)$. For example, if $y(x)$ is bounded, uniformly continuous, of bounded total variation, etc., then $y^{(\rho)}(x)$, $0 \leq \rho \leq r$, is also bounded, uniformly continuous, of bounded total variation, etc. If $f(x) = O(|x|^n)$, then the same valuation holds also for the $y^{(\rho)}(x)$. If $f(x)$ is almost periodic, so also is the $y^{(\rho)}(x)$. — The same observations can also be made for the more general equation (7), but we shall not pursue this subject further.

If restriction (8) is not satisfied by (7), but the transform of the given function $f(x)$ is equivalent to zero outside of a finite interval (a, b), then again at least a solution exists. But the arbitrary solution of the homogeneous equation cannot be described as simply as in Theorem 45, unless one limits oneself to such solutions whose transforms are likewise equivalent to zero outside of (a, b).

3. The statement of Theorem 45 is valid, for fixed $\lambda \neq 0$, also for the integral equation

$$\lambda y(x) - \frac{1}{2\pi}\int y(\xi)K(x - \xi)\, d\xi = f(x) \quad ,$$

provided $K(\xi)$ satisfies (4) and $G(\alpha)$ has only finitely many zeros τ_ν with well determined multiplicity ℓ_ν; conditions which are fulfilled, for example, if

$$|K(\xi)| \leq Ae^{-a|\xi|} \quad , \qquad A > 0,\ a > 0 \quad .$$

A well known example is [84]

$$K(t) = 2\pi e^{-|t|} \quad ,$$

in which case

$$\lambda y(x) - \int e^{-|x-\xi|} y(\xi)\, d\xi = f(x) \quad . \tag{9}$$

In this case

$$\lambda(\alpha) = \frac{2}{1 + \alpha^2} \ , \qquad G(\alpha) = \lambda - \frac{2}{1 + \alpha^2}$$

The numbers $\lambda \leq 2$ are eigenvalues. The equation

$$\lambda - \frac{2}{1 + \alpha^2} = 0$$

yields

$$\alpha = \pm\sqrt{\frac{2 - \lambda}{\lambda}}$$

and these zeros are both simple. The totality of eigenfunctions (out of the totality of functions of $\mathfrak{F}$) consists therefore of the functions

$$c_1 e\left(\sqrt{\frac{2-\lambda}{\lambda}}\, x\right) + c_2 e\left(-\sqrt{\frac{2-\lambda}{\lambda}}\, x\right) \quad .$$

If $f(x)$ belongs to $\mathfrak{F}$, the non-homogeneous equation (9) always has a solution (in $\mathfrak{F}$). The solution $y(x)$ must belong to the lowest class $\mathfrak{F}_k$ in which $f(x)$ is already contained, except when λ is an eigenvalue. If $f(x)$ is bounded, $y(x)$ need not likewise be bounded.

4. We shall not, in the present chapter, go into the system of functional equations. But in the next chapter, we shall have the opportunity to examine certain special systems in a concrete connection.

CHAPTER VII

ANALYTIC AND HARMONIC FUNCTIONS

§35. Laplace Integrals [85]

1. We shall examine analytic functions of a complex variable. As frequently done, we denote the complex variable by

$$s = \sigma + it, \quad \sigma = \Re(s), \quad t = \Im(s) \quad .$$

By a strip $[\lambda, \mu]$, we mean all the points of the complex plane for which $\lambda < \sigma < \mu$. It is also admissible to have $\lambda = -\infty$ or $\mu = \infty$, in which case a left half or a right half plane or the whole plane is involved. By (λ_1, μ_1) we shall always mean a real closed sub-strip of $[\lambda, \mu]$, $(\lambda <)\lambda_1 \leqq \sigma \leqq \mu_1 (< \mu)$.

2. In $0 < \alpha < \infty$, let $E(\alpha)$ be a given function integrable in every finite interval. If the integral

$$(1) \qquad \int_0 e^{\sigma\alpha}|E(\alpha)|\, d\alpha$$

converges for a certain σ_0, then since for $\sigma \leqq \sigma_0 : e^{\sigma\alpha} \leqq e^{\sigma_0\alpha}$, the integral

$$(2) \qquad f(s) = \int_0 e^{s\alpha}E(\alpha)\, d\alpha$$

converges absolutely and uniformly for $\Re(s) \leqq \sigma_0$. It is easy to show that the function

$$f_n(s) = \int_0^n e^{s\alpha}E(\alpha)\, d\alpha , \qquad n = 1, 2, 3, \dots ,$$

is everywhere differentiable. Therefore it is analytic, and indeed

$$f_n'(s) = \int_0^n \alpha e^{s\alpha} E(\alpha)\, d\alpha \quad .$$

Since the sequence of analytic functions $f_n(s)$ converges uniformly as $n \longrightarrow \infty$ to the function $f(s)$ in the open region $\Re(s) < \sigma_0$, it follows by a general theorem that the latter function is likewise analytic there. Because of

$$\lim_{\alpha \to 0} \alpha^r e^{\sigma\alpha - \sigma_0\alpha} = 0, \qquad \sigma < \sigma_0, \quad r = 1, 2, 3, \ldots,$$

the integral

$$\int_0 \alpha e^{s\alpha} E(\alpha)\, d\alpha$$

is absolutely and uniformly convergent in each partial half plane $\Re(s) \leq \sigma_1 (< \sigma_0)$. It is the limit of the sequence $f_n'(s)$ and by a general theorem, it has therefore the value $f'(s)$. More generally

$$(3) \qquad f^{(r)}(s) = \int_0 \alpha^r e^{s\alpha} E(\alpha)\, d\alpha \quad , \qquad r = 1, 2, 3, \ldots \quad ,$$

where the integral on the right converges absolutely and uniformly for each r in each partial half plane $\Re(s) \leq \sigma_1$.

<u>We call the integral</u> (2) <u>a Laplace integral (in the narrower sense</u>).

It is possible to give a simple formula for the upper limit μ of all numbers σ_0 for which (1) is finite. And indeed μ has the value

$$\overline{\lim_{\alpha \to \infty}} \frac{1}{\alpha} \int_0^\alpha |E(\alpha)|\, d\alpha \quad \text{or} \quad \overline{\lim_{\alpha \to \infty}} \frac{1}{\alpha} \int_\alpha |E(\alpha)|\, d\alpha \quad ,$$

according as $\mu < 0$ or $\mu > 0$. In what follows, we shall not discuss questions of function theory of this kind.

3. For any $\sigma < \min.(\sigma_0, 0)$, consider the absolutely integrable functions

$$H(\alpha) = e^{\sigma\alpha} E(\alpha) \quad \text{for } \alpha \geq 0, \quad = 0 \quad \text{for } \alpha < 0$$

$$K(\alpha) = e^{\sigma\alpha} \qquad \text{for } \alpha \geq 0, \quad = 0 \quad \text{for } \alpha < 0 \quad .$$

Their Faltung, cf. §13, 3, is zero for $\alpha < 0$ and

$$\frac{1}{2\pi}\int_0^\alpha e^{\sigma\beta}E(\beta)e^{\sigma(\alpha-\beta)}\,d\beta = \frac{1}{2\pi}\,e^{\sigma\alpha}\int_0^\alpha E(\beta)\,d\beta$$

for $\alpha > 0$. And since the Faltung is likewise absolutely integrable, cf. §13, 3, it follows that

$$\int_0 e^{\sigma\alpha}|E(\alpha, 1)|\,d\alpha \tag{4}$$

is finite, where

$$E(\alpha, 1) = \int_0^\alpha E(\beta)\,d\beta \quad . \tag{5}$$

From

$$\int_0^\alpha e^{\sigma\alpha}E(\alpha)\,d\alpha = e^{\sigma\alpha}E(\alpha, 1) - \sigma\int_0^\alpha e^{\sigma\alpha}E(\alpha, 1)\,d\alpha \quad ,$$

it follows, since both integrals converge as $\alpha \longrightarrow \infty$, that $e^{\sigma\alpha}E(\alpha, 1)$ also converges as $\alpha \longrightarrow \infty$. By (4), this limit must vanish, i.e.,

$$\lim_{\alpha \to \infty} e^{\sigma\alpha}E(\alpha, 1) = 0 \quad . \tag{6}$$

With the aid of (6) and (4), we have by partial integration of (2), the Laplace integral

$$-\frac{f(s)}{s} = \int_0 e^{\sigma\alpha}E(\alpha, 1)\,d\alpha \quad , \qquad \Re(s) < \min(\sigma_0, 0) \quad .$$

By induction one obtains for $r = 1, 2, 3, \cdots$

$$(-1)^r\,\frac{f(s)}{s} = \int_0 e^{s\alpha}E(\alpha, r)\,d\alpha \quad , \qquad \Re(s) < \min(\sigma_0, 0) \tag{7}$$

where

$$E(\alpha, r+1) = \int_0^\alpha E(\beta, r)\,d\beta \quad , \qquad r = 1, 2, 3 \quad .$$

Conversely, if $E(\alpha)$ is differentiable, then the formula

$$- sf(s) = \int_0 e^{s\alpha}E'(\alpha)\,d\alpha + E(0) \quad , \qquad \Re(s) < \min(\sigma_0, 0) \quad ,$$

is valid, provided the integral on the right converges absolutely. Inductively one finds that if the integrals

$$\int_0 e^{s\alpha}E^{(\sigma)}(\alpha)\,d\alpha \quad , \qquad \Re(s) < \min(\sigma_0, 0) \, , \qquad \sigma = 1, 2, \ldots, r \quad ,$$

converge absolutely, then if $\Re(s) < \min(\sigma_0, 0)$

(7') $$(-1)^r[s^r f(s) + s^{r-1}E(0) - s^{r-2}E'(0) + \ldots + (-1)^{r-1}E^{(r-1)}(0)]$$

$$= \int_0 e^{s\alpha}E^{(r)}(\alpha)\,d\alpha \quad .$$

4. For the integrals

$$\int^0 e^{s\alpha}E(\alpha)\,d\alpha \quad ,$$

there are analogous observations to those made thus far: instead of the half plane $[-\infty, \sigma_0]$, substitute the half plane $[\sigma_0, \infty]$, and replace $\min(\sigma_0, 0)$ by $\max(\sigma_0, 0)$.

Through the combination of both kinds, we obtain Laplace integrals (in the wider sense) of the form

(7") $$f(s) = \int e^{s\alpha}E(\alpha)\,d\alpha \quad .$$

If the integral (7") converges absolutely for $\sigma = \lambda$, and $\sigma = \mu$, $\lambda < \mu$, then by the decomposition

$$\int = \int^0 + \int_0 \quad ,$$

one recognizes that it converges absolutely and uniformly in $[\lambda, \mu]$, and represents an analytic function. For arbitrary r we have

$$f^{(r)}(s) = \int \alpha^r e^{s\alpha}E(\alpha)\,d\alpha \quad ,$$

where the integral on the right converges absolutely and uniformly in each sub strip (λ_1, μ_1).

In the sense of §13,

$$(8) \qquad e^{c\alpha}e(\alpha) \sim \frac{1}{2\pi}\int f(c + it)e(-\alpha t)\,dt$$

is valid for each c in $\lambda < c < \mu$. Therefore, if two Laplace integrals of an analytic function $f(s)$ are known, whose admissible strips have common interior points, then the function $E(\alpha)$ must be the same both times (by Theorem 14). It is different, however, if the strips lie outside one another. Thus for example, the function $f(s) = s^{-1}$ has in the strips $[-\infty, 0]$, $[0, \infty]$, the different Laplace integrals

$$-\int_0 e^{s\alpha}d\alpha\ , \qquad \int^0 e^{s\alpha}d\alpha\ .$$

If one considers that $E(\alpha)$ and $e^{\sigma\alpha}E(\alpha)$ are simultaneously of bounded variation, then for each point α in whose neighborhood $E(\alpha)$ is of bounded variation, and for $\lambda < c < \mu$, one finds the inverse formulas

$$(9) \qquad f(s) = \int e^{s\alpha}E(\alpha)\,d\alpha\ , \qquad \frac{E(\alpha+0) + E(\alpha-0)}{2} = \frac{1}{2\pi i}\int_{c-i\infty}^{c+i\infty} f(s)e^{-s\alpha}\,d\alpha\ ,$$

where the last integral is to be taken as the principal value of

$$\lim_{A \to \infty}\int_{c-iA}^{c+iA}\ .$$

If $x = e^{\alpha}$, the pair of formulas (9) becomes

$$(10) \qquad f(s) = \int_0 \varphi(x)x^{s-1}\,dx\ , \qquad \frac{\varphi(x+0) + \varphi(x-0)}{2} = \frac{1}{2\pi i}\int_{c-i\infty}^{c+i\infty} f(s)x^{-s}\,ds\ .$$

Here $x > 0$, $x^{-s} = e^{-s \log s}$, and $\log x$ is real. These inverse formulas are named after Mellin [85]. The convergence assumption of $\varphi(x)$ states that for $s = \lambda$ and $s = \mu$ with $\lambda < c < \mu$, the integral

$$\int_0 \varphi(x)x^{s-1}\,dx$$

converges absolutely. This occurs, for example, if $\varphi(x)$, as $x \longrightarrow \infty$, has the order of magnitude $O(x^{-\mu-\epsilon})$, and as $x \longrightarrow 0$, the order of magnitude $O(x^{-\lambda+\epsilon})$, $\epsilon > 0$.

5. Examples of the inverse formula

1. The best known is $[\Re(s) > 0,\ c > 0,\ x > 0]$

$$\Gamma(s) = \int_0^\infty e^{-x}x^{s-1}\,dx\,, \qquad e^{-x} = \frac{1}{2\pi i}\int_{c-i\infty}^{c+i\infty} \Gamma(s)x^{-s}\,ds \quad .$$

The second formula, after having been established for $x > 0$, may also be claimed for $\Re(x) > 0$.

2. $\varphi(x) = x^p(1 + x)^{-q}$, $0 < p < q$. The first integral in (10) converges for $-p < \sigma < q - p$, and indeed

$$f(s) = \int_0 x^p(1 + x)^{-q}x^{s-1}\,dx = \int_0 x^{p+s-1}(1 + x)^{-q}\,dx = \frac{1}{\Gamma(q)}\,\Gamma(s + p)\Gamma(q - p - s).$$

Hence

$$(11)\quad \Gamma(q)x^p(1 + x)^{-q} = \frac{1}{2\pi i}\int_{c-i\infty}^{c+i\infty} \Gamma(s + p)\Gamma(q - p - s)x^{-s}\,ds, \quad -p < c < q - p.$$

In particular $q = 2p$, $c = 0$ gives

$$\Gamma(2p)x^p(1 + x)^{-2p} = \frac{1}{2\pi}\int_0 |\Gamma(p + it)|^2\cos(t\log x)\,dt \quad ,$$

using the fact that $\Gamma(p - it) = \overline{\Gamma(p + it)}$.

3. $\varphi(x) = (1 - x)^{q-1}$ for $0 < x < 1$, and $= 0$ for $x > 1$; $q > 0$.

$$f(s) = \int_0^1 (1 - x)^{q-1}x^{s-1}\,dx = \frac{\Gamma(q)\Gamma(s)}{\Gamma(s+q)} \quad .$$

Hence for $c > 0$

$$(12)\quad \frac{1}{2\pi i}\int_{c-i\infty}^{c+i\infty} \frac{\Gamma(s)}{\Gamma(s+q)}\,x^{-s}\,ds = \begin{cases} \frac{1}{\Gamma(q)}\,(1 - x)^{q-1} & \text{for } 0 < x < 1 \\ 0 & \text{for } 1 < x < \infty \; . \end{cases}$$

If $q = 1$, one obtains the important formula $[c > 0]$

$$\frac{1}{2\pi i}\int_{c-i\infty}^{c+i\infty} \frac{x^{-s}}{s}\,ds = \begin{cases} 1 & \text{for } 0 < x < 1 \\ 1/2 & \text{for } x = 1 \\ 0 & \text{for } 1 < x < \infty \end{cases} .$$

If q is a positive integer > 1, we have

$$\frac{1}{2\pi i}\int_{c-i\infty}^{c+i\infty} \frac{x^{-s}\,ds}{s(s+1)\ldots(s+q-1)} = \begin{cases} \frac{(1-x)^{q-1}}{(q-1)!} & \text{for } 0 < x \leqq 1 \\ 0 & \text{for } 1 \leqq x < \infty \end{cases} .$$

4. $\varphi(x) = \left(\log\frac{1}{x}\right)^p$ for $0 < x \leqq 1$, and $= 0$ for $x > 1$; $p > -1$. Hence

$$f(s) = \int_0^1 x^{s-1}\left(\log\frac{1}{x}\right)^p dx = \int_0 e^{-y(s-1)}y^p e^{-y}\,dy = \int_0 e^{-ys}y^p\,dy .$$

But for $\Re(s) > 0$, this is

$$\frac{\Gamma(p+1)}{s^{p+1}} .$$

Hence for $p > -1$, $c > 0$, we have

$$\frac{1}{2\pi i}\int_{c-i\infty}^{c+i\infty} \frac{x^{-s}}{s^{p+1}}\,ds = \begin{cases} \frac{\left(\log\frac{1}{x}\right)^p}{\Gamma(p+1)} & \text{for } 0 < x < 1 \\ 0 & \text{for } 1 < x < \infty \end{cases} .$$

6. We now make, without proof, a few assertions concerning ordinary (not necessarily absolute) convergence of a Laplace integral

$$f(s) = \int_0 e^{s\alpha}E(\alpha)\,d\alpha . \tag{13}$$

If the integral on the right converges at a point $s = \sigma_0$, or only oscillates between finite limits, then it is uniformly convergent in each bounded, closed point set of the half plane $\Re(s) < \sigma_0$, and therefore represents an analytic function there. The relations (3), (7) and (7') retain validity, and the analogue to the above uniqueness theorem again holds [85].

The integral (13) is a special case of the integral

$$(14) \qquad \int_0 e^{s\alpha} d\Phi(\alpha) \quad ,$$

where $\Phi(\alpha)$ is of bounded variation in each finite interval $(0, A)$, and the integral is to be taken in the Stieltjes sense [87]. The integrals (14) include the Dirichlet series which arise if the function $\Phi(\alpha)$ is piecewise constant and has only isolated jump points. By an extension, the above assertions about the integral (13) are valid for the integral (14), and include the corresponding assertions for Dirichlet series. Many other statments can be carried over from Dirichlet series to the integral (14). For example, there is the theorem that for monotone $\Phi(\alpha)$ the point $\sigma = \mu$ of the convergence abscissa is a singular point of the represented function. In addition the summation theories of M. Riesz, the theorems about the order of magnitude of the represented function, etc.

We mention that theorems of Abelian and Tauberian character have been carried over from power series to Laplace integrals (13) [88], and also that frequently the distribution of the (complex) zeros of functions has been investigated which can be represented by integrals of the form

$$\int e(s\alpha)E(\alpha)\, d\alpha \quad , \qquad \int {\cos \atop \sin} s\alpha E(\alpha)\, d\alpha \quad ,$$

and by correspondingly more general Stieltjes integrals [89]. We observe further that under suitable assumptions concerning $E(\alpha)$, asymptotic developments [90] can be set up for Laplace integrals (13) which are of importance for various questions of analysis and in the theory of probability.

We point out in closing, that as a rule we shall understand by Laplace integrals only the above absolutely convergent integrals, treated in extenso.

§36. Union of Laplace Integrals [91]

1. If two Laplace integrals

$$f_\nu(s) = \int e^{s\alpha} E_\nu(\alpha)\, d\alpha \quad , \qquad \nu = 1, 2 \quad ,$$

have a strip $[\sigma_1, \sigma_2]$ in common, then

$$f_1(s)f_2(s) = \int e^{s\alpha} E(\alpha)\, d\alpha$$

is valid there, where by Theorem 13

$$e^{\sigma\alpha}E(\alpha) = \int e^{\sigma\beta}E_1(\beta)e^{\sigma(\alpha-\beta)}E_2(\alpha - \beta)\, d\beta \quad .$$

Hence

$$E(\alpha) = \int E_1(\beta)E_2(\alpha - \beta)\, d\beta \quad , \tag{1}$$

and it is part of the assertion that the last integral is convergent (for almost all α). If f_1 and f_2 are Laplace integrals in the narrower sense, then the same holds for their Faltung, i.e.,

$$f_1(s)f_2(s) = \int_0 e^{s\alpha}E(\alpha)\, d\alpha$$

where

$$E(\alpha) = \int_0^\alpha E_1(\beta)E_2(\alpha - \beta)\, d\beta \quad . \tag{2}$$

Example [92]. For the Bessel function $J_\mu(\alpha)$, we have

$$\int_0 e^{s\alpha}\alpha^\mu J_\mu(\alpha)\, d\alpha = \frac{\Gamma(\mu+1/2)}{\sqrt{\pi}} \cdot \frac{2^\mu}{(1+s^2)^{\mu+1/2}} \tag{3}$$

if $\sigma < 0$, $\mu > -1/2$. From this it follows if $\mu > -1/2$, $\nu > -1/2$, $\sigma < 0$, that

$$\int_0^\alpha \beta^\mu J_\mu(\beta)(\alpha - \beta)^\nu J_\nu(\alpha - \beta)\, d\beta = \frac{\Gamma(\mu+1/2)\Gamma(\nu+1/2)2^{\mu+\nu}}{\sqrt{\pi^2}} \cdot \frac{1}{2\pi i}\int_{c-i\infty}^{c+i\infty} \frac{e^{-s\alpha}ds}{(1+s^2)^{\mu+\nu+1}}$$

$$= \frac{1}{\sqrt{2\pi}}\,\frac{\Gamma(\mu+1/2)\Gamma(\nu+1/2)}{\Gamma(\mu+\nu+1)}\,\alpha^{\mu+\nu+1/2}J_{\mu+\nu+1/2}(\alpha) \; .$$

If $\mu = \nu = 0$, we have

$$\int_0^\alpha J_0(\beta)J_0(\alpha - \beta)\, d\beta = \sqrt{\frac{\pi}{2}}\,\alpha^{1/2}J_{1/2}(\alpha) = \sin\alpha \quad . \tag{4}$$

2. In the Mellin manner of writing, §35, (10), the Faltung rule (1) corresponds to the formula

$$\varphi(x) = \int \varphi_1(y)\varphi_2\left(\frac{x}{y}\right)\frac{dy}{y} \; . \tag{5}$$

EXAMPLES.

1. $$\varphi_1(x) = \frac{x^{-\alpha}}{(1+x)^{\gamma-\alpha}}, \qquad \varphi_2(x) = \frac{x^{-\beta}}{(1+x)^{\delta-\beta}},$$

gives by example 2 of §35, in the strip $[\alpha, \gamma]$ or $[\beta, \delta]$ with $\alpha < 0 < \gamma$, $\beta < 0 < \delta$,

$$f_1(s) = \frac{\Gamma(s-\alpha)\Gamma(\gamma-s)}{\Gamma(\gamma-\alpha)}, \qquad f_2(s) = \frac{\Gamma(s-\beta)\Gamma(\delta-s)}{\Gamma(\delta-\beta)} .$$

Now in the common strip $[\text{Max}(\alpha, \beta), \text{Min}(\gamma, \delta)]$

$$(6) \qquad \varphi(x) = \int_0 \frac{y^{-\alpha}}{(1+y)^{\gamma-\alpha}} \cdot \frac{x^{-\beta}}{y^{-\beta}(1+x/y)^{\delta-\beta}} \frac{dy}{y} = x^{-\beta} \int_0 \frac{y^{\delta-\alpha-1}\, dy}{(1+y)^{\gamma-\alpha}(x+y)^{\delta-\beta}}$$

and

$$(7) \qquad f(s) = f_1(s)f_2(s) = \frac{\Gamma(s-\alpha)\Gamma(s-\beta)\Gamma(\gamma-s)\Gamma(\delta-s)}{\Gamma(\gamma-\alpha)\Gamma(\delta-\beta)} .$$

These strips contain the straight line $\sigma = 0$. As is easily established, $\varphi(x)$ is a differentiable function, and hence

$$\varphi(1) = \int_0 \frac{y^{\delta-\alpha-1}\, dy}{(1+y)^{\gamma+\delta-\alpha-\beta}} = \frac{\Gamma(\delta-\alpha)\Gamma(\gamma-\beta)}{\Gamma(\gamma+\delta-\alpha-\beta)} ,$$

$$f(it) = \frac{\Gamma(it-\alpha)\Gamma(it-\beta)\Gamma(\gamma-it)\Gamma(\delta-it)}{\Gamma(\gamma-\alpha)\Gamma(\delta-\beta)} .$$

By §35, (10), we obtain for $x = 1$, $c = 0$ [93]

$$(8) \quad \frac{1}{2\pi} \int \Gamma(it - \alpha)\Gamma(it - \beta)\Gamma(\gamma - it)\Gamma(\delta - it)\, dt = \frac{\Gamma(\gamma-\alpha)\Gamma(\delta-\alpha)\Gamma(\gamma-\beta)\Gamma(\delta-\beta)}{\Gamma(\gamma-\delta-\alpha-\beta)}$$

2. Starting with

$$\varphi_1(x) = x^{-\alpha}e^{-x}, \qquad \varphi_2(x) = x^{-\beta}e^{-x} ,$$

one finds for $c > \text{Max}(\alpha, \beta)$

$$(9) \qquad \frac{1}{2\pi i} \int_{c-i\infty}^{c+i\infty} \Gamma(s - \alpha)\Gamma(s - \beta)x^{-s}\, ds = x^{-\beta} \int_0 y^{\beta-\alpha-1}e^{-y-x/y}\, dy .$$

If especially $\alpha = \beta = 0$, $c > 0$, we obtain

$$(10) \qquad \frac{1}{2\pi}\int \Gamma(c + it)^2 x^{-c-it}\, dt = \int_0 e^{-y-x/y}\, \frac{dy}{y} \quad .$$

3. Different Faltung formulas are obtained if the product $f_1(c + it)f_2(c - it)$ is formed in place of the product $f_1(c + it)f_2(c + it)$. From

$$f_1(c + it) = \int e(\alpha t)e^{\alpha c}E_1(\alpha)\, d\alpha, \qquad f_2(c - it) = \int e(\alpha t)e^{-\alpha c}E_2(-\alpha)\, d\alpha \quad ,$$

there results by Faltung

$$(11) \quad e^{-\alpha c}\int e^{2\beta c}E_1(\beta)E_2(\beta - \alpha)\, d\beta \sim \frac{1}{2\pi}\int f_1(c + it)f_2(c - it)e(-t\alpha)\, dt \quad .$$

If the function on the left is by chance differentiable in α, then there results, if α is now replaced by 2α, and β by $\beta + \alpha$

$$(12) \quad \frac{1}{2\pi}\int f_1(c + it)f_2(c - it)e(-2t\alpha)\, dt = \int e^{2\beta c}E_1(\beta + \alpha)E_2(\beta - \alpha)\, d\beta$$

$$= \int_0 x^{2c-1}\varphi_1(xe^{\alpha})\varphi_2(xe^{-\alpha})\, dx \quad .$$

4. For $t > 0$, consider the theta function

$$(13) \qquad \vartheta(t) = \sum_{n=-\infty}^{\infty} e^{-\pi^2 n^2 t} \quad .$$

Starting with

$$\int_0 e^{st}e^{-\pi^2 n^2 t}\, dt = -\frac{1}{s - \pi^2 n^2}, \qquad \Re(s) < 0 \quad ,$$

we find for the partial sums

$$\vartheta_n(t) = \sum_{\nu=-n}^{n} e^{-\pi^2 \nu^2 t} \quad ,$$

the relation

$$\int_0^{\infty} e^{st}\vartheta_n(t)\,dt = -\sum_{-n}^{n} \frac{1}{s - \pi^2\nu^2}, \qquad \Re(s) < 0 \quad .$$

Let $s = \sigma < 0$. As $n \longrightarrow \infty$, the series on the right on the one hand, converges to a finite number, and on the other hand the positive integrand $e^{\sigma t}\vartheta_n(t)$ converges monotonically increasing to $e^{\sigma t}\vartheta(t)$. Hence by a general theorem [Appendix 7, 9], the latter function is also integrable in $[0, \infty]$, and

$$\int_0^{\infty} e^{st}\vartheta(t)\,dt = -\sum_{n=-\infty}^{\infty} \frac{1}{s - \pi^2 n^2} \quad .$$

As is well known, the series on the right has the value

$$f(s) = -\frac{\operatorname{ctg}\sqrt{s}}{\sqrt{s}} \tag{14}$$

and we have therefore found the Laplace integral for this function

$$f(s) = \int_0^{\infty} e^{st}\vartheta(t)\,dt, \qquad \Re(s) < 0 \quad . \tag{15}$$

But now (14) satisfies the differential equation

$$f(s)^2 - 2f'(s) - \frac{1}{s}f(s) + \frac{1}{s} = 0 \quad . \tag{16}$$

If we make use of the formula

$$\frac{1}{s} = -\int_0^{\infty} e^{st}\,dt, \qquad \Re(s) < 0 \quad ,$$

and apply foregoing statements then from the Laplace integral (15) we can deduce a Laplace integral for the left side of (16), valid in $\Re(s) < 0$. By the uniqueness theorem, the integrand of the Laplace integral must vanish and this leads to the relation

$$\int_0^{t} \vartheta(\tau)\vartheta(t-\tau)\,d\tau - 2t\vartheta(t) + \vartheta'(t) - 1 = 0 \quad .$$

This surprising relation for the theta function (13), which can also be verified directly, has been discovered strangely enough only recently [94], and although it had been found in another way, it was soon afterwards verified in the manner here described. It is not an isolated formula. If the Laplace integral

$$f(s) = \int_0 e^{s\alpha} E(\alpha)\, d\alpha$$

satisfies an algebraic differential equation

$$\mathfrak{P}(s, f, f', f'', \ldots) = 0 \quad ,$$

where $\mathfrak{P}(s, f, f', f'', \ldots)$ is a polynomial of the arguments $s, f, f',$ $f'', \ldots,$ then the function $E(\alpha)$ satisfies an integral differential equation of the above type [95].

§37. Representation of Given Functions by Laplace Integrals

1. A function $f(s)$, analytic in $[\lambda, \mu]$, shall belong to class $\mathfrak{F}_k$, $k = 0, 1, 2, \ldots,$ if for each sub-strip $[\lambda_1, \mu_1]$, there is a constant $K = K(\lambda_1, \mu_1)$ such that

$$(1) \qquad \int |f(\sigma + it)| p_k(t)\, dt \leqq K \quad ,$$

where, as in the previous chapter,

$$(2) \qquad p_k(t) = \frac{1}{1 + |t|^k} \quad .$$

We require, therefore, that the function $f(\sigma + it)$, as a function of t, "uniformly in σ", belong to $\mathfrak{F}_k$.

THEOREM 46. In $[\lambda, \mu]$, let $f(s)$ be a function of $\mathfrak{F}_k$. In each sub-strip (λ_1, μ_1), there is, for each ϵ, a $T > 0$, such that

$$(3) \qquad |f(s)| \leqq \epsilon |t|^k \quad \text{for} \quad |t| > T \quad ,$$

and

$$(4) \qquad \left(\int^{-T} + \int_T \right) |f(\sigma + it)| p_k(t)\, dt \leqq \epsilon \quad .$$

PROOF. For a function $g(z)$ of the variable $z = x + iy$, analytic in $|x| < r$, $|y| < r$, let

$$\int_{-r}^{r}\int_{-r}^{r} |g(z)|\, dx\, dy \leqq G \quad .$$

Then

$$|g(0)| \leqq \frac{G}{\pi r^2} \quad .$$

For from

$$2\pi g(0) = \int_0^{2\pi} g(\rho e^{i\varphi})\, d\varphi \quad , \qquad 0 < \rho < r \quad ,$$

we have

$$2\pi\rho |g(0)| \leqq \int_0^{2\pi} |g(\rho e^{i\varphi})| \rho\, d\varphi \quad , \qquad 0 < \rho < r \quad ,$$

and from this, by integration with respect to ρ between 0 and r, we obtain

$$\pi r^2 |g(0)| \leqq \int_0^r \int_0^{2\pi} |g(re^{i\rho})| \rho\, d\rho\, d\varphi \leqq G \quad .$$

We introduce a sub-strip $(\lambda_1 - r, \mu_1 + r)$ of $[\lambda, \mu]$ enclosing the sub-strip (λ_1, μ_1). By hypothesis, we have for this sub-strip

$$\int |f(\sigma + it)| p_k(t)\, dt \leqq K_1 \quad .$$

For the function

$$\chi(t) = \int_{\lambda_1 - r}^{\mu_1 + r} |f(\sigma + it)|\, d\sigma \quad , \tag{5}$$

we obtain

$$\int \chi(t) p_k(t)\, dt \leqq K_2 \quad .$$

Hence there is, for each ϵ, a $T_0 > 0$, such that

$$\int_{T_0} \frac{\chi(t)}{t^k}\, dt \leqq \pi r^2 \epsilon \quad . \tag{6}$$

If r is sufficiently small, it follows for $t > T_0 + r$, that

$$\int_{t-r}^{t+r} x(t)\, dt \leqq \pi r^2 \epsilon t^k \quad . \tag{7}$$

Therefore for $t > T_0 + r$ and σ in (λ_1, μ_1)

$$\int_{-r}^{r}\int_{-r}^{r} |f(\sigma + it + \xi + i\eta|\, d\xi\, d\eta \leqq \pi r^2 \epsilon t^k \quad . \tag{8}$$

Because

$$\pi r^2 |f(\sigma + it)| \leqq \int_{-r}^{r}\int_{-r}^{r} |f(\sigma + it + \xi + i\eta|\, d\xi\, d\eta \quad , \tag{9}$$

it follows that

$$|f(s)| \leqq \epsilon t^k \quad \text{for} \quad t > T_0 + r \quad .$$

The inequality (3) has thus been proved for the upper half plane. The proof proceeds analogously for the lower half plane.

By (8), in conjunction with (5), and for sufficiently small r, it follows that

$$\pi r^2 |f(\sigma + it)| p_k(t) \leqq p_k(t) \int_{-r}^{r}\int_{-r}^{r} |f(\sigma + it + \xi + i\eta)|\, d\xi\, d\eta$$

$$\leqq p_k(t) \int_{-r}^{r} x(t + \eta)\, d\eta \leqq 2 \int_{-r}^{r} x(t + \eta) p_k(t + \eta)\, d\eta \quad .$$

From this in conjunction with (6), we obtain

$$\int_{T_0+r} |f(\sigma + it)| p_k(t)\, dt \leqq \frac{2}{\pi r^2} \int_{-r}^{r} d\eta \int_{T_0} x(\tau) p_k(\tau)\, d\tau = \epsilon' \quad ,$$

where for fixed r, ϵ' becomes arbitrarily small along with ϵ. From this and from the corresponding observations for the lower half plane, (4) follows.

DEDUCTION. To each closed rectangle lying in $[\lambda, \mu]$

$$\lambda_1 \leqq \sigma \leqq \mu_1, \quad |t| \leqq T \quad ,$$

there is for each ϵ, a δ such that

$$\int |f(s + i\tau) - f(s' + i\tau)| p_k(\tau)\, d\tau \leqq \epsilon$$

for every two points s, s' of the rectangle for which $|s - s'| \leqq \delta$.

2. THEOREM 47. In $[\lambda, \mu]$, let $f(s)$ be a function of $\mathfrak{F}_0$. The function

$$E(\alpha) = \frac{1}{2\pi i} \int_{c-i\infty}^{c+i\infty} f(s) e^{-\alpha s}\, ds, \qquad \lambda < c < \mu \tag{10}$$

is independent of c, and for all s in $[\lambda, \mu]$, the inverse

$$f(s) = \int e^{s\alpha} E(\alpha)\, d\alpha \tag{11}$$

holds. The integral (11) is absolutely convergent.

PROOF. Since

$$\int_{c-i\infty}^{c+i\infty} f(s) e^{-\alpha s}\, ds = i e^{-\alpha c} \int f(c + it) e(-\alpha t)\, dt \quad ,$$

the integral (10) is absolutely convergent for each c. The independence of c results from the Cauchy theorem for

$$\int_{c_2 - iT}^{c_2 + iT} f(s) e^{-\alpha s}\, ds - \int_{c_1 - iT}^{c_1 + iT} = \int_{c_1 + iT}^{c_2 + iT} - \int_{c_1 - iT}^{c_2 - iT} \quad ,$$

and by (3), the two integrals on the right are convergent to zero as $T \longrightarrow \infty$, provided c_1 and c_2 are fixed. — The function $f(\sigma + it)$ has a 0-transform $E(\sigma, \alpha)$ for each σ. Hence

$$E(\sigma, \alpha) = e^{\sigma\alpha} E(\alpha) \quad .$$

It is therefore natural to write, in the sense of §13,

$$f(s) \sim \int e^{s\alpha} E(\alpha)\, d\alpha$$

and to speak of $E(\alpha)$ as the 0-transform of the analytic function $f(s)$. Since $f(s)$, as a function of t, is differentiable, it follows from Theorem 11b that we have equality (11) in a literal sense. It is easy to see that (11) is absolutely convergent. We prove it for $\sigma = \sigma_1$. Choose a σ_2 in the interval $\sigma_1 < \sigma_2 < \mu$. Then

$$E(\sigma_1, \alpha) = e^{(\sigma_1-\sigma_2)\alpha}E(\sigma_2, \alpha)$$

and since $E(\sigma_2, \alpha)$, as a function of $\mathfrak{z}_0$, is bounded, $E(\sigma_1, \alpha)$ is absolutely integrable in $[0, \infty]$. A similar statement is valid in $[-\infty, 0]$, Q.E.D.

In order that the integral (11) be a Laplace integral in the narrower sense:

$$(12) \qquad E(\alpha) = 0 \quad \text{for} \quad \alpha < 0 \quad ,$$

it is necessary that the quantity

$$(13) \qquad J(\sigma) = \int |f(\sigma + it)| \, dt$$

be bounded as $\sigma \longrightarrow -\infty$. It is already sufficient that for each $\epsilon > 0$

$$(14) \qquad J(\sigma) = O(e^{-\epsilon\sigma}) \quad , \qquad \sigma \longrightarrow -\infty \quad .$$

For example, if (14) is satisfied, then it is at once seen from the relation

$$E(\alpha) = \frac{e^{-\sigma\alpha}}{2\pi}\int f(\sigma + it)e(-\alpha t)\, dt \quad ,$$

that (12) holds. Conversely, if (12) holds, then if $\sigma < \sigma_1 < \mu$, $\delta = \sigma_1 - \sigma$, one writes

$$f(\sigma + it) = \int_0 e^{-\delta|\alpha|}E(\alpha)e^{\sigma_1\alpha}e(\alpha t)\, d\alpha \quad .$$

Hence, by the Faltung rule,

$$f(\sigma + it) = \frac{1}{2\pi}\int f(\sigma_1 + i\tau)\frac{2\delta}{\delta^2 + (t-\tau)^2}\, d\tau \quad ,$$

from which follows

$$2\pi J(\sigma) \leqq \int |f(\sigma_1 + i\tau)|\, d\tau \int \frac{2\delta d\tau}{\delta^2 + \tau^2} = 2\pi J(\sigma_1) \quad .$$

Therefore $J(\sigma)$ is also a monotonically decreasing function as $\sigma \to -\infty$.

3. THEOREM 48. In $[\lambda, \mu]$, let $f(s)$ be a function of $\mathfrak{F}_k$. Denote the k-transforms belonging to the functions $f(\sigma + it)$, $\lambda < \sigma < \mu$, by $E(\sigma, \alpha) \equiv E(\sigma, \alpha, k)$. There is a function $E(\alpha) = E(\alpha, k)$ which we shall call the k-transform of $f(s)$ (in $[\lambda, \mu]$), such that

$$d^k E(\sigma, \alpha) = e^{\sigma\alpha} d^k E(\alpha) \quad , \qquad (\lambda < \sigma < \mu) \quad . \tag{15}$$

According to this, one can write

$$f(s) \sim \int e^{s\alpha} d^k E(\alpha) \quad .$$

Two functions of the same class in $[\lambda, \mu]$ whose transforms are equivalent, are identical.

PROOF. It can be assumed for the proof that $[\lambda, \mu]$ is not the whole plane $[-\infty, \infty]$ (otherwise the theorem results from the fact that the whole plane can be covered by two overlapping half planes), and that $[\lambda, \mu]$ does not contain the point $\sigma = 0$ (for example, if the function $f_1(s) = f(s + a)$, for any real a, satisfies the theorem in $[\lambda + a, \mu + a]$, then $f(s)$ satisfies it in the strip $[\lambda, \mu]$). Under these assumptions the function

$$\frac{f(s)}{s^k} \tag{16}$$

is contained in $\mathfrak{F}_0$, because then a valuation

$$|s| = \sqrt{\sigma^2 + t^2} \geqq A(1 + |t|)$$

exists in $\lambda_1 \leqq \Re(s) \leqq \mu_1$. We now set

$$E_T(\sigma, \alpha) = \frac{1}{2\pi} \int_{-T}^{T} \frac{f(\sigma + it)}{(-it)^k} \Big[e(-\alpha t) - L_k(\alpha, t) \Big] dt \quad ,$$

$$H_T(\sigma, \alpha) = \frac{(-1)^k}{2\pi} \int_{-T}^{T} \frac{f(\sigma + it)}{(\sigma + it)^k} e^{-(\sigma + it)\alpha} dt \quad .$$

Then

$$\frac{d^k E_T(\sigma, \alpha)}{d\alpha^k} = e^{\sigma\alpha} \frac{d^k H_T(\sigma, \alpha)}{d\alpha^k} \quad ,$$

for which we can also write

$$d^k E_T(\sigma, \alpha) = e^{\sigma\alpha} d^k H_T(\sigma, \alpha) \quad .$$

If $T \longrightarrow \infty$, the functions $E_T(\sigma, \alpha)$ and $H_Y(\sigma, \alpha)$ are uniformly convergent in each finite α-interval; the first indeed to $E(\sigma, \alpha)$, the second to a function which apart from sign, agrees with the 0-transform of (16). There actually is, therefore, a function $E(\alpha)$ for which (15) holds, Q.E.D.

4. By suitable extension, many properties of the k-transform established in the previous chapter are also valid in the case of analytic functions. Thus, for example, $E(\alpha + \lambda)$ is the k-transform of $e^{-\lambda s}f(s)$ – λ real –, and the k-transform $\Phi(\alpha)$ of $f(s + \lambda)$ satisfies the relation

$$d^k \Phi(\alpha) = e^{\lambda\alpha} d^k E(\alpha) \quad .$$

By a trivial function of $\mathfrak{F}_k$, we mean a function (with real τ_ν)[1]

$$\sum_{\mu=0}^{k-2} \sum_{\nu=1}^{n} c_{\nu\mu} e^{\tau_\nu s} s^\mu \quad . \tag{17}$$

Its k-transform amounts to

$$\frac{1}{2} \sum_{\mu=0}^{k-2} \sum_{\nu=1}^{n} c_{\nu\mu} \frac{(-1)^\mu}{(k-\mu-1)!} |\overline{\alpha - \tau_\nu}|^{k-\mu-1} \quad ,$$

and it is again characterized by having, after the omission of the points $\tau_1, \ldots, \tau_n$ of the α-axis, a vanishing k-th derivative.

If outside of a finite interval (a_0, b_0), $E(\alpha)$ is equivalent to zero, then again for $a < a_0$, $b_0 < b$, we have

$$f(s) = e^{bs} \sum_{\mu=0}^{k-1} (-1)^\mu E^{(k-\mu-1)}(b) s^\mu - e^{as} \sum_{\mu=0}^{k-1} (-1)^\mu E^{(k-\mu-1)}(a) s^\mu$$

$$+ (-s)^k \int_a^b e^{s\alpha} E(\alpha)\, d\alpha \quad .$$

[1] The trivial functions now defined agree for $\sigma = 0$ with those defined in §28, 7.

5. If $E(\alpha)$ is equivalent to zero on a half line $[-\infty, a]$, and if the function $f(s)$ was given originally in an interval $[\lambda, \mu]$, with $\lambda > -\infty$, then there exists an analytic continuation of $f(s)$ in $[-\infty, \mu]$ which likewise belongs to $\mathfrak{F}_k$. It can be assumed for the proof that $a \geqq 0$ — otherwise we consider $e^{-sa}f(s)$ instead of $f(s)$ — and that $\lambda < 0$ — otherwise we consider $f(s + \lambda + \epsilon)$, $\epsilon > 0$, instead of $f(s)$ —. By the proof of Theorem 48, we obtained the result that the k-transform $E(\alpha)$ of $f(s)$ in a suitable normalization, equalled in $[\lambda, 0]$, apart from sign, the 0-transform of the function (16). Assuming this normalization, then

$$\lim_{\alpha \to -\infty} e^{\sigma\alpha}E(\alpha) = 0$$

for $\lambda < \sigma < 0$, and since $E(\alpha)$ in $[-\infty, 0]$ is a polynomial and, therefore $E(\alpha)$ must vanish there. Hence the function (16) exists in $[-\infty, \mu]$, and is there a function of $\mathfrak{F}_0$. Our assertion concerning $f(s)$ itself follows from this.

If $a \geqq 0$ at the outset, then in $[-\infty < \sigma \leqq \mu)$, we have by 2

$$\int \frac{|f(\sigma+it)|}{|\sigma+it|^k}\, dt \leqq K(\mu_1) \quad .$$

From this it easily follows that the function

$$J(\sigma) = \int |f(\sigma + it)| p_k(t)\, dt$$

satisfies the valuation

$$J(\sigma) = O(|\sigma|^k) \quad , \qquad \sigma \longrightarrow -\infty \quad . \tag{19}$$

Conversely, if $f(s)$ belongs to $\mathfrak{F}_k$ in $[-\infty, \mu]$, and for each $\epsilon > 0$ satisfies the valuation [more general than (19)]

$$J(\sigma) = O(e^{-\epsilon\sigma}) \quad , \qquad \sigma \longrightarrow -\infty \quad ,$$

then, as can be recognized with the help of (16), $d^kE(\alpha) = 0$ for $\alpha < 0$.

If in $[-\infty, \infty]$, i.e., in the whole plane, a function $f(s)$ is analytic and belongs to $\mathfrak{F}_k$, and if it has the valuation, for each $\epsilon > 0$,

$$J(\sigma) = O(e^{\epsilon|\sigma|})$$

not only as $\sigma \longrightarrow \infty$ but also as $\sigma \longrightarrow -\infty$, then $d^kE(\alpha) = 0$ for

$\alpha < 0$ and $0 < \alpha$. Hence $f(s)$ is identically zero for $k = 0$ and $k = 1$ and is a polynomial of (at most) degree $(k - 2)$ for $k \geqq 2$. This theorem is a generalization of the basic theorem of Liouville that an everywhere analytic and bounded function is a constant. Since a bounded entire function $f(s)$ is contained in $\mathfrak{F}_2$, and

$$\int |f(\sigma + it)| p_2(t)\, dt \leqq K \quad ,$$

it follows by our general theorem that $f(s)$ is a polynomial of degree 0, i.e., a constant.

§38. Continuation. Harmonic Functions

1. We now consider harmonic functions

$$u(s) = u(\sigma, t)$$

which are regular in a strip $[\lambda, u]$. It will be profitable to fix the harmonic functions as complex valued, i.e., as functions

$$u(s) = u_1(\sigma, t) + iu_2(\sigma, t)$$

where from the outset, the real components $u_1(\sigma, t)$ and $u_2(\sigma, t)$ have no analytical connection with one another.

2. A function $u(s)$ in $[\lambda, \mu]$ shall belong to class $\mathfrak{F}_k$, if to each sub-strip (λ_1, μ_1) there is a constant $K = K(\lambda_1, \mu_1)$, such that

$$(1) \qquad \int |u(\sigma, t)| p_k(t)\, dt < K \quad .$$

Here again the proof is the same as in §37. There is a T such that for a given (λ_1, μ_1) and for each ϵ,

$$(2) \qquad |u(s)| \leq \epsilon |t|^k \quad , \qquad |t| > T \quad ,$$

and

$$(3) \qquad \int^{-T} + \int_T |u(\sigma)| p_k(t)\, dt \leqq \epsilon \quad .$$

From this it follows again, for a given rectangle

$$(4) \qquad \lambda_1 \leqq \sigma \leqq \mu_1 \; , \qquad |t| \leqq T \quad ,$$

that for each ϵ there exists a δ such that for every two points of the rectangle for which $|\sigma' - \sigma| \leqq \delta$, $|t' - t| \leqq \delta$

$$\int |u(\sigma', t' + \tau) - u(\sigma, t + \tau)| p_k(\tau)\, d\tau \leqq \epsilon \quad .$$

We say of this last property that $u(s)$ is <u>k-continuous</u> in the interior of $[\lambda, \mu]$.

3. Let the left end point $s = \lambda$ be a finite number, and let a boundary function $u(t)$ of the class $\mathfrak{F}_k$ be given, to which the function $u(\sigma, t)$ k-converges as $\sigma \longrightarrow \lambda$:

$$\int |u(t) - u(\sigma, t)| p_k(t)\, dt \longrightarrow 0 \quad \text{as} \quad \sigma \longrightarrow \lambda \quad .$$

We say that the function $u(\sigma, t)$ <u>joins itself k-continuously to the boundary value</u> $u(t)$ <u>as</u> $\sigma \longrightarrow \lambda$. In order that such a boundary function exist, it is necessary and sufficient that the functions $u(\sigma, t)$ be k-convergent as $\sigma \longrightarrow \lambda$, i.e., cf. §29, 1,

$$\lim_{\substack{\sigma_1 \to \lambda \\ \sigma_2 \to \lambda}} \int |u(\sigma_1, t) - u(\sigma_2, t)| p_k(t)\, dt = 0 \quad .$$

Whenever such a boundary function $u(t)$ exists, if we extend the function $u(\sigma, t)$ on the straight line $\sigma = \lambda$ by putting $u(\lambda, t) = u(t)$, then the <u>k-continuity</u> as defined above will also hold in a strip which includes the left boundary line

$$\lambda \leq \sigma \leq \mu_1 \ (< \mu) \ , \qquad - \infty < t < \infty \quad .$$

If μ is finite, similar considerations naturally apply to $\sigma \longrightarrow \mu$.

4. For an harmonic function $u(x, y)$ in $|x| \leq r$, $|y| \leq r$, we have

$$r^2 \frac{\partial u(0,0)}{\partial x} = \frac{1}{\pi} \int_0^{2\pi} \rho u(\rho \cos \varphi, \rho \sin \varphi) \cos \varphi \, d\varphi, \qquad 0 < \rho \leq r \quad .$$

Integrating with respect to ρ between 0 and r, we obtain

$$(5) \qquad \left| \frac{\partial u(0,0)}{\partial x} \right| \leqq \frac{3}{\pi r^3} \int_{-r}^{r} \int_{-r}^{r} |u(x, y)|\, dx\, dy \quad .$$

Now let $u(\sigma, t)$ be a function of $\mathfrak{F}_k$ in $[\lambda, \mu]$. To each sub-strip (λ_1, μ_1), it is possible to specify a larger sub-strip $(\lambda_1 - r, \mu_1 + r)$. Then for all (σ, t) in (λ_1, μ_1)

$$\left|\frac{\partial u(\sigma,t)}{\partial \sigma}\right| \leqq \frac{3}{\pi r^3}\int_{-r}^{r}\int_{-r}^{r} |u(\sigma + \xi, t + \eta)|\, d\xi\, d\eta \quad .$$

With the aid of the function $\chi(t)$ found in the proof of Theorem 46, the valuation

$$\int \left|\frac{\partial u(\sigma,t)}{\partial \sigma}\right| p_k(t)\, dt \leqq K_1 , \qquad \lambda_1 \leqq \sigma \leqq \mu_1$$

is found without difficulty. That is, the derivative of $u(\sigma, t)$ with respect to σ, is likewise a function of $\mathfrak{F}_k$ in $[\lambda, \mu]$. Since the letters x and y can be interchanged in (5), the same assertions for the derivative of $u(\sigma, t)$ with respect to t can also be deduced. But since a partial derivative of any higher order arises by a succession of simple differentiations, it follows that all partial derivatives of $u(\sigma, t)$ with respect to σ and t, likewise belong to $\mathfrak{F}_k$ in $[\lambda, \mu]$.

This applies in particular to an analytic function

$$f(s) = u(s) + iv(s)$$

which belongs to $\mathfrak{F}_k$ in $[\lambda, \mu]$. Its first derivative

$$\frac{\partial f}{\partial s} = - i\left(\frac{\partial u}{\partial t} + i\,\frac{\partial v}{\partial t}\right) \quad , \tag{6}$$

also belongs to $\mathfrak{F}_k$ there and therefore also each higher derivative. The transforms of the derivatives of $f(s)$ can be determined at once by (6) with the use of Theorem 44, 1. Hence

$$f^{(r)}(s) \sim \int e^{s\alpha}\alpha^r d^k E(\alpha) \quad .$$

5. On the other hand, the integral

$$F(s) = \int_{s_0}^{s} f(s)\, ds$$

(where s_0 is a point of $[\lambda, \mu]$) need not belong to $\mathfrak{F}_k$. However for $k \geqq 1$, it belongs to $\mathfrak{F}_{k+1}$ and for $k = 0$ to $\mathfrak{F}_{k+2}$. For, on putting $s_0 = \sigma_0$, that is s_0 real, we have

$$F(\sigma + it) = \int_{\sigma_0}^{\sigma} f(s)\, ds + i \int_0^t f(\sigma + it)\, dt = A(\sigma) + \varphi_\sigma(t) \quad .$$

By §29, 8, the function $\varphi_\sigma(t)$ belongs to $\mathfrak{R}_{k+1}$ or $\mathfrak{R}_{k+2}$, and for $x = k + 1$ or $= k + 2$, satisfies the valuation

$$\int |\varphi_\sigma(t)| p_x(t)\, dt \leqq K \text{ in } (\lambda_1, \mu_1) \quad ,$$

and the function $A(\sigma)$ is bounded in $\lambda_1 \leqq \sigma \leqq \mu_1$.

The following is additionally of interest. If $F(\sigma + it)$ belongs to $\mathfrak{R}_k$ on a single interior[1] straight line $\sigma = \sigma_0$, then it belongs to $\mathfrak{R}_k$ in the whole strip $[\lambda, \mu]$. For example, if one sets

$$F(\sigma + it) = F(\sigma_0 + it) + \psi(\sigma, t)$$

for σ in $\lambda_1 \leqq \sigma \leqq \mu_1$, then it is sufficient to show that

$$\int |\psi(\sigma, t)| p_k(t)\, dt \leqq K(\lambda_1, \mu_1) \quad .$$

But this follows from

$$|\psi(\sigma, t)| \leqq \int_{\sigma_0}^{\sigma} |f(\xi + it)|\, d\xi \quad .$$

By Theorem 44, 4), the following assertion results immediately. If inside an interval (a, b) which contains the zero point: $a < 0 < b$, the k-transform $E(\alpha)$ of $f(s)$ is equivalent to zero, then in $[\lambda, \mu]$, $f(s)$ is integrable arbitrarily often in $\mathfrak{R}_k$.

6. To each harmonic function $u(\sigma, t)$, there is a conjugate in each simply connected region, and therefore in particular in each strip $[\lambda, \mu]$. It is determined, to within a real additive constant, as a solution of the equations

$$\frac{\partial v}{\partial t} = \frac{\partial u}{\partial \sigma}, \qquad \frac{\partial v}{\partial \sigma} = -\frac{\partial u}{\partial t} \quad .$$

---

[1] The assertion is also valid for a straight boundary. If, by chance, $u(\sigma, t)$ of $\mathfrak{R}_k$ joins itself k-continuously to a boundary value $u(t)$ as $\sigma \longrightarrow \lambda$, and if $u(t)$ is integrable in $\mathfrak{R}_k$, then the integral of $f(s)$ in a suitable normalization of the additive constants is contained in $\mathfrak{R}_k$, and as $\sigma \longrightarrow \lambda$, likewise joins itself k-continuously to a boundary value.

If (σ_0, t_0) is a point of the strip, then

$$v(\sigma, t) = v(\sigma, t_0) + \int_{t_0}^{t} \frac{\partial u(\sigma, \tau)}{\partial \sigma} d\tau .$$

One recognizes from this, just as in the case of the integral of an analytic function, that $v(\sigma, t)$ belongs in any case to $\mathfrak{F}_{k+1}$ or $\mathfrak{F}_{k+2}$. And from

$$v(\sigma, t) = v(\sigma_0, t) + \int_{\sigma_0}^{\sigma} \frac{\partial u}{\partial t} (\xi, t)\, d\xi ,$$

one recognizes that $v(\sigma, t)$ belongs to $\mathfrak{F}_k$ in $[\lambda, \mu]$ provided it belongs to $\mathfrak{F}_k$ on a single straight line $\sigma = \sigma_0$.

If $u(\sigma, t)$ is real, and belongs to $\mathfrak{F}_k$ in $[\lambda, \mu]$, then $u(\sigma, t)$ is, in $[\lambda, \mu]$, the real part of an analytic function $f(s)$ which belongs at least to $\mathfrak{F}_{k+2}$. Writing, if need be, k for $k + 2$, we can fix $u(\sigma, t)$ as the real part of a function $f(s)$ of $\mathfrak{F}_k$. From

$$(7) \qquad f(s) \sim \int e^{s\alpha} d^k E(\alpha) = \int e(t\alpha) e^{\sigma\alpha} d^k E(\alpha) ,$$

follows

$$\overline{f(s)} \sim (-1)^k \int e(t\alpha) e^{-\sigma\alpha} d^k \overline{E(-\alpha)} .$$

Hence if we put

$$(7') \qquad u(\sigma, t) \sim \int e(t\alpha) d^k E(\sigma, \alpha) ,$$

then

$$(8) \qquad d^k E(\sigma, \alpha) = e^{\sigma\alpha} d^k E^+(\alpha) + e^{-\sigma\alpha} d^k E^-(\alpha) ,$$

where

$$(9) \qquad 2E^+(\alpha) = E(\alpha), \qquad 2E^-(\alpha) = (-1)^k \overline{E(-\alpha)} .$$

In particular, if $k = 0$

$$E(\sigma, \alpha) = e^{\sigma\alpha} E^+(\alpha) + e^{-\sigma\alpha} E^-(\alpha) .$$

If for $k \geq 2$, a function

$$g(s) \sim \int e^{\sigma\alpha} d^k \Phi(\alpha)$$

is introduced which differs from $f(s)$ by an imaginary constant, and therefore has the same real part, then analogous to (8)

$$(10) \qquad d^k E(\sigma, \alpha) = e^{\sigma\alpha} d^k \Phi^+(\alpha) + e^{-\sigma\alpha} d^k \Phi^-(\alpha) \quad .$$

And since $\Phi(\alpha) - E(\alpha)$ is the k-transform of a constant, we obtain

$$(11) \qquad \Phi^+(\alpha) \asymp E^+(\alpha) + a\overline{|\alpha|}^{k-1}, \qquad \Phi^-(\alpha) \asymp E^-(\alpha) + b\overline{|\alpha|}^{k-1} \quad ,$$

and herein the numbers a and b are certain constants. According to (8), the representation of the function $E(\sigma, \alpha)$ by two functions $E^+(\alpha)$ and $E^-(\alpha)$ independent of σ is therefore not entirely unique. But we assert that if there is a representation (10) besides the representation (8), a connection (11) necessarily exists. Setting

$$D^+(\alpha) = \Phi^+(\alpha) - E^+(\alpha), \qquad D^-(\alpha) = \Phi^-(\alpha) - E^-(\alpha) \quad ,$$

it then follows from (8) and (10) that

$$e^{2\sigma\alpha} d^k D^+(\alpha) = - d^k D^-(\alpha) \quad ,$$

and therefore

$$(12) \qquad \left(e^{2\sigma_1\alpha} - e^{2\sigma_2\alpha} \right) d^k D^+(\alpha) = 0$$

for any two numbers σ_1, σ_2 in $[\lambda, \mu]$. If $\sigma_1 \neq \sigma_2$, $e^{2\sigma_1\alpha} - e^{2\sigma_2\alpha}$ has a simple zero at the point $\alpha = 0$. Therefore (12) implies

$$i\alpha d^k D^+(\alpha) = 0 \quad .$$

Hence $D^+(\alpha)$ is a trivial function of $\mathfrak{T}_k$ and indeed it is the transform of a function whose derivative vanishes, therefore the transform of a constant. A similar result is valid for $D^-(\alpha)$. Therefore

$$D^+(\alpha) \asymp a\overline{|\alpha|}^{k-1}, \qquad D^-(\alpha) \asymp b\overline{|\alpha|}^{k-1} \quad ,$$

Q.E.D.

7. Taking (8) into consideration, the transforms of the partial derivatives of $u(\sigma, t)$ are obtained by formal partial differentiation of (7'). This follows from the fact that every partial derivative of $u(\sigma, t)$ can be expressed by the real or imaginary part of a certain derivative of

$f(s)$, and that the transform of every derivative of $f(s)$ results from formal differentiation of (7).

8. For the imaginary part of $f(s)$, that is for the conjugate $v(\sigma, t)$ of $u(\sigma, t)$, we have

$$d^k E(\sigma, \alpha) = - ie^{\sigma\alpha} d^k E^+(\alpha) + ie^{-\sigma\alpha} d^k E^-(\alpha) \quad .$$

In particular, if $\lambda < 0 < \mu$, then according to (8) and (9)

$$E(0, \alpha) \asymp E^+(\alpha) + E^-(\alpha) \asymp \frac{1}{2} [E(\alpha) + (-1)^k \overline{E(-\alpha)}]$$

$$E(0, \alpha) \asymp - iE^+(\alpha) + iE^-(\alpha) \asymp \frac{1}{2} [- iE(\alpha) + i(-1)^k \overline{E(-\alpha)}] \quad .$$

Two real variable functions $u(t)$ and $v(t)$ of $\mathfrak{F}_k$ ought to be denoted as <u>conjugate</u>, if their k-transforms $\Phi(\alpha)$ and $\Psi(\alpha)$, with the aid of a suitable function $E(\alpha)$, stand to one another in the relation

$$2\Phi(\alpha) = E(\alpha) + (-1)^k \overline{E(-\alpha)}, \qquad 2\Psi(\alpha) = - iE(\alpha) + i(-1)^k \overline{E(-\alpha)} \quad .$$

We shall not proceed further in the study of conjugate functions [97].

9. If the function $u(\sigma, t)$ joins itself k-continuously to a function $u(\lambda, t)$ as $\sigma \longrightarrow \lambda$, then the transform $E(\sigma, \alpha)$ goes over to the transform $E(\lambda, \alpha)$ continuously. By §30, 9, (8) is therefore also valid for $\sigma = \lambda$. An analogous remark is valid as $\sigma \longrightarrow \mu$.

If the analytic function $f(s)$ joins itself k-continuously to a function $f_\lambda(t)$ as $\sigma \longrightarrow \lambda$, then

$$f_\lambda(t) \sim \int e^{(\lambda+it)\alpha} d^k E(\alpha) \quad .$$

<u>§39. Boundary Value Problems for Harmonic Functions</u>

1. In a strip $[\lambda, \mu]$, let a real harmonic function $u(\sigma, t)$ be given, where λ and μ, until further notice, are finite numbers. We say that the function $u(\sigma, t)$ belongs to the function class $\mathfrak{F}$ in $[\lambda, \mu]$, if it belongs to $\mathfrak{F}_k$ for a certain k. If for sufficiently large k, $u(\sigma, t)$ joins itself k-continuously to the boundary value $u(\lambda, t)$ as $\sigma \longrightarrow \lambda$, then we say that $u(\sigma, t)$ belongs to $\mathfrak{F}$ in $(\lambda, \mu]$. An analogous remark is valid for $\sigma = \mu$. The assumption that $u(\sigma, t)$ belongs to $\mathfrak{F}$ in (λ, μ) is supposed to signify that for sufficiently large k, a k-continuous joining takes place at the boundary value not only as $\sigma \longrightarrow \lambda$ but also as $\sigma \longrightarrow \mu$. In this sense, for example, the statement that the function

$$\frac{\partial^2(\sigma,t)}{\partial\sigma\partial t} \tag{1}$$

likewise belongs to $\mathfrak{F}$ in (λ, μ) is to be understood as follows. According to §38, 7, not only is the function (1) contained in $\mathfrak{F}_k$ in $[\lambda, \mu]$, but also a suitable $k = k'$ exists which if necessary can be larger than the k needed until now, such that (1) by approximation of σ to λ and μ, is k-convergent to two limit functions. We shall denote them purely symbollically by

$$\frac{\partial^2 u(\lambda,t)}{\partial\sigma\partial t}, \qquad \frac{\partial^2 u(\mu,t)}{\partial\sigma\partial t} \quad .$$

For what follows, there will be no restriction of generalness if we set $\lambda = 0$, $\mu = 1$.

THEOREM 49. If the harmonic function $u(\sigma, t)$ of $\mathfrak{F}$ joins itself k-continuously to the value zero at both boundary lines, then it is identically zero.

PROOF. Since

$$d^kE(\sigma, \alpha) = e^{\sigma\alpha}d^kE^+(\alpha) + e^{-\sigma\alpha}d^kE^-(\alpha) \tag{2}$$

is also valid for the voundary value $\sigma = 0$, therefore

$$d^kE^+(\alpha) + d^kE^-(\alpha) = 0 \tag{3}$$

$$e^{+\alpha}d^kE^+(\alpha) + e^{-\alpha}d^kE^-(\alpha) = 0 \quad . \tag{4}$$

From this it follows that

$$(1 - e^{2\alpha})d^kE^+(\alpha) = 0 \quad . \tag{5}$$

Hence

$$\alpha d^kE^+(\alpha) = 0 \quad . \tag{6}$$

Because

$$2E^+(\alpha) = E(\alpha), \qquad 2E^-(\alpha) = (-1)^k\overline{E(-\alpha)} \quad , \tag{7}$$

we obtain

$$\alpha d^kE(\alpha) = 0 \quad . \tag{8}$$

But $E(\alpha)$ was the k-transform of the function

$$f(s) = u(\sigma, t) + iv(\sigma, t) \quad , \tag{9}$$

and (8) says that the derivative of $f(s)$ vanishes. Hence $f(s)$ is a constant. Therefore $u(\sigma, t)$ is a constant, and because of the k-continuous approximation of $u(\sigma, t)$ to the value zero, this constant vanishes, Q.E.D.

A generalization of Theorem 49 is the following.

THEOREM 50. Let

$$\Lambda_0 u = \sum_{\mu,\nu=0}^{m} a_{\mu\nu} \frac{\partial^{\mu+\nu} u}{\partial\sigma^\mu \partial t^\nu}, \qquad \Lambda_1 u = \sum_{\mu,\nu=0}^{m} b_{\mu\nu} \frac{\partial^{\mu+\nu} u}{\partial\sigma^\mu \partial t^\nu} \tag{10}$$

be two given functionals with real constant coefficients $a_{\mu\nu}$, $b_{\mu\nu}$. Let it be known of a function of $\mathfrak{F}$, harmonic in $[0, 1]$, that it displays the following behavior at the boundary.

1. The harmonic function

$$u_0(\sigma, t) = \Lambda_0 u(\sigma, t)$$

is contained in $\mathfrak{F}$ even in $(0, 1]$, and what is more

$$u_0(0, t) = 0 \tag{11}$$

2. The harmonic function

$$u_1(\sigma, t) = \Lambda_1 u(\sigma, t)$$

is contained in $\mathfrak{F}$ even in $[0, 1]$, and what is more

$$u_1(1, t) = 0 \quad . \tag{12}$$

Then $u(\sigma, t)$ is the real part of a trivial function, i.e., of a function

$$\sum_{\nu=1}^{n} \sum_{\mu=0}^{\ell_\nu - 1} c_{\mu\nu} s^\mu e^{\tau_\nu s}, \qquad \tau_\nu \text{ real} \quad . \tag{13}$$

REMARK. The proof will suggest that the given functionals

$\Lambda_0 u$, $\Lambda_1 u$ can be fixed in a considerably more general form than that of (10). However we shall not discuss this matter further.

PROOF. For $0 < \sigma < 1$, we write

$$u(\sigma, t) \sim \int e(t\alpha) d^k E(\sigma, \alpha)$$

where (2) and (7) are valid. The function $\Lambda_0 u$ has the "differential"

$$e^{\sigma\alpha} P^+(\alpha) d^k E^+(\alpha) + e^{-\sigma\alpha} P^-(\alpha) d^k E^-(\alpha) \quad ,$$

where

$$P^+(\alpha) = \sum_{\mu,\nu} a_{\mu\nu} i^\nu \alpha^{\mu+\nu}, \qquad P^-(\alpha) = \sum_{\mu,\nu} (-1)^\mu a_{\mu\nu} i^\nu \alpha^{\mu+\nu} \quad .$$

By (11) we have therefore

$$P^+(\alpha) d^k E^+(\alpha) + P^-(\alpha) d^k E^-(\alpha) = 0 \quad . \tag{14}$$

Correspondingly, it follows from (12)

$$e^{\alpha} Q^+(\alpha) d^k E^+(\alpha) + e^{-\alpha} Q^-(\alpha) d^k E^-(\alpha) = 0 \quad , \tag{15}$$

where the polynomial Q results from the polynomial P by replacing $a_{\mu\nu}$ by $b_{\mu\nu}$. Again using (7), it follows from (14) and (15) that

$$G(\alpha) d^k E(\alpha) = 0 \tag{16}$$

where

$$G(\alpha) = e^{\alpha} Q^+(\alpha) P^-(\alpha) - e^{-\alpha} Q(-\alpha) P^+(\alpha) \quad .$$

It is easy to see that $G(\alpha)$ has at most a finite number of zeros $\tau_1, \ldots, \tau_n$ with well determined multiplicities $\ell_1, \ldots, \ell_n$. It therefore follows from (16) that $u(\sigma, t)$ is the real part of (13), Q.E.D.

We have not proved and do not claim that each function (13) actually gives a solution. For which values of the constants $c_{\mu\nu}$ a solution actually results is best established directly, for given concrete functions (10), by verifying the boundary conditions (11) and (12) in the expression (13).

EXAMPLE. In a certain problem of hydronamics [98]

$$\Lambda_0 u \equiv u, \qquad \Lambda_1 u \equiv qu - \frac{\partial u}{\partial \sigma} \quad .$$

Then $P^+(\alpha) = 1$, $P^-(\alpha) = 1$, $Q^+(\alpha) = q - \alpha$, $Q^-(\alpha) = q + \alpha$. Hence

$$G(\alpha) = e^{\alpha}(q - \alpha) - e^{-\alpha}(q + \alpha) = q(e^{\alpha} - e^{-\alpha}) - \alpha(e^{\alpha} + e^{-\alpha}) \quad .$$

This function always has the zero $\tau = 0$ which is simple if $q \neq 1$ and threefold if $q = 1$. Its contribution to the function (13) has the form c_0 if $q \neq 1$, and because of the first boundary condition, this real part must vanish. Hence there results no contribution at all to the function u itself. For $q = 1$ on the contrary, the contribution to (13) has the value

$$c_0 + c_1 s + c_2 s^2 \quad .$$

If for complex constants c_0, c_1, c_2, the real part is taken and both boundary conditions are verified for it, there results the contribution

$$A\sigma + B\sigma t$$

to the solution u, where A and B are arbitrary real constants. — In order to discover further zeros of $G(\alpha)$, we write the equation $G(\alpha) = 0$ in the form

$$\frac{\tan h\alpha}{\alpha} = \frac{1}{q} \quad . \tag{17}$$

The function on the left has the value 1 for $\alpha = 0$ and converges to zero as $\alpha \longrightarrow \infty$. In $[0, \infty)$ it is monotonic. For where its derivative vanished, we would have

$$4\alpha = e^{2\alpha} - e^{-2\alpha} \quad .$$

But for $\alpha > 0$, this equality cannot be true since $4\alpha < e^{2\alpha} - e^{-2\alpha}$. Further the function on the left of (17) is even. We have therefore the result that for $q \leq 1$, $G(\alpha)$ has no additional zeros; for $q > 1$, it has two symmetrical zeros

$$\tau_2 = \tau > 0, \qquad \tau_3 = -\tau < 0$$

both of which are single. For the real part of

$$c_3 e^{\tau s} + c_4 e^{-\tau s} \quad ,$$

there results an expression

$$e^{\tau\sigma}(a_1 \cos \tau t + b_1 \sin \tau t) + e^{-\tau\sigma}(a_2 \cos \tau t + b_2 \sin \tau t) \quad .$$

Condition (11) reads

$$(a_1 + a_2) \cos \tau t + (b_1 + b_2) \sin \tau t = 0 \quad .$$

From this it follows that $a_2 = - a_1 = - a$, $b_2 = - b_1 = - b$. Hence

$$u = (e^{\tau\sigma} - e^{-\tau\sigma})(a \cos \tau t + b \sin \tau t) \quad .$$

This function satisfies condition (12) for any a and b, and thus all solutions of our problem have been found.

3. The result of Theorem 50 brings the question closer whether for given functions $\varphi_0(t)$, $\varphi_1(t)$ of $\mathfrak{F}$, there is an harmonic function $u(\sigma, t)$ of $\mathfrak{F}$ in $[0, 1]$ which satisfies the non-homogeneous boundary conditions

$$\Lambda_0 u(0, t) = \varphi_0(t)$$

$$\Lambda_1 u(1, t) = \varphi_1(t) \quad .$$

Since a discussion of this question is somewhat lengthy we shall disregard it.

4. Consider the real part $u(\sigma, t)$ of a function $f(s)$ which belongs to $\mathfrak{F}$ in $[- \infty, 0]$, and has a k-transform which for $\alpha < 0$ is equivalent to zero.[1] Let it now be required of a function $u(\sigma, t)$ that $\Lambda_0 u$ belong to $\mathfrak{F}$ in $[- \infty, 0)$, and that for $\sigma = 0$, the boundary condition $\Lambda_0 u = 0$ be satisfied. Here $\Lambda_0 u$ is a functional as in Theorem 50. Again we obtain the relation

$$P^+(\alpha)d^k E^+(\alpha) + P^-(\alpha)d^k E^-(\alpha) = 0 \quad .$$

And if (7) and the special assumption concerning $E(\alpha)$ are considered, there results

$$P^+(\alpha)d^k E(\alpha) = 0 \quad \text{for} \quad \alpha > 0$$

therefore again only the real part of an expression (13) is eligible, where now the τ_ν are the non-negative zeros of the polynomial $P^+(\alpha)$.

---

[1] Such an harmonic function $u(\sigma, t)$ can be characterized "directly" by requiring that it belong to $\mathfrak{F}$ in $[- \infty, 0]$, and that for sufficiently large k, the function

$$J(\sigma) = \int |u(\sigma, t)| p_k(t) \, dt$$

satisfy for each $\epsilon > 0$, the valuation

$$J(\sigma) = o(e^{-\epsilon\sigma}), \qquad \sigma \longrightarrow - \infty \quad .$$

CHAPTER VIII

QUADRATIC INTEGRABILITY

§40. The Parseval Equation

1. By the class $\mathfrak{F}_k$, $k = 0, 1, 2, \ldots,$ we understood the totality of those functions $f(x)$ for which the function

$$\frac{f(x)}{1 + |x|^k}$$

is absolutely integrable in $[-\infty, \infty]$. For $k = 0, 1, 2,$ we shall have occasion to consider the totality of those functions $f(x)$ for which

$$\frac{f(x)^2}{1 + |x|^{2k}}$$

is absolutely integrable in $[-\infty, \infty]$. We shall denote this function class by $\mathfrak{F}_k^2$, and for accurate distinction we shall also write $\mathfrak{F}_k^1$ instead of $\mathfrak{F}_k$. <u>In particular,</u> $\mathfrak{F}_0^2$ <u>is the totality of those functions for which</u>

$$\int |f(x)|^2 \, dx$$

<u>is finite</u>.

If $|f(x)|^2$ is integrable over a certain finite interval, then $|f(x)|$ is also integrable there, but not conversely. This fact is illustrated by the Schwarz inequality

$$(b - a)\left(\int_a^b |f(x)| \, dx\right)^2 \leqq \int_a^b |f(x)|^2 \, dx \quad .$$

But nevertheless, each function of $\mathfrak{F}_k^2$ is not therefore already contained in $\mathfrak{F}_k^1$ as the following counter example demonstrates. For $n = 1, 2, 3, \ldots$ let

$$f(x) = n^{-1} \quad \text{in} \quad n < x < n + 1 \quad ,$$

and $= 0$ for $x \leqq 1$. This function is contained in $\mathfrak{F}_0^2$ but not in $\mathfrak{F}_0^1$.

By $\mathfrak{F}_k^{12}$ we mean the totality of those functions which belong not only to $\mathfrak{F}_k^1$ but also to $\mathfrak{F}_k^2$.

For functions of $\mathfrak{F}_0^2$ we shall make use of the following facts, of which the basic ones will have to be stated without proof [99].

2. The sum of two functions of $\mathfrak{F}_0^2$ is again a function of $\mathfrak{F}_0^2$. The product of two functions of $\mathfrak{F}_0^2$ is absolutely integrable in $[-\infty, \infty]$, and what is more, the Schwarz inequality

$$\left|\int f(x)g(x)\, dx\right|^2 \leqq \int |f(x)|^2\, dx \cdot \int |g(x)|^2\, dx$$

holds. Hence

$$\int |f(x) + g(x)|^2\, dx \leqq \int (|f|^2 + |g|^2 + 2|fg|)\, dx \leqq 2\int (|f|^2 + |g|^2)\, dx \quad .$$

In particular, if

$$\int |f(x) - g(x)|^2\, dx \leqq \epsilon, \qquad \int |g(x) - h(x)|^2\, dx \leqq \epsilon \quad ,$$

then

$$\int |f(x) - h(x)|^2\, dx \leqq 4\epsilon \quad .$$

3. From

$$\text{(1)} \qquad \lim_{n \to \infty} \int |f_n(x) - f(x)|^2\, dx = 0 \quad ,$$

it always follows that

$$\text{(2)} \qquad \lim_{\substack{m \to \infty \\ n \to \infty}} \int |f_m(x) - f_n(x)|^2\, dx = 0 \quad .$$

Conversely, if a sequence of functions $f_n(x)$ of $\mathfrak{F}_0^2$ converges in the quadratic mean, i.e., satisfies (2), then there is a function $f(x)$ of $\mathfrak{F}_0^2$ to which it converges in the quadratic mean, i.e., for which (1) is satisfied. This function $f(x)$ (apart from a null set) is unique.

Let $f_n(x)$, $\varphi_n(x)$ be two given sequences of functions of $\mathfrak{F}_0^2$ which converge to the functions $f(x)$, $\varphi(x)$ in the quadratic mean. In order that the functions $f(x)$ and $\varphi(x)$ agree, it is necessary and sufficient that

$$\lim_{n \to \infty} \int |f_n(x) - \varphi_n(x)|^2 \, dx = 0 \quad .$$

If (1) holds, and if the sequence $f_n(x)$ converges almost everywhere, in the usual sense, to a function $\varphi(x)$, then $f(x) = \varphi(x)$.

From (1) we obtain

$$\lim_{n \to \infty} \int |f_n(x)|^2 \, dx = \int |f(x)|^2 \, dx \quad .$$

4. To each function $f(x)$ of $\mathfrak{F}_0^2$, there is a sequence of functions $f_n(x)$ of $\mathfrak{F}_0^{12}$ for which

$$\lim_{n \to \infty} \int |f_n(x) - f(x)|^2 \, dx = 0 \quad .$$

One sets, say

$$f_n(x) = \begin{cases} f(x) & \text{for} \quad |x| \leq n \\ 0 & \text{for} \quad |x| > n \quad . \end{cases}$$

The "approximation functions" $f_n(x)$ of $\mathfrak{F}_0^{12}$, as just constructed, are each zero outside of a finite interval. But still further "regularity properties" can be required of the approximation functions. We shall use the fact that there are functions $f_n(x)$, each of which has, in a finite number of finite intervals, a constant value in each and otherwise vanishes ("step function"). Such a function is eo ipso bounded.

5. For each function of $\mathfrak{F}_0^2$, we have

$$\lim_{\xi \to 0} \int |f(x) - f(x + \xi)|^2 \, dx = 0 \quad . \tag{3}$$

If $f(x)$ belongs to $\mathfrak{F}_0^{12}$ and is bounded: $f(x) \leq G$, then (3) follows from $|f(x) - f(x + \xi)|^2 = |f(x) - f(x + \xi)| \cdot |f(x) - f(x + \xi)| \leq 2G|f(x) - f(x + \xi)|$ in conjunction with

$$\lim_{\xi \to 0} \int |f(x) - f(x + \xi)| \, dx = 0 \quad .$$

cf. the beginning of the proof of Theorem 39.

For an arbitrary function of $\mathfrak{F}_0^2$, (3) now follows by the approximation property established in 4, through the valuation

$$\frac{1}{3}\,|f(x) - f(x+\xi)|^2 \leqq |f_n(x) - f(x)|^2$$

$$+\,|f_n(x+\xi) - f(x+\xi)|^2 + |f_n(x) - f_n(x+\xi)|^2 \quad .$$

6. The object of this paragraph is the following theorem.

THEOREM 51 [100]. For each function

$$f(x) \sim \int e(x\alpha)E(\alpha)\, d\alpha$$

of $\mathfrak{F}_0^{12}$ (The Parseval equation)

$$\int |E(\alpha)|^2\, d\alpha = \frac{1}{2\pi}\int |f(x)|^2\, dx \tag{4}$$

is valid.

PROOF. For each function $f(x)$ of $\mathfrak{F}_0^2$, the function $f(x)\overline{f(x-y)}$ with y held fixed, is by 2. absolutely integrable in $-\infty < x < \infty$. We now consider the function

$$g(x) = \frac{1}{2\pi}\int f(\xi)\overline{f(\xi - x)}\, d\xi$$

for all (and not only for almost all) values x in $[-\infty, \infty]$. It is bounded

$$|g(x)|^2 \leqq \frac{1}{4\pi^2}\int |f(\xi)|^2\, d\xi \cdot \int |f(\xi)|^2\, d\xi \quad ,$$

and continuous. The latter follows from

$$4\pi^2|g(x) - g(x+y)|^2 \leqq \left(\int |f(\xi)|\; |f(\xi - x) - f(\xi - x - y)|\, d\xi\right)^2$$

$$\leqq \int |f(\xi)|^2\, d\xi \cdot \int |f(\xi) - f(\xi - y)|^2\, d\xi \quad ,$$

in conjunction with (3).

If $f(x)$ is contained in $\mathfrak{F}_0^{12}$, then by the Faltung rule, $g(x)$ is a function of $\mathfrak{F}_0^1$ and its transform amounts to $|E(\alpha)|^2$. But we have already proved in §20, 3 (at the very end) that if the transform of a bounded function of $\mathfrak{F}_0^1$ is not negative, then it is absolutely integrable. Therefore, by Theorem 15, 1), since $g(x)$ is continuous for all x,

$$g(x) = \int |E(\alpha)|^2 e(x\alpha)\, d\alpha \quad .$$

From this (4) results for $x = 0$. Q.E.D.

7. If two functions $f_1(x)$ and $f_2(x)$ of $\mathfrak{F}_0^{12}$ are considered, one obtains, by applying (4) to $\lambda f_1 + f_2$ and $\lambda f_1 - f_2$ and subtracting

$$\lambda \int E_1(\alpha)\overline{E_2(\alpha)}\, d\alpha + \overline{\lambda} \int \overline{E_1(\alpha)}E_2(\alpha)\, d\alpha$$

$$= \frac{\lambda}{2\pi} \int f_1(x)\overline{f_2(x)}\, dx + \frac{\overline{\lambda}}{2\pi} \int \overline{f_1(x)}f_2(x)\, dx \quad .$$

By means of the specifications $\lambda = 1$ and $\lambda = i$, one finally obtains

$$(5) \qquad \int E_1(\alpha)\overline{E_2(\alpha)}\, dx = \frac{1}{2\pi} \int f_1(x)\overline{f_2(x)}\, dx \quad .$$

If $\overline{f_2(x)}$ is replaced by $f_2(y - x)$, there results

$$(6) \qquad \int E_1(\alpha)E_2(\alpha)e(y\alpha)\, d\alpha = \frac{1}{2\pi} \int f_1(x)f_2(y - x)\, dx \quad ;$$

in particular

$$(7) \qquad \int E_1(\alpha)E_2(\alpha)\, d\alpha = \frac{1}{2\pi} \int f_1(x)f_2(- x)\, dx \quad .$$

8. Let $f(x)$ be a given function of $\mathfrak{F}_1^{12}$. For its 1-transform $E(\alpha) = E(\alpha, 1)$, we have

$$E(\alpha + \epsilon) - E(\alpha - \epsilon) = \frac{1}{2\pi} \int f(x)\, \frac{2 \sin \epsilon x}{x}\, e(- \alpha x)\, dx \quad ,$$

and therefore by Theorem 51

$$\int |E(\alpha + \epsilon) - E(\alpha - \epsilon)|^2\, d\alpha = \frac{2\epsilon}{\pi} \int |f(x)|^2\, \frac{\sin^2 \epsilon x}{\epsilon x^2}\, dx \quad .$$

If therefore a "mean value" exists for the function $|f(x)|^2$ in the sense of §9, 2, then

$$(8) \qquad \lim_{\epsilon \to 0} \frac{1}{2\epsilon} \int |E(\alpha + \epsilon) - E(\alpha - \epsilon)|^2\, d\alpha = \mathfrak{M}\left\{|f(x)|^2\right\} \quad .$$

If $f(x)$ belongs to $\mathfrak{F}_2^{12}$, then for the 2-transform $E(\alpha) = E(\alpha, 2)$ of $f(x)$, one obtains in a similar manner

$$\int |E(\alpha + \epsilon) - 2E(\alpha) + E(\alpha - \epsilon)|^2\, d\alpha = \frac{1}{2\pi} \int |f(x)|^2 \left(\frac{\sin \frac{\epsilon x}{2}}{\frac{x}{2}} \right)^4 dx \quad .$$

With the aid of §4, (18), we obtain from this, insofar as the "mean value" of $|f(x)|^2$ exists in the sense of §9, 2,

$$\lim_{\epsilon \to 0} \frac{1}{\epsilon^3} \int |E(\alpha + \epsilon) - 2E(\alpha) + E(\alpha - \epsilon)|^2 \, d\alpha = \frac{2}{3} \mathfrak{M}\left\{|f(x)|^2\right\} .$$

§41. The Theorem of Plancherel

1. THEOREM 52. Each function $f(x)$ of $\mathfrak{F}_0^2$ is attached to a function $F(\alpha)$ of $\mathfrak{F}_0^2$ in a reversibly unique manner with the following properties.

If $f(x)$ belongs to $\mathfrak{F}_0^{12}$, then $F(\alpha)$ is the 0-transform of $f(x)$. If $f(x)$ is an arbitrary function of $\mathfrak{F}_0^2$, and if the functions

$$f_n(x) \sim \int F_n(\alpha) e(x\alpha) \, d\alpha \tag{1}$$

of $\mathfrak{F}_0^{12}$ satisfy the relation

$$\int |f_n(x) - f(x)|^2 \, dx \longrightarrow 0 , \qquad (n \longrightarrow \infty) , \tag{2}$$

then

$$\int |F_n(\alpha) - F(\alpha)|^2 \, d\alpha \longrightarrow 0 , \qquad (n \longrightarrow \infty) . \tag{3}$$

Hence

$$\int |F(\alpha)|^2 \, d\alpha = \frac{1}{2\pi} \int |f(x)|^2 \, dx . \tag{4}$$

PROOF. Let $f(x)$ be an arbitrary function of $\mathfrak{F}_0^2$. Consider any functions (1) whatever of $\mathfrak{F}_0^{12}$ for which (2) holds. By Theorem 51

$$2\pi \int |F_m(\alpha) - F_n(\alpha)|^2 \, d\alpha = \int |f_m(x) - f_n(x)|^2 \, dx .$$

Hence

$$\int |F_m(\alpha) - F_n(\alpha)|^2 \, d\alpha \longrightarrow 0, \qquad (m, n \longrightarrow \infty) . \tag{5}$$

By (5) there is a function $F(\alpha)$ of $\mathfrak{F}_0^2$ for which

$$\int |F_n(\alpha) - F(\alpha)|^2 \, d\alpha \longrightarrow 0, \qquad (n \longrightarrow \infty) . \tag{6}$$

From

$$2\pi \int |F_n(\alpha)|^2 \, d\alpha = \int |f_n(x)|^2 \, dx$$

and

$$\int |F_n(\alpha)|^2 \, d\alpha \longrightarrow \int |F(\alpha)|^2 \, d\alpha, \qquad \int |f_n(x)|^2 \, dx \longrightarrow \int |f(x)|^2 \, dx$$

(4) now follows.

We still have to show that the function $F(\alpha)$ is independent of the special approximation sequence (2), and that the functions $F(\alpha)$ which belong to two different functions $f(x)$, differ from one another.

Consider an approximation sequence

$$g_n(x) \sim \int G_n(\alpha) e(x\alpha) \, d\alpha$$

along with (1). From

$$(7) \qquad 2\pi \int |F_n(\alpha) - G_n(\alpha)|^2 \, d\alpha = \int |f_n(x) - g_n(x)|^2 \, dx \longrightarrow 0, \quad (m, n, \to \infty)$$

it follows that the "limit function" of the sequence $G_n(\alpha)$ actually agrees with $F(\alpha)$. Thus the independence of the function $F(\alpha)$ of the special approximation sequence has been shown.

In this way, each function of $\mathfrak{F}_0^2$ is attached to a well determined function $F(\alpha)$ which we call the Plancherel transform of $f(x)$.

In particular, if one sets

$$f_a(x) = \begin{cases} f(x) & \text{for } |x| < a \\ 0 & \text{for } |x| > a \end{cases},$$

then the functions group $f_a(x)$ is convergent to $f(x)$ in the quadratic mean as $a \longrightarrow \infty$. Hence the functions

$$F_a(\alpha) = \frac{1}{2\pi} \int_{-a}^{a} f(x) e(-x\alpha) \, dx$$

are convergent to $F(\alpha)$ in the quadratic mean as $a \longrightarrow \infty$. We signify this fact by the relation

$$(8) \qquad F(\alpha) = \ell.\text{m.} \frac{1}{2\pi} \int f(x) e(-x\alpha) \, dx \quad .$$

In particular, if the expression

$$\frac{1}{2\pi} \int_{-a}^{a} f(x) e(-x\alpha) \, dx$$

converges as $a \longrightarrow \infty$, at almost all points of an α-interval, then the limit function is identical with $F(\alpha)$. The attaching of the Plancherel transform to the functions of $\mathfrak{F}_0^2$ is an additive one. In particular

$$2\pi \int |F_n(\alpha) - F(\alpha)|^2 \, d\alpha = \int |f_n(x) - f(x)|^2 \, dx$$

is valid for any two functions $f(x)$ and $f_n(x)$ of $\mathfrak{F}_0^2$.

2. It follows from this that if a sequence of functions $f_n(x)$ of $\mathfrak{F}_0^2$ is convergent to $f(x)$ in the quadratic mean, then the sequence of Plancherel transforms belonging to them

$$(9) \qquad F_n(\alpha) = \text{l.m.} \frac{1}{2\pi} \int f_n(x) e(-x\alpha) \, dx$$

is also convergent to the function (8).

3. If $f(x)$ and $g(x)$ have the same transform $F(\alpha)$, then by (4)

$$\int |f(x) - g(x)|^2 \, dx = 2\pi \int |F(\alpha) - F(\alpha)|^2 \, d\alpha = 0 \quad ,$$

and therefore $f(x) = g(x)$. Hence different transforms belong to different functions. Thus Theorem 52 has been proved completely.

4. THEOREM 53. Each function $F(\alpha)$ of $\mathfrak{F}_0^2$ is a Plancherel transform, and the function of which it is the transform is in turn

$$(10) \qquad f(x) = \text{l.m.} \int F(\alpha) e(\alpha \bar{x}) \, d\alpha \; .$$

The relations (8) and (10) therefore are each an inverse of the other [101].

REMARK. By multiplying $F(\alpha)$ by $\sqrt{2\pi}$, the pair of formulas (8) and (10) become

$$(11) \qquad \begin{aligned} g(\alpha) &= \text{l.m.} \frac{1}{\sqrt{2\pi}} \int f(x) e(\alpha x) \, dx \\ f(x) &= \text{l.m.} \frac{1}{\sqrt{2\pi}} \int g(x) e(-x\alpha) \, d\alpha \; . \end{aligned}$$

Also for $x > 0$, $\alpha > 0$, we obtain the pair of formulas

$$(12) \quad c(\alpha) = \text{l.m.} \sqrt{\frac{2}{\pi}} \int_0 f(x) \cos \alpha x \, dx, \quad f(x) = \text{l.m.} \sqrt{\frac{2}{\pi}} \int_0 c(\alpha) \cos x\alpha \, d\alpha \; ,$$

(13) $s(\alpha) = \ell.\text{m.} \sqrt{\frac{2}{\pi}} \int_0 f(x) \sin \alpha x \, dx, \quad f(x) = \ell.\text{m.} \sqrt{\frac{2}{\pi}} \int_0 s(\alpha) \sin x\alpha \, d\alpha$

PROOF. We must show that (8) is a consequence of (10). Since by an interchange of the functions $2\pi F(-\alpha)$, $f(x)$, the pair of formulas (8), (10) become the pair (10), (8), it is sufficient to prove that (10) is a consequence of (8). If $f(x)$ is a step function $f_n(x)$ in accordance with §40, 4, then actually by Theorem 12 (apart from finitely many values x)

$$f_n(x) = \int F_n(\alpha) e(\alpha x) \, d\alpha = \ell.\text{m.} \int F_n(\alpha) e(\alpha x) \, d\alpha \quad .$$

From this (10) follows by allowing $n \longrightarrow \infty$, and by considering the statement in 2. with the functions $2\pi F(-\alpha)$, $f(x)$ interchanged, Q.E.D.

5. As in §40, 7, the following relations are obtained

$$\int F_1(\alpha) \overline{F_2(\alpha)} \, d\alpha = \frac{1}{2\pi} \int f_1(x) \overline{f_2(x)} \, dx$$

$$\int F_1(\alpha) F_2(\alpha) e(y\alpha) \, d\alpha = \frac{1}{2\pi} \int f_1(x) f_2(y - x) \, dx \quad .$$

As an application, we determine the solutions which are contained in $\mathfrak{F}_0^2$, of the equation [102]

$$f(x) = \frac{1}{\pi} \int_0 \frac{\sin t}{t} [f(x + t) + f(x - t)] \, dt \quad .$$

Since $\frac{\sin t}{t}$ also belongs to $\mathfrak{F}_0^2$, we obtain from the Plancherel transform $F(\alpha)$ of $f(x)$, the necessary and sufficient condition

$$F(\alpha) = \frac{F(\alpha)}{\pi} \int_0 \frac{\sin t}{t} [e(\alpha t) + e(-\alpha t)] \, d\alpha = F(\alpha) \frac{2}{\pi} \int_0 \frac{\sin t \cos \alpha t}{t} \, dt \quad .$$

Therefore

$$F(\alpha)\delta(\alpha) = 0$$

where $\delta(\alpha) = 0$ for $|\alpha| < 1$ and $\delta(\alpha) = 1$ for $|\alpha| > 1$. Hence the general solution reads

$$f(x) = \int_{-1}^{1} F(\alpha) e(\alpha x) \, d\alpha \quad ,$$

where $F(\alpha)$ is any quadratic integrable function. The real solutions are

$$f(x) = \int_0^1 \lambda(\alpha) \cos \alpha x \, d\alpha + \int_0^1 \mu(\alpha) \sin \alpha x \, d\alpha \quad ,$$

where $\lambda(\alpha)$ and $\mu(\alpha)$ are any real quadratic integrable functions.

6. Each function of $\mathfrak{F}_0^2$ belongs to $\mathfrak{F}_1^1$. For the function

$$f(x) \cdot \frac{1}{1 + |x|}$$

is the product of two functions of $\mathfrak{F}_0^2$ and is therefore absolutely integrable. If $f(x)$ belongs to $\mathfrak{F}_0^2$, then the functions

$$f_n(x) = \begin{cases} f(x) & \text{for} \quad |x| < n \\ 0 & \text{for} \quad |x| > n \end{cases}$$

are on the one hand convergent to $f(x)$ in the quadratic mean, and on the other hand 1-convergent to $f(x)$. From the latter, it follows for the 1-transforms of the functions $f_n(x)$, $f(x)$, that

$$E_n(\alpha, 1) \longrightarrow E(\alpha, 1) \tag{14}$$

and because of (6) it follows from the first that

$$\int_0^\alpha F_n(\beta)\, d\beta \longrightarrow \int_0^\alpha F(\beta)\, d\beta \quad . \tag{15}$$

Since $f_n(x)$ belongs to $\mathfrak{F}_0^{12}$

$$E_n(\alpha, 1) = E_n(0, 1) + \int_0^\alpha F_n(\beta)\, d\beta \quad ,$$

and from (14) and (15) follows

$$E(\alpha, 1) = E(0, 1) + \int_0^\alpha F(\beta)\, d\beta \quad .$$

Since if $f(x)$ belongs to $\mathfrak{F}_0^{12}$, the integrals

$$\left(\int^{-1} + \int_1 \right) f(x) \frac{-1}{-ix}\, dx$$

converge, therefore

$$2\pi E(\alpha, 1) \asymp \int f(x) \frac{e(-\alpha x) - L_1(\alpha, x)}{-ix}\, dx \asymp \int f(x) \frac{e(-\alpha x) - 1}{-ix}\, dx \quad .$$

From this, the following theorem results.

THEOREM 54. For the Plancherel transform of a function $f(x)$ of $\mathfrak{F}_0^2$, we have the relation [103]

(15')
$$F(\alpha) = \frac{d}{d\alpha}\frac{1}{2\pi}\int f(x)\,\frac{1 - e(-\alpha x)}{ix}\,dx \quad .$$

If $f(x)$ is even or odd, then one can also write for (15')

$$F(\alpha) = \frac{d}{d\alpha}\frac{1}{\pi}\int_0 f(x)\,\frac{\sin\alpha x}{x}\,dx$$

or

$$F(\alpha) = \frac{d}{d\alpha}\frac{1}{\pi i}\int_0 f(x)\,\frac{1 - \cos\alpha x}{x}\,dx \quad .$$

7. A partial generalization of the theorems of this paragraph reads as follows [104].

Consider a function $f(x)$ for which

$$\int |f(x)|^p\,dx$$

is finite. If $1 < p \leq 2$, then the derivative

(16)
$$F(\alpha) = \frac{1}{2\pi}\frac{d}{d\alpha}\int f(x)\,\frac{1 - e(-\alpha x)}{ix}\,dx$$

exists and has a finite integral

(17)
$$\int |F(\alpha)|^{\frac{p}{p-1}}\,d\alpha \quad .$$

Hence

$$f(x) = \frac{d}{dx}\int F(\alpha)\,\frac{e(\alpha x) - 1}{ix}\,d\alpha \quad .$$

For two functions $f(x)$ with equal p, the following is also valid [105],

$$\int f_1(t)\overline{F_2(t)}\,dt = \int f_2(t)\overline{F_1(t)}\,dt \quad .$$

But it is to be observed that for $1 < p < 2$, not every function $F(\alpha)$ for which (17) is finite occurs as a transform, i.e., stands in the relation (16) to a suitable function $f(x)$.

§42. Hankel Transform

In the present paragraph, all functions are real and defined on the half line $[0, \infty]$. We count them as belonging to $\mathfrak{F}_0^2$ if their square is integrable in $[0, \infty]$.

1. Let $g(t)$ be a function of $\mathfrak{F}_0^2$ and let $g(t) \geq 0$. The

function

$$\varphi_g(y) = \frac{1}{y}\int_0^y g(t)\,dt \tag{1}$$

is also contained in $\mathfrak{F}_0^2$ and satisfies

$$\int_0 \varphi_g(y)^2\,dy \leqq 4\int_0 g(t)^2\,dt \quad . \tag{2}$$

To prove this, let us consider for fixed $a > 0$, the functions

$$F_a(y) = \int_a^y g(t)\,dt, \qquad \Phi_a(y) = \frac{1}{y}F_a(y) \quad .$$

If $A > a$,

$$\int_a^A \Phi_a(y)^2\,dy = -\int_a^A F_a(y)^2 d\,\frac{1}{y} = -\frac{1}{A}F_a(A)^2 + 2\int_a^A \Phi_a(y)g(y)\,dy$$

$$\leqq 2\int_a^A \Phi_a(y)g(y)\,dy \leqq 2\left[\int_a^A \Phi_a(y)^2\,dy\right]^{1/2}\left[\int_a^A g(y)^2\,dy\right]^{1/2} .$$

From this it follows, by squaring the outermost terms of the inequality, and then removing a common factor, that

$$\int_a^A \Phi_a(y)^2\,dy \leqq 4\int_a^A g(y)^2\,dy \leqq 4\int_0 g(t)^2\,dt \quad .$$

Let b and B be two fixed numbers with $B > b > 0$. Then for $0 < a < b$

$$\int_b^B \Phi_a(y)^2\,dy \leqq 4\int_0 g(t)^2\,dt \quad . \tag{3}$$

But in each fixed interval (b, B), the function $\Phi_a(y)$ is convergent to $\varphi_g(y)$ as $a \longrightarrow 0$. Hence by (3)

$$\int_b^B \varphi_g(y)^2\,dy \leqq 4\int_0 g(t)^2\,dt \quad .$$

By letting $b \longrightarrow 0$, $B \longrightarrow \infty$, (2) results, Q.E.D.

2. Now consider the function

$$\psi_g(a) = \int_0^{1/a} g(t)\, dt \quad .$$

Hence

$$\frac{1}{y}\, \psi_g\left(\frac{1}{y}\right) = \frac{1}{y}\int_0^y g(t)\, dt = \varphi_g(y) \quad .$$

From this, with the aid of the variable transformation $y = \frac{1}{\alpha}$, one finds

$$\int_0 \frac{1}{y^2}\, \psi_g\left(\frac{1}{y}\right)^2 dy = \int_0 \psi_g(\alpha)^2\, d\alpha \leqq 4\int_0 g(t)^2\, dt \quad .$$

Finally by means of the variable transformation $t = \frac{1}{\tau}$, one obtains for the function

$$\chi_g(\alpha) = \frac{1}{\alpha}\int_{1/\alpha} \frac{g(t)}{t}\, dt \quad ,$$

the relation

$$\chi_g(\alpha) = \frac{1}{\alpha}\int_0^{\alpha} \frac{1}{\tau}\, g\left(\frac{1}{\tau}\right) d\tau \quad .$$

Hence again by (2)

$$\int_0 \chi_g(\alpha)^2\, d\alpha \leqq 4\int_0 \frac{1}{\tau^2}\, g\left(\frac{1}{\tau}\right)^2 d\tau = 4\int_0 g(t)^2\, dt \quad .$$

3. THEOREM 55 [106]. For $\Re(\nu) \geqq -\frac{1}{2}$, form with the Bessel function $J_\nu(t)$, the function

$$S_\nu(t) = t^{1/2} J_\nu(t) \quad ,$$

and hold ν fixed.

Then the reciprocity formulas

$$F(\alpha) = \int_0 S_\nu(\alpha x) f(x)\, dx, \qquad f(x) = \int_0 S_\nu(x\alpha) F(\alpha)\, d\alpha \tag{4}$$

hold, and are to be understood in the following sense.

1) For an arbitrary function $f(x)$ of $\mathfrak{F}_0^2$, the function

$$F_n(\alpha) = \int_0^n S_\nu(\alpha x) f(x)\, dx \tag{5}$$

is again contained in $\mathfrak{F}_0^2$ provided $n > 0$, and the sequence $F_n(\alpha)$ is convergent in the quadratic mean as $n \longrightarrow \infty$. The limit function

$$(6) \qquad F(\alpha) = \ell.m. \int_0 S_\nu(\alpha x) f(x)\, dx$$

is naturally again a function of $\mathfrak{F}_0^2$.

2). The relation

$$(7) \qquad \int_0 F(\alpha)^2\, d\alpha = \int_0 f(x)^2\, dx$$

holds.

3). The reciprocal relation

$$(8) \qquad f(x) = \ell.m. \int_0 S_\nu(x\alpha) F(\alpha)\, d\alpha$$

holds.

REMARK. If $\nu = -\frac{1}{2}$, then $S_\nu(t) = \sqrt{\frac{2}{\pi}} \cos t$, and (4) agrees with (12) of §41.

PROOF of 1). Together with the proof, we shall show the existence of a constant c, independent of $f(x)$, such that

$$(9) \qquad \int_0 F(\alpha)^2\, d\alpha \leqq c \int_0 f(x)^2\, dx \quad .$$

Because $\Re(\nu) \geqq -\frac{1}{2}$, the function $S_\nu(t)$ is bounded in finite intervals. And because of the asymptotic behavior of $J_\nu(t)$, we can set

$$S_\nu(t) = \sqrt{\frac{2}{\pi}} \cos\left(t - \frac{\pi}{4} - \frac{\nu\pi}{2}\right) + R(t) \quad ,$$

where

$$|R(t)| \leqq A \quad \text{for} \quad t \leqq 1$$

and

$$|R(t)| \leqq \frac{B}{t} \quad \text{for} \quad t > 1 \quad .$$

We now introduce the three functions

$$(10) \qquad \Phi_1(\alpha) = \ell.m. \sqrt{\frac{2}{\pi}} \int_0 \cos\left(\alpha x - \frac{\pi}{4} - \frac{\nu\pi}{2}\right) f(x)\, dx \quad ,$$

$$\Phi_2(\alpha) = \int_0^{1/\alpha} R(\alpha x) f(x)\, dx, \quad \Phi_3(\alpha) = \int_{1/\alpha} R(\alpha x) f(x)\, dx \quad .$$

Because of

$$\cos\left(\alpha x - \frac{\pi}{4} - \frac{\nu\pi}{2}\right) = \cos \alpha x \cos\left(\frac{\pi}{4} + \frac{\nu\pi}{2}\right) + \sin \alpha x \sin\left(\frac{\pi}{4} + \frac{\nu\pi}{2}\right) ,$$

and by the results of the previous paragraphs, the right integral in (10) actually exists, and therefore

$$(11) \qquad \int_0 \Phi_1(\alpha)^2 \, d\alpha \leqq c_1 \int_0 f(x)^2 \, dx \quad .$$

Further

$$|\Phi_2(\alpha)| \leqq A \int_0^{1/\alpha} |f(x)| \, dx, \qquad |\Phi_3(\alpha)| \leqq \frac{B}{\alpha} \int_{1/\alpha} \frac{1}{x} |f(x)| \, dx \quad ,$$

and by the auxilliary considerations in 2., the functions $\Phi_2(\alpha)$ and $\Phi_3(\alpha)$ are functions of $\mathfrak{F}_0^2$. They satisfy certain estimates

$$(12) \qquad \int_0 \Phi_\rho(\alpha)^2 \, d\alpha \leqq C_\rho \int_0 f(x)^2 \, dx , \qquad (\rho = 1, 2) \quad .$$

But now

$$\Phi_1(\alpha) + \Phi_2(\alpha) + \Phi_3(\alpha) = \ell.\text{m.} \int_0 S_\nu(\alpha x) f(x) \, dx \quad ,$$

and we have thus proved that the integral (6) exists. By (11) and (12), a valuation (9) holds for the function

$$F(\alpha) = \Phi_1(\alpha) + \Phi_2(\alpha) + \Phi_3(\alpha) \quad .$$

The function $F(\alpha)$ is called the (ν-th) Hankel transform of the function $f(x)$.

PROOF of 3). Let $f(x)$, $f_n(x)$, $n = 1, 2, 3, \ldots$ be given functions of $\mathfrak{F}_0^2$. Consider their Hankel transforms

$$F_n(\alpha) = \ell.\text{m.} \int_0 S_\nu(\alpha x) f_n(x) \, dx , \qquad F(\alpha) = \ell.\text{m.} \int_0 S_\nu(\alpha x) f(x) \, dx \quad ,$$

and with them, form the functions

$$g_n(x) = \ell.\text{m.} \int_0 S_\nu(x\alpha) F_n(\alpha) \, d\alpha, \qquad g(x) = \ell.\text{m.} \int_0 S_\nu(x\alpha) F(\alpha) \, d\alpha \quad .$$

By (9), it follows that

$$\int_0 |F(\alpha) - F_n(\alpha)|^2 \, d\alpha \leqq c \int_0 |f(x) - f_n(x)|^2 \, dx$$

$$\int_0 |g(x) - g_n(x)|^2 \, dx \leqq c \int_0 |F(\alpha) - F_n(\alpha)|^2 \, d\alpha \quad .$$

Therefore if $g_n(x) = f_n(x)$, then

$$\int_0 |g(x) - f_n(x)|^2 \, dx \leqq c^2 \int_0 |f(x) - f_n(x)|^2 \, dx \quad ,$$

from which the following immediately results. If for a sequence of functions the inverse formula (8) holds, and if these functions converge in the quadratic mean, then the inverse formula also holds for the limit function.

On the other hand, we find directly that the inverse formula is valid for the linear combination $c_1 f_1 + c_2 f_2$, and in particular for the difference $f_1 - f_2$, provided it holds for f_1 and f_2.

The inverse formula is valid for the function

$$f_A(x) = \begin{cases} x^{\nu+1/2} & \text{for } 0 < x < A \\ 0 & \text{for } A < x < \infty \quad . \end{cases}$$

In fact

$$\alpha^{\nu+1/2} F(\alpha) = \int_0^A (\alpha x)^{\nu+1} J_\nu(\alpha x) \, dx = \frac{1}{\alpha} \int_0^{\alpha A} \xi^{\nu+1} J_\nu(\xi) \, d\xi = \frac{1}{\alpha} \xi^{\nu+1} J_{\nu+1}(\xi) \Big|_0^{\alpha A}$$

$$= \alpha^\nu A^{\nu+1} J_{\nu+1}(\alpha A) \quad .$$

Hence

$$\int_0 S_\nu(\alpha x) F(\alpha) \, d\alpha = \int_0 (\alpha x)^{1/2} J_\nu(\alpha x) A^{-1/2} A^{\nu+1} J_{\nu+1}(\alpha A) \, d\alpha$$

$$= x^{1/2} A^{\nu+1} \int_0 J_\nu(\alpha x) J_{\nu+1}(\alpha A) \, d\alpha \quad ,$$

which, by a formula about Bessel functions [107], is actually $f_A(x)$.

The inverse formula is therefore also valid for the function

$f_b(x) - f_a(x)$ which has the value $x^{\nu+1/2}$ in the interval (a, b) and vanishes otherwise. Consequently, it is also valid for the function which has the value $c_p x^{\nu+1/2}$ in n consecutive intervals $x_p < x < x_{p+1}$, $p = 0, 1, \ldots, n - 1$, — the c_p are constants —, and vanishes otherwise. Each function $\varphi(x)$ which has the value 1 in an interval (a, b) and vanishes otherwise, can now be approximated as accurately as one wishes through the last functions considered. For the interval (a, b) can be subdivided in sufficiently many equal parts (x_p, x_{p+1}), and c_p can be defined by the restriction $c_p x_p^{\nu+1/2} = 1$. Hence the converse function is valid also for $\varphi(x)$; hence also for each step function; hence also for each function of $\mathfrak{F}_0^2$.

PROOF of 2). It can be assumed that $f(x)$ vanishes outside of a finite interval $[0, a]$. (The general case follows easily from this by extending the limit.) Hence

$$(13) \quad \int_0 f(x)^2\,dx = \int_0^a f(x)^2\,dx = \int_0^a f(x)\left[\ \ell.\mathrm{m.} \int_0 S_\nu(\alpha x)F(\alpha)\,d\alpha\ \right] dx \ .$$

If the functions $\varphi_n(x)$ of $\mathfrak{F}_0^2$ converge to $\varphi(x)$ in the quadratic mean, then

$$\int_0^a f(x)\varphi(x)\,dx = \lim_{n\to\infty} \int_0^a f(x)\varphi_n(x)\,dx \ .$$

Hence the last integral in (13) has the value

$$\lim_{n\to\infty} \int_0^n F(\alpha)\left[\int_0^a S_\nu(\alpha x)f(x)\,dx\right] d\alpha = \lim_{n\to\infty} \int_0^n F(\alpha)^2\,d\alpha \ ,$$

Q.E.D.

CHAPTER IX

FUNCTIONS OF SEVERAL VARIABLES

§43. Trigonometric Integrals in Several Variables

1. We shall consider in the present chapter (complex valued) functions of several (real) variables. The number of variables will always be denoted by k. By an interval, we shall always mean a point set

$$(1) \qquad a_\chi \leq x_\chi \leq b_\chi , \qquad \chi = 1, 2, \ldots, k .$$

For each function occurring under a k-fold integral, we again tacitly assume that in each finite interval of its range of definition, it is integrable in the sense of Lebesgue (the interval is called finite if the numbers a_χ, b_χ are all finite). For brevity, in place of

$$(2) \qquad \int_{a_1}^{b_1} \cdots \int_{a_k}^{b_k} F(x_1, \ldots, x_k) dx_1 \ldots dx_k ,$$

we shall also write

$$(3) \qquad \int_a^b F(x_1, \ldots, x_k)\, dx ,$$

and correspondingly for

$$\int_{-\infty}^{\infty} \cdots \int_{-\infty}^{\infty} F(x_1, \ldots, x_k) dx_1 \ldots dx_k \quad \text{or} \quad \int_0^{\infty} \cdots \int_0^{\infty} F(x_1, \ldots, x_k) dx_1 \ldots dx_k ,$$

also write

$$(4) \qquad \int F(x_1, \ldots, x_k)\, dx \qquad \text{or} \qquad \int_0 F(x_1, \ldots, x_k)\, dx .$$

Other letters, as for example $\alpha_1, \ldots, \alpha_k, \alpha$ instead of the letters $x_1, \ldots, x_k, x$, will appear in an obvious manner.

We call the integral (4) convergent if the (2k)-fold or the k-fold limit of the integral

$$(5) \qquad \int_a^b F(x_1, \ldots, x_k)\, dx \quad \text{or} \quad \int_0^c F(x_1, \ldots, x_k)\, dx$$

exists as $a_\chi \longrightarrow -\infty$ and $b_\chi \longrightarrow +\infty$ or as $c_\chi \longrightarrow +\infty$. We shall denote this limit as the value of the integral in question. We call the integral (4) absolutely convergent, if the integral in question converges for the function $|F(x_1, \ldots, x_k)|$. We shall say that the integral

$$\int F(x_1, \ldots, x_k)\, dx$$

exists as a Cauchy principal value, if the k-fold limit of the integral

$$\int_{-A}^{A} F(x_1, \ldots, x_k)\, dx$$

exists as $A \longrightarrow +\infty$.

If the function F can be written in the form

$$F(x_1, \ldots, x_k) = \prod_{\chi=1}^{k} F_\chi(x_\chi) \quad ,$$

then the integral (4) is convergent, absolutely convergent, or convergent as a Cauchy principal value, if for each individual χ, the (simple) integral

$$\int F_\chi(x)\, dx \quad \text{or} \quad \int_0 F_\chi(x)\, dx$$

is convergent, absolutely convergent, or convergent as a Cauchy principal value.

2. In the main we will consider a function $f(x_1, \ldots, x_k)$ which is defined and absolutely integrable in the whole space. Then the trigonometric integral

$$E(\alpha_1, \ldots, \alpha_k) = \frac{1}{(2\pi)^k} \int f(x_1, \ldots, x_k) e\left(-\sum_\chi \alpha_\chi x_\chi\right)\, dx$$

exists for all α. We shall denote it as the (Fourier) transform of f.

It is very easy to see that the transform is bounded and continuous. Further it is convergent to zero as

$$|\alpha_1| + \dots + |\alpha_k| \longrightarrow \infty \quad .$$

However we shall not use this fact

3. If

$$f(x_1, \dots, x_k) = \prod_\chi f_\chi(x_\chi)$$

and each individual factor $f_\chi(x)$ is absolutely integrable in $[-\infty, \infty]$, then

$$E(\alpha_1, \dots, \alpha_k) = \prod_\chi E_\chi(\alpha_\chi)$$

where

$$E_\chi(\alpha) = \frac{1}{2\pi}\int f_\chi(x) e(-\alpha x)\, dx \quad .$$

For example, in the interval $0 \leqq |\alpha_\chi| < n_\chi$,

$$(6) \qquad \frac{1}{(2\pi)^k}\int \prod_\chi \left(\frac{\sin \frac{n_\chi x_\chi}{2}}{\frac{x_\chi}{2}} \right)^2 \frac{e(-\alpha_\chi x_\chi)}{n_\chi}\, dx = \prod_\chi \left(1 - \frac{|\alpha_\chi|}{n_\chi} \right) ,$$

and for other values $(\alpha_1, \dots, \alpha_k)$ the integral is 0. Further if $a_\chi > 0$

$$(7) \qquad \int e^{-\Sigma_\chi a_\chi x_\chi{}^2 - i\Sigma_\chi \alpha_\chi x_\chi}\, dx = \frac{\pi^{k/2}}{\prod_\chi \sqrt{a_\chi}}\, e^{-1/4 \Sigma_\chi \frac{\alpha_\chi{}^2}{a_\chi}}$$

4. More generally the integral

$$J(a, \alpha) = \int e^{-\Sigma_{\chi,\lambda} a_{\chi\lambda} x_\chi x_\lambda - i\Sigma_\chi \alpha_\chi x_\chi} dx$$

can be computed from the above provided the quadratic form in the $a_{\chi\lambda}$ is symmetric and positive definite. Indeed we shall show that it has the value [108]

$$W(a, \alpha) = \frac{\pi^{k/2}}{\sqrt{D}}\, e^{\rho/4} \quad ,$$

where

$$D(a) = \begin{vmatrix} a_{11} & \cdots & a_{1k} \\ \cdot & \cdots & \cdot \\ \cdot & \cdots & \cdot \\ a_{k1} & \cdots & a_{kk} \end{vmatrix} , \qquad D(a)\rho(a, \alpha) = \begin{vmatrix} 0 & \alpha_1 & \cdots & \alpha_k \\ \alpha_1 & \alpha_{11} & \cdots & \alpha_{1k} \\ \vdots & \vdots & \vdots & \vdots \\ \alpha_k & \alpha_{k1} & \cdots & \alpha_{kk} \end{vmatrix}$$

Thus we claim the identity

$$(8) \qquad J(a, \alpha) = W(a, \alpha) \quad .$$

Because

$$(9) \qquad \sum_{\chi,\lambda} a_{\chi\lambda} x_\chi x_\lambda \geqq A \sum_\chi x_\chi^2 \quad , \qquad A > 0 \quad ,$$

the integral $J(a, \alpha)$ is absolutely integrable. Hence one can apply to it arbitrary coordinate transformations with continuous partial derivatives. The affine transformation

$$(10) \qquad x_\chi = \sum_\lambda \gamma_{\chi\lambda} y_\lambda$$

with positive determinant

$$\Delta = | \gamma_{\chi\lambda} |$$

gives on the one hand

$$\sum_{\chi\lambda} a_{\chi\lambda} x_\chi x_\lambda + i \sum_\chi \alpha_\chi x_\chi = \sum_{\chi\lambda} b_{\chi\lambda} y_\chi y_\lambda + i \sum_\chi \beta_\chi y_\chi$$

where

$$b_{\chi\lambda} = \sum_{\mu\nu} a_{\mu\nu} \gamma_{\mu\chi} \gamma_{\nu\lambda} \qquad \beta_\chi = \sum_\mu \alpha_\mu \gamma_{\mu\chi} \quad ,$$

and on the other hand

$$(11) \qquad J(a, \alpha) = J(b, \beta) \cdot \Delta \quad .$$

But now by formulas in the theory of determinants, we have

$$D(b) = D(a) \cdot \Delta^2, \qquad D(b)\rho(b, \beta) = D(a)\rho(a, \alpha) \cdot \Delta^2 \quad ,$$

and therefore

$$(12) \qquad W(a, \alpha) = W(b, \beta) \cdot \Delta \quad .$$

But the transformation (10) can now be determined in such a way that

$$\sum_{\chi\lambda} b_{\chi\lambda} x_\chi x_\lambda = \sum_\chi y_\chi^2 \quad .$$

In this special case, however, the relation

$$J(b, \beta) = W(b, \beta)$$

holds because of (7). By (11) and (12), (8) now also follows.

5. THEOREM 56 [109]. If the absolutely integrable function $f(x_1, \ldots, x_k)$ depends only on the quantity $r = \sqrt{x_1^2 + \ldots + x_k^2}$, i.e.,

$$f(x_1, \ldots, x_k) = \varphi\left(\sqrt{x_1^2 + \ldots + x_k^2}\right) \quad , \tag{13}$$

then the function

$$J(\alpha_1, \ldots, \alpha_k) = \int f(x_1, \ldots, x_k) e\left(-\sum_\chi \alpha_\chi x_\chi\right) dx \quad , \tag{14}$$

in the same way depends only on the quantity $\alpha = \sqrt{\alpha_1^2 + \ldots + \alpha_k^2}$. The k-fold integral (14) can be expressed by a single integral

$$J(\alpha) = \frac{(2\pi)^{k/2}}{\alpha^{(k-2)/2}} \int_0 \varphi(\rho)\rho^{k/2} J_{(k-2)/2}(\alpha\rho)\, d\rho \quad , \tag{15}$$

where $J_\mu(t)$ denotes the Bessel function of the μ-th order.

If in addition the quantity

$$s = \alpha^2 = \alpha_1^2 + \ldots + \alpha_k^2 \tag{16}$$

is introduced, then one finds [110]

1. For $k = 2m + 2$

$$J(\alpha) = (-1)^m \pi^{m+1} 2^{2m+1} \frac{d^m}{ds^m} \int_0 \varphi(\rho)\rho J_0(\sqrt{s}\rho)\, d\rho \quad , \tag{17}$$

and

2. For $k = 2m + 1$

(18) $$J(\alpha) = (-1)^m \pi^m 2^{2m+1} \frac{d^m}{ds^m} \int_0^\infty \varphi(\rho) \cos(\sqrt{s}\rho)\, d\rho \quad .$$

PROOF. Hold α_χ fixed and consider an orthogonal transformation

$$y_\chi = \sum_\lambda o_{\chi\lambda} x_\lambda$$

in the integral (14) with the determinant $+1$, in which

$$o_{11} : o_{12} : \dots : o_{1k} = \alpha_1 : \alpha_2 : \dots : \alpha_k \quad .$$

Then in particular

$$\sum_\chi x_\chi^2 = \sum_\chi y_\chi^2, \qquad \sum_\chi \alpha_\chi x_\chi = \sqrt{\sum_\chi \alpha_\chi^2} \cdot y_1 \quad .$$

If after the transformation we write x_κ instead of y_κ, we obtain

(19) $$J(\alpha_1, \dots, \alpha_k) = \int \varphi\left(\sqrt{x_1^2 + \dots + x_k^2}\right) e(-\alpha x_1)\, dx \quad ,$$

so that J actually depends only on α.

Let $F(x_1, \dots, x_k)$ be dependent only on the distance r. If the integral

$$\int F(x_1, \dots, x_k)\, dx$$

is evaluated by first integrating for fixed r, over the $(k-1)$ dimensional sphere $x_1^2 + \dots + x_k^2 = r^2$, and then between 0 and $+\infty$ with respect to r, we obtain the value

$$\omega_k \int_0^\infty F(r) r^{k-1}\, dr \quad ,$$

where ω_k denotes the $(k-1)$ dimensional content of the unit sphere

$$x_1^2 + \dots + x_k^2 = 1 \quad .$$

Therefore

(20) $$\omega_k = \frac{2\pi^{k/2}}{\Gamma(k/2)} \quad ,$$

a relation which is valid also for $k = 1$. In this case the assumption in regard to $F(x_1)$ says that it is an even function.

Now let $k \geq 2$. The integrand of the integral (19) depends only on x_1 and

$$\rho = \sqrt{x_2^2 + \dots + x_k^2} \quad .$$

For fixed x_1, the other variables run through the whole $(k-1)$ dimensional space. Hence by the observation just set forth

$$J(\alpha) = \omega_{k-1} \int dx_1 \int_0 \omega\left(\sqrt{x_1^2 + \rho^2}\right) e(-\alpha x_1) \rho^{k-2} \, d\rho \quad .$$

In the (x_1, ρ) plane, the integral is to extend over the upper half plane $\rho \geq 0$. In polar coordinates

$$x_1 = r \cos \theta, \qquad \rho = r \sin \theta \; ,$$

r lies between the limits 0 and ∞, and θ between 0 and π. Hence

$$J(\alpha) = \omega_{k-1} \int_0 \varphi(r) r^{k-1} K(\alpha r) \, dr$$

where

$$K(t) = \int_0^\pi e(-t \cos \theta)(\sin \theta)^{k-2} \, d\theta \quad .$$

But now, cf. §16, (1)

$$J_\nu(t) = \frac{(1/2t)^\nu}{\Gamma(\nu+1/2)\Gamma(1/2)} \int_0^\pi e(-t \cos \theta)(\sin \theta)^{2\nu} \, d\theta \quad .$$

From this and (20), (15) results, which is also valid for $k = 1$. Then

$$J_{-1/2}(t) = \sqrt{\frac{2}{\pi t}} \cos t \quad .$$

One gains (17) and (18) from (15), if in (15) the expression

$$\frac{J_p(\alpha\rho)}{\alpha^p} = \left(\frac{-2}{\rho}\right)^m \frac{d^m}{ds^m} \left(\frac{J_{p-m}(\sqrt{s}\rho)}{\sqrt{s}^{p-m}} \right)$$

is inserted for $p = \frac{k-2}{2}$. This expression results from the substitution of $x = \sqrt{s\rho}$ in the formula

$$\frac{J_p(x)}{x^p} = -\frac{1}{x}\frac{d}{dx}\left(\frac{J_{p-1}(x)}{x^{p-1}}\right) .$$

EXAMPLE. For odd k, $u > 0$,

$$(21)\quad \frac{1}{(2\pi)^k}\int e^{-u\sqrt{x_1^2+\dots+x_k^2} - i(\alpha_1 x_1+\dots+\alpha_k x_k)}\,dx = \frac{\pi^{-\frac{k+1}{2}}\,\Gamma(\frac{k+1}{2})u}{(u^2+\alpha_1^2+\dots+\alpha_k^2)^{(k+1)/2}}$$

is obtained very easily from (18). The same relation is obtained from (17) for even k by means of the formula

$$\int_0 e^{-u\rho}\rho J_o(a\rho)\,d\rho = \frac{u}{\sqrt{u^2+a^2}^{\,3}}$$

which results from differentiation of [111]

$$\int_0 e^{-u\rho}J_o(a\rho)\,d\rho = \frac{1}{\sqrt{u^2+a^2}} .$$

6. Again the <u>Faltung rule</u> holds. Let

$$f_1(x_1, \dots, x_k), \qquad f_2(x_1, \dots, x_k)$$

be two given absolutely integrable functions with an equal number of variables. Then there exists the k-fold integral

$$\frac{1}{(2\pi)^k}\int f_1(y_1, \dots, y_k)f_2(x_1 - y_1, \dots, x_k - y_k)\,dy$$

for almost all points of the k-dimensional x-space, and the resulting function (the "faltung")

$$f(x_1, \dots, x_k)$$

is again absolutely integrable. The proof proceeds just as it did in the case $k = 1$, cf. §13, 3, because the theorem of Fubini used in the case $k = 1$, in regard to the interchange of the order of integration in a two-fold integral is still correct if each of the variables x and y runs through not a linear interval but a k-fold interval (cf. Appendix 7, 10). In a similar way the proof carries over that <u>the Faltung of absolutely integrable functions corresponds to the multiplication of the transforms</u>

$$E(\alpha_1, \ldots, \alpha_k) = E_1(\alpha_1, \ldots, \alpha_k)E_2(\alpha_1, \ldots, \alpha_k) \quad .$$

Thus for example, the transform of the function

$$(22) \qquad \frac{1}{(2\pi)^k}\int f(y_1, \ldots, y_k) \prod_{\chi} \frac{1}{n_\chi} \left(\frac{\sin \frac{n_\chi}{2}(x_\chi - y_\chi)}{\frac{1}{2}(x_\chi - y_\chi)} \right)^2 dy$$

has in the interval $0 \leqq |\alpha_\chi| < n_\chi$, the value

$$(22') \qquad E(\alpha_1, \ldots, \alpha_k) \prod_{\chi} \left(1 - \frac{|\alpha_\chi|}{n_\chi} \right)$$

and the value 0 for other $(\alpha_1, \ldots, \alpha_k)$; while the transform of the function

$$(23) \qquad \int f(y_1, \ldots, y_k) e^{-\Sigma_\chi n_\chi (x_\chi - y_\chi)^2} dy$$

has the value

$$\frac{\pi^{k/2}}{\sqrt{n_1 \ldots n_k}} E(\alpha_1, \ldots, \alpha_k) e^{-1/4 \Sigma_\chi \frac{\alpha_\chi^2}{n_\chi}} \quad .$$

§44. The Fourier Integral Theorem

1. Let $\varphi(\xi_1, \ldots, \xi_k)$ be a given bounded function, and $K(\xi_1, \ldots, \xi_k)$ a given absolutely integrable function in the "octant"

$$(1) \qquad 0 < \xi_\chi < \infty , \qquad \chi = 1, 2, \ldots, k \quad .$$

For simplification we assume

$$(2) \qquad \int_0 K(\xi_1, \ldots, \xi_k)\, d\xi = 1 \quad .$$

For each system of positive numbers

$$n = (n_1, \ldots, n_k), \qquad n_\chi > 0 \quad ,$$

there exists the quantity

$$(3) \qquad A_n = \int_0 \varphi\left(\frac{\xi_1}{n_1}, \ldots, \frac{\xi_k}{n_k} \right) K(\xi_1, \ldots, \xi_k)\, d\xi \quad .$$

If now, the (k-fold) limit exists for the function φ,

$$\varphi(+0, \ldots, +0) = \lim_{\xi_\chi \to 0} \varphi(\xi_1, \ldots, \xi_k) \quad , \tag{4}$$

then the (k-fold) limit

$$\lim_{n_\chi \to \infty} A_n \tag{5}$$

also exists, and its value amounts to (4). For its proof, we can restrict ourselves to the case

$$\varphi(+0, \ldots, +0) = 0 \quad . \tag{6}$$

We divide the octant (1) into a finite interval

$$0 < \xi_\chi < a_\chi \quad , \tag{7}$$

and its exterior. Correspondingly, we divide the integral (3) into the sum

$$P_n + Q_n \quad .$$

If we denote a bound of φ by M, then $|Q_n| \leqq M$ times the integral of $|K|$ over the exterior. Hence Q_n, uniformly in n, is arbitrarily small if only the a_χ are sufficiently large. Further $|P_n| \leqq$ the upper limit of the function $\varphi(t_1, \ldots, t_k)$ in the interval

$$0 < t_\chi < \frac{a_\chi}{n_\chi}$$

times the integral

$$\int |K(\xi_1, \ldots, \xi_k)| \, d\xi \quad .$$

But because of (4), this upper limit becomes arbitrarily small for fixed a_χ as $n_\chi \longrightarrow \infty$, and from this it follows that the limit (5) exists and has the value (6).

From this the following generalization of Theorem 3, a) is found without difficulty.

THEOREM 57 [112]. Let $f(x_1, \ldots, x_k)$ be a given bounded function, and $K(\xi_1, \ldots, \xi_k)$ be a given absolutely integrable function in the whole space, for which

$$\int K(\xi_1, \ldots, \xi_k) \, d\xi = 1 \quad . \tag{8}$$

As $n_\kappa \longrightarrow \infty$ the function

$$\int f\left(x_1 + \frac{\xi_1}{n_1}, \dots, x_k + \frac{\xi_k}{n_k}\right) K(\xi_1, \dots, \xi_k)\, d\xi$$

(9)

$$= n_1 \dots n_k \int f(\xi_1, \dots, \xi_k) K(n_1(x_1 - \xi_1), \dots, n_k(x_k - \xi_k))\, d\xi$$

is convergent to f at each point where f is continuous. More generally, it converges at each point at which the 2^k limits

$$f(x_1 \pm 0, \dots, x_k \pm 0) \tag{10}$$

exist, to the "mean value"

$$\frac{1}{2^k} \sum f(x_1 \pm 0, \dots, x_k \pm 0) \quad . \tag{11}$$

2. For the relation

$$E(\alpha_1, \dots, \alpha_k) = \frac{1}{(2\pi)^k} \int f(x_1, \dots, x_k) e\left(-\sum_{\chi} \alpha_\chi x_\chi\right) dx \quad , \tag{12}$$

the inverse formula

$$f(x_1, \dots, x_k) = \int E(\alpha_1, \dots, \alpha_k) e\left(\sum_{\chi} \alpha_\chi x_\chi\right) d\alpha \tag{13}$$

holds. We shall establish criteria for its validity. If for absolutely integrable f, the integral (12) is substituted in

$$\int_{-n}^{n} E(\alpha_1, \dots, \alpha_k)\left(\sum_{\chi} \alpha_\chi x_\chi\right) d\alpha \tag{14}$$

and the order of integration interchanged, we obtain

$$\frac{1}{\pi^k} \int f(\xi_1, \dots, \xi_k) \prod_{\chi} \frac{\sin n_\chi (x_\chi - \xi_\chi)}{x_\chi - \xi_\chi}\, d\xi \quad .$$

The question arises under what conditions this integral converges to $f(x_1, \dots, x_k)$ as $n_\chi \longrightarrow \infty$. However we shall leave the examination of this question for the next paragraph, and shall now make another statement.

Let us form with the absolutely integrable function f, the expression (22) of §43, and denote it by f_n. According to (22'), the integral (13), formed with the transform of f_n, amounts to

$$(15) \qquad \int_{-n}^{n} E(\alpha_1, \ldots, \alpha_k) \prod_{\chi} \left(1 - \frac{|\alpha_\chi|}{n_\chi}\right) e\left(\sum_{\chi} \alpha_\chi x_\chi\right) d\alpha .$$

Substituting the integral (12) herein, and interchanging the order of integration, the function f_n results after a slight adjustment. Hence the inverse formula (13) is valid for f_n. By Theorem 57, we now obtain the following theorem.

THEOREM 58. If $f(x_1, \ldots, x_k)$ is absolutely integrable and bounded, then the inverse formula (13) holds at each point at which f is continuous (or more generally has the limit (11)), provided the integral (13) is interpreted as the limit of the integral (15) as $n_\chi \longrightarrow \infty$, i.e., provided the integral (13) is summed by the method of the k-dimensional arithmetic mean.

If the limit of (15) exists, and at the same time the integral (13) exists as a Cauchy principal value, then by a general theorem of summation related to the method of arithmetic mean, the two values are equal to one another [Appendix 18]. Therefore from Theorem 58, we obtain the following.

THEOREM 59. If $f(x_1, \ldots, x_k)$ is absolutely integrable and bounded, then the inverse formula (13) holds at each point at which the integral (13) exists as a Cauchy principal value and $f(x_1, \ldots, x_k)$ is continuous (or more generally has the limit (11)).

3. The following theorem is of a totally different kind.

THEOREM 60. If $f(x_1, \ldots, x_k)$ and $E(\alpha_1, \ldots, \alpha_k)$ are both absolutely integrable, then the inverse formula (13) holds for almost all x.

Since for absolutely integrable E, the integral (13) represents a continuous function, therefore f, after correction if need be on a point set of measure zero, is a continuous function, and for continuous f, (13) is valid everywhere.

If f is absolutely integrable and bounded, then, at any rate, E is absolutely integrable if it is of uniform sign, $E \geq 0$.

PROOF. Let f and E be both absolutely integrable. The function f_n considered in 2. has the value (15). Since E is absolutely integrable, for fixed x, the integral (15) is convergent to the integral (13). On the other hand, by Theorem 57, f_n is convergent to f, if f is known to be continuous and bounded. Thus, for absolutely integrable E, the inverse formula certainly holds if f is also continuous and bounded. In the general case where only absolute integrability is known of $f(x)$, we introduce the function, for fixed $h_\chi > 0$

$$(16) \qquad F_h(x_1, \ldots, x_k) = \frac{1}{h_1 \ldots h_k} \int_0^h f(x_1 + t_1, \ldots, x_k + t_k)\, dt \quad .$$

It is known from the theory of integration that the function (16) is continuous (Appendix 9). Moreover it is bounded

$$h_1 \ldots h_k |F_h| \leqq \int |f(\xi_1, \ldots, \xi_k)|\, d\xi \quad .$$

If a constant factor is disregarded, (16) is the Faltung of the function f with that function which has the value 1 in the interval $-h_\chi < t_\chi < 0$ and vanishes otherwise. The function, F_h, therefore, is also absolutely integrable. Its transform computes easily to

$$E_h(\alpha_1, \ldots, \alpha_k) = E(\alpha_1, \ldots, \alpha_k)\delta_h(\alpha_1, \ldots, \alpha_k) \quad ,$$

where

$$\delta_h(\alpha_1, \ldots, \alpha_k) = \prod_\chi \frac{e(\alpha_\chi h_\chi) - 1}{i\alpha_\chi h_\chi} \quad .$$

Because

$$(17) \qquad |\delta_h(\alpha_1, \ldots, \alpha_k)| \leqq 1 \quad ,$$

the transform is likewise absolutely integrable. Hence by the special case already established we have

$$(18) \qquad F_h(x_1, \ldots, x_k) = \int E_h(\alpha_1, \ldots, \alpha_k) e\left(\sum_\kappa \alpha_\kappa x_\kappa\right) d\alpha$$

for all x. Relation (13) in the general case will be deduced from this by a passage to a limit. We let $h_\kappa \longrightarrow 0$, for which δ_h converges to 1. Moreover since (17) is valid and because of the absolute integrability of E, the right side of (18) is convergent to the right side of (13) as $h_\chi \longrightarrow 0$. On the other hand, we know from the theory of integration if in (16) the numbers h_χ are say, equal to one another,

$$h_1 = h_2 = \dots = h_k = h \quad ,$$

and the common value h is allowed to decrease to zero, that then the function F_h is convergent to f for almost all x. Hence (13) holds for almost all x.

Now let f be absolutely integrable and bounded; $|f| \leqq G$, and let $E \geqq 0$. We gather immediately from (22) of §43, that likewise

$$|f_n| \leqq G$$

for all $n_\chi > 0$. By the observations in 2, the inverse formula is now valid for f_n at all points. If it is applied to the origin $x_\chi = 0$, we obtain

$$\int_{-n}^{n} E(\alpha_1, \dots, \alpha_k) \prod_\chi \left(1 - \frac{|\alpha_\chi|}{n_\chi}\right) d\alpha \leqq G \quad .$$

If fixed numbers $a_\chi > 0$ are now taken, then because $E \geqq 0$,

$$\int_{-a}^{a} E(\alpha_1, \dots, \alpha_k) \prod_\chi \left(1 - \frac{|\alpha_\chi|}{n_\chi}\right) d\alpha \leqq G$$

is valid for $n_\chi > a_\chi$. By allowing $n_\chi \longrightarrow \infty$, we obtain

$$\int_{-a}^{a} E(\alpha_1, \dots, \alpha_k) \, d\alpha \leqq G \quad .$$

And since the numbers a_χ can be arbitrarily large, this says that E is absolutely integrable, Q.E.D.

4. This presents the question of how to recognize for a given function f whether its transform is absolutely integrable. Delicate criteria do not seem to be known. But a rather obvious criterion [113] is the following. The function f has the 3^k derivatives

$$(19) \qquad \frac{\partial^{p_1 + \dots + p_k} f}{\partial x_1^{p_1} \dots \partial x_k^{p_k}} \quad , \qquad\qquad 0 \leqq p_\chi \leqq 2$$

and these are continuous, absolutely integrable, and convergent to zero as

$$|x_1| + \dots + |x_k| \longrightarrow \infty \quad .$$

(For the highest derivatives: $p_\chi = 2$, the decreasing to zero part can be abandoned, and certain discontinuities can also be allowed.) — For if the expression for the transform

$$E_{p_1 \cdots p_k}$$

of the function (19) is formed and integrated partially, there results apart from a constant factor, the function

$$\alpha_1^{p_1} \cdots \alpha_k^{p_k} E \quad . \tag{20}$$

Since the functions (19) are absolutely integrable, the functions (20) are bounded. Hence the function

$$(1 + |\alpha_1|^2) \cdots (1 + |\alpha_k|^2) \cdot E$$

is also bounded. Therefore E is absolutely integrable.

5. An application [114]. By the inverse of (21), §43, we obtain for $u > 0$

$$e^{-u\sqrt{\alpha_1^2+\cdots+\alpha_k^2}} = \frac{\Gamma\left(\frac{k+1}{2}\right)u}{\pi^{\frac{k+1}{2}}} \int \frac{e(\Sigma_\chi \alpha_\chi x_\chi)\, dx}{(u^2+x_1^2+\cdots+x_k^2)^{\frac{k+1}{2}}} \quad . \tag{21}$$

By a Faltung, we find for $u > 0$, $u' > 0$

$$e^{-(u+u')\sqrt{\alpha_1^2+\cdots+\alpha_k^2}} = \frac{\Gamma\left(\frac{k+1}{2}\right)^2 uu'}{\pi^{k+1}} \int F(x_1, \ldots, x_k) e(\Sigma_\chi a_\chi x_\chi)\, dx \quad , \tag{22}$$

where

$$F(x_1, \ldots, x_k) = \int \frac{dy}{\{[u^2+y_1^2+\cdots+y_k^2] \cdot [u'^2+(x_1-y_1)^2+\cdots+(x_k-y_k)^2]\}^{(k+1)/2}} \quad .$$

By the combination of (21) and (22), with the aid of Theorem 61 which follows, there results

$$\frac{u + u'}{[(u+u')^2+x_1^2+\cdots+x_k^2]^{(k+1)/2}} = \frac{\Gamma\left(\frac{k+1}{2}\right) uu'}{\pi^{(k+1)/2}} F(x_1, \ldots, x_k) \quad .$$

Integrating with respect to u' from u' to ∞, we obtain

$$\frac{1}{[(u+u')^2+x_1^2+\dots+x_k^2]^{(k-1)/2}}$$

(23)

$$= \frac{\Gamma\left(\frac{k+1}{2}\right) u}{\pi^{(k+1)/2}} \int \frac{dy}{[u^2+y_1^2+\dots+y_k^2]^{(k+1)/2}\,[u'^2+(x_1-y_1)^2+\dots+(x_k-y_k)^2]^{(k-1)/2}} .$$

6. THEOREM 61. If two absolutely integrable functions have the same transform, then they are identical (almost everywhere).

PROOF. The difference of both functions has the null transform, and by Theorem 60, therefore, this difference likewise has the value zero, Q.E.D.

7. In Theorem 58, we have summed the inverse integral by the method of the arithmetic mean. A rather meaningful convergence criterion is arrived at, if one sums the inverse integral in the way that one forms the "spherical" partial sums

$$f_R(x_1, \dots, x_k) = \int_{K_R} E(\alpha_1, \dots, \alpha_k) e(\Sigma_\chi \alpha_\chi x_\chi)\, d\alpha \quad . \tag{24}$$

Here K_R means the volume

$$\alpha_1^2 + \dots + \alpha_k^2 \leqq R^2 \quad .$$

Inserting (12), there results

$$f_R = c_1 \int f(x_1 + \xi_1, \dots, x_k + \xi_k) H_R(\xi_1, \dots, \xi_k)\, d\xi \quad , \tag{25}$$

where

$$H_R = \int_{K_R} e(-\Sigma_\chi \xi_\chi \alpha_\chi)\, d\alpha \quad ,$$

and c_1 (later c_2, c_3, ...) denotes a constant dependent only on the dimension k.

If for brevity we set

$$\xi = \sqrt{\xi_1^2 + \ldots + \xi_k^2}$$

and

$$\mu = \frac{k-2}{2} \quad ,$$

we obtain by Theorem 56

$$H_R = \frac{c_2}{\xi^\mu} \int_0^R \gamma^{k/2} J_\mu(\xi\gamma)\, d\gamma = c_2 \xi^{-k} \int_0^{R\xi} \gamma^{\mu+1} J_\mu(\gamma)\, d\gamma$$

$$= c_2 \xi^{-k} \int_0^{R\xi} \frac{d}{d\gamma}\left[\gamma^{\mu+1} J_{\mu+1}(\gamma)\right] = c_2 \left(\frac{R}{\xi}\right)^{\mu+1} J_{\mu+1}(\xi R) \quad .$$

For fixed x, we now introduce the function of one variable

(26) $$F(\xi) = \frac{1}{\omega_{k-1}} \int_S f(x_1 + \xi\lambda_1, \ldots, x_k + \xi\lambda_k)\, d\omega \quad ,$$

where S means the unit sphere $\lambda_1^2 + \ldots + \lambda_k^2 = 1$, and $d\omega$ denotes the $(k-1)$ dimensional element of volume of this sphere, and ω_{k-1} its volume. $F(\xi)$ is therefore the mean value of f on the sphere of radius ξ around the fixed point $(x_1, \ldots, x_k)$. For example, for $k = 2$

$$F(\xi) = \frac{1}{2\pi} \int_0^{2\pi} f(x_1 + \xi \cos\varphi, x_2 + \xi \sin\varphi)\, d\varphi \quad .$$

If in (25), we integrate for absolutely integrable f, first over the sphere of radius t, and then over the variable t from 0 to ∞, we obtain

(27) $$f_R(x_1, \ldots, x_k) = c_3 \int F(t) R^{\mu+1} t^\mu J_{\mu+1}(Rt)\, dt$$

$$= c_3 \int_0 F\left(\frac{t}{R}\right) t^\mu J_{\mu+1}(t)\, dt \quad .$$

THEOREM 62. In order that the partial integral

(28) $$f_R(x_1, \ldots, x_k)$$

converge, for absolutely integrable f, to a

finite number as $R \longrightarrow \infty$, it is sufficient that the spherical mean value $F(t)$ defined by (26) be of the following character. If we denote by λ the quantity $(k-2)/2$ when k is even, and the quantity $(k-1)/2$ when k is odd, then the function $F(t)$ is λ-times continuously differentiable in $(0, \infty]$, and each of the functions

$$g(t) = t^{\lambda}F(t),\ g'(t),\ g''(t),\ \ldots,\ g^{(\lambda)}(t)$$

has an absolutely integrable derivative in $[0, \infty]$. The limit of (28) has the value

$$F(0) = \lim_{\xi \to 0} F(\xi) \quad .$$

REMARK. The following is to be noticed in regard to the existence of the function $F(\xi)$. The function $f(\xi_1, \ldots, \xi_k)$ is by assumption integrable everywhere in finite intervals. Introducing spherical polar coordinates around the point $(x_1, \ldots, x_k)$, there results by the theorem of Fubini (Appendix 7, 10), that the integral (26) exists for almost all positive ξ, and represents an integrable function in finite intervals. And our convergence condition demands that the function $F(\xi)$, after suitable modification on a null set, fulfill the given assumptions.

PROOF. By a suitable application of Theorem 6 to (27), we obtain the result that the limit of (28) exists and has the value

$$c_4 F(0) \quad .$$

That the constant c_4, which is independent of f, has the value 1 is obtainable from Theorem 60, since for instance for the function

$$f = e^{-(x_1^2 + \ldots + x_k^2)}$$

the corresponding value $F(0)$ must come out.

8. We record without details, the following theorem.

THEOREM 63. Theorems 52 and 53 and the definitions underlying them are also valid for functions of k-variables [115].

We need only replace dx everywhere by $dx_1 \ldots dx_k$. The particulars to Theorems 52 and 53 were so set up that they can be carried over to the k-dimensional case by the results of the present chapter.

§45. The Dirichlet Integral

1. In the closed "octant"

$$(1) \qquad 0 \leqq x_\chi < \infty, \qquad \chi = 1, 2, \ldots, k \;,$$

let a non-negative function $f(x_1, \ldots, x_k)$ be defined at all points. We shall define by recurrence what we mean when we denote this function as monotonically decreasing by proceeding from $k - 1$ to k.

For $k = 1$, we apply the usual definition

$$f(\xi_1) - f(\xi_1') \geqq 0 \quad \text{if} \quad \xi_1 \leqq \xi_1' \;.$$

Assume the definition is already known for functions of $k - 1$ variables. Then we call the function $f(x_1, \ldots, x_k)$ monotonically decreasing, if for each value $x_k = \xi_k$, the function

$$(2) \qquad f(x_1, \ldots, x_{k-1}, \xi_k)$$

and for every two values $\xi_k < \xi_k'$, the function

$$(3) \quad g(x_1, \ldots, x_{k-1}) = f(x_1, \ldots, x_{k-1}, \xi_k) - f(x_1, \ldots, x_{k-1}, \xi_k')$$

is non-negative and montonically decreasing[1] in the octant

$$0 \leqq x_\chi < \infty \;, \qquad \chi = 1, \ldots, k - 1 \;.$$

Let a function $\varphi(x_1, \ldots, x_k)$ be $\geqq 0$ in (1) and have a finite integral there. Then the function

$$f(x_1, \ldots, x_k) = \int_X \varphi(y_1, \ldots, y_k)\, dy$$

is monotonically decreasing. For $k = 1$, this can be seen immediately. For $k \geqq 1$, it can be obtained inductively by considering that the function (2) belonging to it has the form

[1] One can raise the objection to this definition, that it depends on the sequence of the variables. Without engaging in a discussion of this definition, we remark that the results which follow would hold all the more if we assumed that the above monotonic definition should be satisfied in every fixed sequence of the variables [116].

$$\int_{x_1} \cdots \int_{x_{k-1}} \left[\int_{\xi_k} \varphi(y_1, \ldots, y_{k-1}, \eta_k)\, d\eta_k \right] dy_1 \ldots dy_{k-1}$$

and the function (3) belonging to it has the form

$$\int_{x_1} \cdots \int_{x_{k-1}} \left[\int_{\xi_k}^{\xi_k'} \varphi(y_1, \ldots, y_{k-1}, \eta_k)\, d\eta_k \right] dy_1 \ldots dy_{k-1} \quad .$$

2. THEOREM 64. Let f be a given non-negative, monotonically decreasing function in (1), and for $\chi = 1, \ldots, k$ in the (one dimensional) interval

$$(0 \leqq)\ a_\chi \leqq \xi_\chi \leqq b_\chi \quad , \tag{4}$$

let $\lambda_\chi(x_\chi)$ be a function such that for all ξ_χ in (4)

$$0 \leqq \int_{a_\chi}^{\xi_\chi} \lambda_\chi(t)\, dt \leqq c_\chi \quad .$$

Then for the integral

$$J = \int_a^b f(x_1, \ldots, x_k) \prod_\chi \lambda_\chi(x_\chi)\, dx \quad , \tag{5}$$

the valuation

$$0 \leqq J \leqq f(a_1, \ldots, a_k) \prod_\chi c_\chi \tag{6}$$

holds.

PROOF. For $k = 1$, this is the second mean value theorem of the integral calculus, cf. §2, 2. We shall now reason from $k - 1$ to k. We set

$$J(x) = \int_{a_1}^{b_1} \cdots \int_{a_{k-1}}^{b_{k-1}} f(x_1, \ldots, x_{k-1}, x) \prod_{\chi=1}^{k-1} \lambda_\chi(x_\chi)\, dx_1 \ldots dx_{k-1} \quad .$$

Then by (5)

$$J = \int_{a_k}^{b_k} J(x) \lambda_k(x)\, dx \quad . \tag{7}$$

Because (2) is monotonic and because we have assumed our theorem as already proved for $k - 1$, therefore

$$(8) \qquad 0 \leqq J(x) \leqq f(a_1, \ldots, a_{k-1}, x) \prod_{\chi=1}^{k-1} c_\chi \quad .$$

Hence $J(x)$ is, in particular, non-negative. On the other hand

$$J(\xi_k) - J(\xi_k') = \int_{a_1}^{b_1} \cdots \int_{a_{k-1}}^{b_{k-1}} g(x_1, \ldots, x_{k-1}) \prod_{\chi=1}^{k-1} \lambda_\chi(x_\chi)\, dx_1 \ldots dx_{k-1} \quad .$$

Since therefore our theorem is already proved for $k - 1$ variables, we have

$$J(\xi_k) \geqq J(\xi_k') \quad \text{for} \quad \xi_k \leqq \xi_k' \quad .$$

Hence $J(x)$ is monotonically decreasing. By (7) therefore

$$0 \leqq J \leqq J(a_k)c_k \quad ,$$

and by (8), (6) follows, Q.E.D.

3. Since the function $\sin x/x$ is integrable in $[0, \infty]$, therefore

$$\int_0^\xi \frac{\sin x}{x}\, dx$$

lies below a bound independent of ξ. Moreover, as one can easily convince oneself, this expression is positive for all ξ. For $a = 2\pi m$, $m = 1, 2, 3, \ldots,$ and $b \geqq a$, we have by 2

$$0 \leqq \int_a^b \frac{\sin x}{x}\, dx \leqq \frac{1}{a} \operatorname*{Max}_\xi \int_a^\xi \sin x\, dx \leqq \frac{2}{a} \quad .$$

There is therefore a numerical constant $B > 0$, such that for

$$a = 2\pi m, \qquad m = 0, 1, 2, 3, \ldots$$

and $\xi \geqq a$

$$0 \leqq \int_a^\xi \frac{\sin x}{x}\, dx \leqq \frac{B}{m + 1} \quad .$$

If therefore the function f in (1) is non-negative and monotonically decreasing, and if

(9) $$a_\chi = 2\pi m_\chi, \qquad m_\chi = 0, 1, 2, 3, \dots ,$$

then by Theorem 64, for the integral

$$J(a, b) = \int_a^b f(x_1, \dots, x_k) \prod_\kappa \frac{\sin x_\kappa}{x_\kappa} dx$$

subject to $b_\kappa \geq a_\kappa$, we have the valuation

(10) $$0 \leq |J(a, b)| \leq \frac{AK}{(m_1+1)\dots(m_k+1)}$$

in which A denotes a numerical constant and K a bound of the function f.

Let $2k$ numbers ρ_χ, σ_χ be given, and non-negative integers m_χ such that

(11) $$\rho_\chi \geqq 2m_\chi\pi, \qquad \sigma_\chi \geqq 2m_\chi\pi, \qquad \chi = 1, 2, \dots, k .$$

In the expression $J(a, b)$, set the value (9) for a_χ, and either ρ_χ or σ_χ for b_χ. In this way there result 2^k integrals, all of which satisfy (10). If these integrals are provided with suitable signs, and then added, we obtain the integral $J(\rho, \sigma)$. By assumption (11), therefore, the valuation

$$|J(\rho, \sigma)| \leqq \frac{2^k AK}{(m_1+1)\dots(m_k+1)}$$

is valid.

From this it is recognized without difficulty that the integral

$$\left(\frac{2}{\pi}\right)^k \int_0 f(x_1, \dots, x_k) \prod_\chi \frac{\sin x_\chi}{x_\chi} dx$$

is convergent for each non-negative, monotonically decreasing function f. And this convergence is uniform for all functions f which satisfy a common valuation

$$|f(x_1, \dots, x_k)| \leqq K .$$

Because of this last assertion, the integral

$$\left(\frac{2}{\pi}\right)^k \int_0 f\left(\frac{x_1}{n_1}, \ldots, \frac{x_k}{n_k}\right) \prod_\chi \frac{\sin x_\chi}{x_\chi}\, dx$$

uniformly in the n_χ, $n_\chi > c$, is approximable by an expression

$$\left(\frac{2}{\pi}\right)^k \int_0^\rho f\left(\frac{x_1}{n_1}, \ldots, \frac{x_k}{n_k}\right) \prod_\chi \frac{\sin x_\chi}{x_\chi}\, dx$$

with sufficiently large ρ_χ. But with ρ_χ held fixed, this expression converges, as $n_\chi \longrightarrow \infty$, to

$$f(+0, \ldots, +0) \prod_\chi \frac{2}{\pi} \int_0^{\rho_\chi} \frac{\sin x_\chi}{x_\chi}\, dx_\chi \quad , \tag{12}$$

where $f(+0, \ldots, +0)$ denotes the limit of the function $f(x_1, \ldots, x_k)$ as $x_\chi \longrightarrow 0$. We assume the existence of this limit without engaging in the question whether it does not follow out of the monotonic assumption. For arbitrarily large ρ_χ, the second factor in (12) differs arbitrarily little from 1. From this we obtain the following theorem.

THEOREM 65 [116]. If in (1) the function $f(x_1, \ldots, x_k)$ is non-negative, bounded and monotonically decreasing, then the relation

$$\lim_{n_\chi \to \infty} \left(\frac{2}{\pi}\right)^k \int_0 f(x_1, \ldots, x_k \prod_\chi \frac{\sin n_\chi x_\chi}{x_\chi}\, dx = f(+0, \ldots, +0)$$

holds.

4. In (1) let the function f be continuous, convergent to zero as

$$|x_1| + |x_2| + \ldots + |x_k| \longrightarrow \infty$$

and have the derivative

$$F = f_{x_1 \ldots x_k}$$

which in (1) is continuous and absolutely integrable. We shall show that in (1), f can then be represented as the difference of two functions, each of which in (1) is non-negative, bounded and monotonically decreasing. We introduce the functions F_1 and F_2, the first of which agrees with

F wherever $F \geqq 0$ and vanishes otherwise, and the second conversely agrees with $|F|$ wherever $F \leqq 0$ and vanishes otherwise. Hence

$$F_1 - F_2 = F \quad .$$

The functions

$$f_\rho = \int_x F_\rho(\xi_1, \ldots, \xi_k)\, d\xi, \qquad \rho = 1, 2 \quad ,$$

are non-negative, bounded and monotonically decreasing, and we wish to show that the function

$$g = f_1 - f_2$$

has the value $(-1)^k f$. In fact,

$$g(a_1, \ldots, a_k) = \int_a F(\xi_1, \ldots, \xi_k)\, d\xi \quad . \tag{13}$$

On the other hand

$$\int_{a_1}^{b_1} \cdots \int_{a_k}^{b_k} f_{x_1 \ldots x_k}(\xi_1, \ldots, \xi_k)\, d\xi_1 \ldots d\xi_k \tag{14}$$

equals a sum

$$\sum \pm f(c_1, \ldots, c_k) \tag{15}$$

where each c_χ has either the value a_χ or the value b_χ, and the term $f(a_1, \ldots, a_k)$ has the coefficient $(-1)^k$. This result is easily obtained if (14) is integrated out. By the assumption concerning f, all components of (15) to within

$$(-1)^k f(a_1, \ldots, a_k)$$

are convergent to zero as $b_\chi \longrightarrow \infty$, and on the other hand (14) is then convergent to (13). From this our assertion follows.

5. THEOREM 66. If the function $f(x_1, \ldots, x_k)$ is continuous in the whole space, is convergent to zero as

$$|x_1| + \ldots + |x_k| \longrightarrow \infty$$

and if it has the derivative

$$f_{x_1 \ldots x_k}$$

which is continuous and absolutely integrable in the whole space, then

$$f(x_1, \ldots, x_k) = \lim_{n_\chi \to \infty} \int f(x_1 + \xi_1, \ldots, x_k + \xi_k) \prod_\chi \frac{\sin n_\chi \xi_\chi}{\pi \xi_\chi} \, d\xi \quad .$$

Moreover, if $f(x_1, \ldots, x_k)$ is absolutely integrable, then for

$$E(\alpha_1, \ldots, \alpha_k) = \frac{1}{(2\pi)^k} \int f(\xi_1, \ldots, \xi_k) e(\Sigma_\chi \alpha_\chi \xi_\chi) \, d\xi$$

the inverse formula

$$(16) \qquad f(x_1, \ldots, x_k) = \int E(\alpha_1, \ldots, \alpha_k) e(\Sigma_\chi x_\chi \alpha_\chi) \, d\alpha$$

holds, where the last integral is to be taken as a Cauchy principal value.

PROOF. The first part of the theorem follows from 4. and Theorem 65. The second part then follows by considering that the value of the integral

$$\int_{-n}^{n} E(\alpha_1, \ldots, \alpha_k) e(\Sigma_\chi \alpha_\chi x_\chi) \, d\alpha$$

amounts to

$$\int f(\xi_1, \ldots, \xi_k) \prod_\chi \frac{\sin n_\chi (x_\chi - \xi_\chi)}{\pi (x_\chi - \xi_\chi)} \, d\xi \quad .$$

§46. The Poisson Summation Formula

1. If the quantity $\varphi(n_1, \ldots, n_k)$ is defined for all "lattice points", i.e., for all combinations of positive and negative integers, then the sum

$$(1) \qquad \sum_n \varphi(n_1, \ldots, n_k)$$

is to extend over all these lattice points. We call the series (1) convergent and denote its sum by ψ, if the partial sums

$$s_{n_1 \ldots n_k} = \sum_{-n_1}^{n_1} \cdots \sum_{-n_k}^{n_k} \varphi(\nu_1, \ldots, \nu_k) \tag{2}$$

converge to ψ as $n_\chi \longrightarrow \infty$.

We shall consider only such functions for which, first of all, the integral

$$\varphi(n_1, \ldots, n_k) = \int f(x_1, \ldots, x_k) e(2\pi \Sigma_\chi n_\chi x_\chi)\, dx \tag{3}$$

exists as a Cauchy principal value. For this it is sufficient but not necessary that f be absolutely integrable.

The Poisson formula in k variables (in its simplest form) reads

$$\sum_m f(m_1, \ldots, m_k) = \sum_n \varphi(n_1, \ldots, n_k) \quad . \tag{4}$$

If $k = 1$, this formula becomes formula (2) of §10.

The question of the validity of (4) embraces the question as to whether the two series occurring in (4) are convergent and as to whether there is equality between their sums. We look at it principally as a question of the equality of both sums, and therefore, as regards convergence of the series, we will have no hesitation to make assumptions explicitly.

THEOREM 67 [117]. In order that (4) hold, the following assumptions are sufficient:

(a) The function $f(x_1, \ldots, x_k)$ be uniquely defined at all points, be bounded and integrable in finite intervals, and be continuous at the lattice points.

(b) The series

$$F(x_1, \ldots, x_k) = \sum_m f(x_1 + m_1, \ldots, x_k + m_k) \tag{5}$$

be uniformly convergent on the base interval J:

$$-\tfrac{1}{2} \leqq x_\chi < \tfrac{1}{2} \quad , \qquad \chi = 1, \ldots, k \quad . \tag{6}$$

— Under the present assumptions, the integral (3) exists and what is more

$$(7) \qquad \varphi(n_1, \ldots, n_k) = \int_J F(\xi_1, \ldots, \xi_k) e(2\pi \Sigma_\chi n_\chi \xi_\chi)\, d\xi \quad .$$

(c) The series

$$(8) \qquad \sum_n \varphi(n_1, \ldots, n_k)$$

be convergent.

REMARK. For example, assumption (a) is then satisfied if f is continuous everywhere, and assumption (b) if there are constants $G > 0$ and $\eta > 0$ such that

$$(9) \qquad |f| \leq G(x_1^2 + \ldots + x_k^2)^{-\frac{k}{2} - \eta} \quad .$$

Concerning assumption (c), it is in a certain sense dispensable. In fact, we shall prove the following. If assumptions (a) and (b) are satisfied, then the series (8) is convergent by arithmetic means, and (4) holds. Under convergence by arithmetic means, we herewith understand that the series

$$(10) \qquad \sigma_{n_1 \ldots n_k} = \frac{1}{n_1 \ldots n_k} \sum_{\nu_1=0}^{n_1-1} \cdots \sum_{\nu_k=0}^{n_k-1} s_{\nu_1 \ldots \nu_k} \quad ,$$

formed with the partial sums (2) of (8), is convergent as $n_\chi \longrightarrow \infty$. — If assumption (c) enters, then the assertion of our theorem results from the following fact (Appendix 18). If a series (8) on the one hand converges, and on the other hand is summable by arithmetic means, then both "sums" are equal to each other.

PROOF. The function $F(x_1, \ldots, x_k)$ is bounded in the interval (6) and continuous at the point $(0, \ldots, 0)$. The first follows because the function f is bounded in finite intervals, and the series (5) is uniformly convergent in (6). The second follows because each term of the series (5) is continuous at the point $(0, \ldots, 0)$ and the series (5) is uniformly convergent.

Because of (7), we obtain from (2) and (10), if use is made of elementary formulas of trigonometric sums [118], that

(11) $$\sigma_{n_1 \ldots n_k} = \int_J F(\xi_1, \ldots, \xi_k) \prod_\chi \frac{\sin^2 \pi n_\chi \xi_\chi}{n_\chi \sin^2 \pi \xi_\chi} d\xi \quad .$$

Owing to

$$F(0, \ldots, 0) = \sum_m f(m_1, \ldots, m_k) \quad ,$$

the proof now results from the following theorem.

THEOREM 68 [119]. If the function $F(\xi_1, \ldots, \xi_k)$ is bounded and continuous at the point $(0, \ldots, 0)$, then the expression (11) is convergent to

$$F(0, \ldots, 0)$$

as $n_\chi \longrightarrow \infty$.

PROOF. Let δ be a sufficiently small number > 0. We divide the interval (6) into the small interval

(12) $$-\delta < \xi_\chi < \delta \, , \qquad \chi = 1, \ldots, k \quad ,$$

and into the "exterior" of this interval which we shall denote by $\mathfrak{A}$. We use the fact that for arbitrary n

(13) $$\int_{-1/2}^{1/2} \frac{\sin^2 \pi n \xi}{n \sin^2 \pi \xi} d\xi = 1 \quad ,$$

and therefore

$$\int_J \prod_\chi \frac{\sin^2 \pi n_\chi \xi_\chi}{n_\chi \sin^2 \pi \xi_\chi} d\xi = 1 \quad .$$

Now

$$\left(\int_{-1/2}^{-\delta} + \int_{\delta}^{1/2} \right) \frac{\sin^2 \pi n \xi}{n \sin^2 \pi \xi} d\xi$$

$$\leq \frac{1}{n \sin^2 \pi \delta} \int_{-1/2}^{1/2} \sin^2 \pi n \xi \, d\xi \leq \frac{1}{n \sin^2 \pi \delta} \quad ,$$

and therefore because of (13)

$$(14) \qquad \lim_{n\to\infty} \int_{-\delta}^{\delta} \frac{\sin^2 \pi n\xi}{n \sin^2 \pi\xi}\, d\xi = 1 \quad .$$

Hence also

$$(15) \qquad \lim_{n_\chi \to\infty} \int_{-\delta}^{\delta} \prod_\chi \frac{\sin^2 \pi n_\chi \xi_\chi}{n_\chi \sin^2 \pi\xi_\chi}\, d\xi = 1 \quad .$$

We now divide the integral (11) into the two terms

$$(16) \qquad \int_{-\delta}^{\delta} + \int_{\mathfrak{R}} \quad .$$

Denoting a bound of F by G, the second term taken absolutely is

$$\leqq G \int_{\mathfrak{R}} \prod_\chi \frac{\sin^2 \pi n_\chi \xi_\chi}{n_\chi \sin^2 \pi\xi_\chi}\, d\xi \quad ,$$

and because of (14) and (15), it is convergent to zero as $n_\chi \longrightarrow \infty$. Because of the continuity of the function $F(\xi_1, \ldots, \xi_k)$ at the point $(0, \ldots, 0)$, and by taking (14) into consideration, the first sum in (16), uniformly in n_χ, differs for sufficiently small δ arbitrarily little from

$$F(0, \ldots, 0) \int_{-\delta}^{\delta} \prod_\chi \frac{\sin^2 \pi n_\chi \xi_\chi}{n_\chi \sin^2 \pi\xi_\chi}\, d\xi \quad ,$$

and because of (15), it is convergent to $F(0, \ldots, 0)$ as $n_\chi \longrightarrow \infty$. Hence our theorem follows.

2. Consider now an arbitrary affine transformation

$$(17) \qquad y_\lambda = \sum_\chi a_{\lambda\chi} x_\chi \quad , \qquad \lambda = 1, \ldots, k \quad ,$$

with positive determinant

$$D = |a_{\lambda\chi}| \quad .$$

If a function $f(x_1, \ldots, x_k)$ is continuous and satisfies (9) for distant points, then (9) is also valid for the function

$$g(x_1, \ldots, x_k) = f(y_1, \ldots, y_k) \quad .$$

There is a connection between the quantities

$$(18) \qquad \varphi(n_1, \ldots, n_k) = \int f(y_1, \ldots, y_k) e\left(2\pi \sum_\chi n_\chi y_\chi\right) dy$$

and

$$\psi(t_1, \ldots, t_k) = \int g(x_1, \ldots, x_k) e\left(2\pi \sum_\chi t_\chi x_\chi\right) dx \quad ,$$

which readily shows if (17) is substituted in (18). In fact

$$\varphi(n_1, \ldots, n_k) = D \cdot \psi\left(\sum_\chi a_{\chi 1} n_\chi, \ldots, \sum_\chi a_{\chi k} n_\chi\right) \cdot$$

If by chance the function $\varphi(t_1, \ldots, t_k)$ vanishes outside of a finite interval, then by Theorem 67

$$D \cdot \sum_n \psi\left(\sum_\chi a_{\chi 1} n_\chi, \ldots, \sum_\chi a_{\chi k} n_\chi\right) = \sum_m f(m_1, \ldots, m_k) \quad .$$

If the inverse transformation to (17)

$$x_\chi = \sum_\lambda A_{\chi\lambda}\, y_\lambda, \qquad \chi = 1, 2, \ldots, k \quad ,$$

is now introduced, there results

$$(19) \qquad D \cdot \sum_n \psi\left(\sum_\chi a_{\chi 1} n_\chi, \ldots, \sum_\chi a_{\chi k} n_\chi\right)$$

$$= \sum_m g\left(\sum_\chi A_{1\chi} m_\chi, \ldots, \sum_\chi A_{k\chi} m_\chi\right) \quad ,$$

which is a generalization [120] of §10, (3).

3. The Poisson summation formula admits numerous applications, and its significance for analytic number theory is constantly rising [120]. A classic application is the deduction of the relations for the theta

functions in several variables, but we shall not discuss it. We bring up only one other application [121], of the formula (19), which is surprising.

If we set

$$g(x_1, \ldots, x_k) = \prod_\chi \left(\frac{\sin \pi x_\chi}{\pi x_\chi} \right)^2$$

then

$$\psi(t_1, \ldots, t_k) = \prod_\chi \left(1 - |t_\chi| \right) \quad ,$$

provided all t_χ belong to the interval

$$0 \leqq |t_\chi| < 1 \quad , \qquad \chi = 1, 2, \ldots, k \ .$$

For all other $(t_1, \ldots, t_k)$, $\psi = 0$. For the special value $m_1 = \ldots = m_k = 0$, the right term of (19) has the value 1. All other terms are in any case non-negative.

Because

$$(20) \qquad g(0, \ldots, 0) = 1 \quad ,$$

the right side of (19) has a value $\geqq 1$. Further

$$(21) \qquad \psi(0, \ldots, 0) = 1 \ .$$

If therefore $0 < D < 1$, then the expression

$$\psi \left(\sum_\chi a_{\chi 1} n_\chi, \ \ldots, \ \sum_\chi a_{\chi k} m_\chi \right)$$

must be different from zero for a system of integers $(n_1, \ldots, n_k)$ which do not vanish completely or what is equivalent to it, that for these n_χ

$$\left| \sum_\chi a_{\chi \rho} n_\chi \right| < 1 \quad , \qquad \rho = 1, 2, \ldots, k \ .$$

But now let D have exactly the value 1. For $1/2 < \lambda < 1$, form the numbers

$$(22) \qquad \begin{aligned} a_{\chi 1}^{(\lambda)} &= \lambda a_{\chi 1} \\ a_{\chi \rho}^{(\lambda)} &= a_{\chi \rho} \quad , \end{aligned} \qquad \rho = 2, 3, \ldots, k \ .$$

The determinant D_λ belonging to it lies in the interval $[1/2, 1]$. Hence by the proofs just given, there are integers $n_\chi^{(\lambda)}$ which do not all vanish, and numbers $\Theta_\rho^{(\lambda)}$ in the interval $(0, 1]$ such that

$$(23) \qquad \sum_\chi a_{\chi\rho}^{(\lambda)} n_\chi^{(\lambda)} = \Theta_\rho^{(\lambda)}, \qquad \rho = 1, 2, \ldots, k \quad .$$

But because for variable λ, the determinant D_λ is bounded below and all coefficients of (23) are bounded above, there is by the theory of linear equations, a bound K independent of λ for which

$$|n_\chi^{(\lambda)}| \leqq K \quad .$$

Since moreover the $n_\chi^{(\lambda)}$ are integers, then by the "bureau drawer" principle, for each sequence of values λ which converge to 1, the corresponding set of number systems $(n_1^{(\lambda)}, \ldots, n_k^{(\lambda)})$ must contain one which occurs infinitely often. We call it $(n_1, \ldots, n_k)$. For this system, we now have

$$(24) \qquad \begin{aligned} \left|\sum_\chi a_{\chi 1} n_\chi\right| &\lneqq 1 \\ \left|\sum_\chi a_{\chi\rho} n_\chi\right| &< 1 \quad , \end{aligned} \qquad \rho = 2, \ldots, k \quad .$$

we have therefore gained the following theorem.

> To each quadratic matrix of real numbers $a_{\chi\rho}$ with the determinant $+1$, there is a system of integers $(n_1, \ldots, n_k)$ which do not all vanish, for which the inequalities (24) hold.

Instead of the index $\rho = 1$, one naturally can designate any index $\rho = 2, 3, \ldots, k$ in the system (24).

A matrix $|a_{\chi\rho}|$ is called an exceptional matrix, if

$$\left|\sum_\chi a_{\chi\rho} n_\chi\right| \geqq 1 \quad ,$$

for each system of integers n_χ not simultaneously vanishing for at least one $\rho = \rho(n_\chi)$.

We consider the relation (19) for the matrix $|a_{\chi\rho}|$ itself, and not for the auxiliary matrix $|a_{\chi\rho}^{(\lambda)}|$; we then set $D = 1$. Our matrix is then and only then an exceptional matrix, if on the left of (19) the member (21) only is different from zero. And for this it is again necessary and sufficient that only the term (20) be different from zero on the right of (19). And for this it is again necessary and sufficient that for each system of integers n_χ which do not all vanish, there be a $\rho = \rho(n_\chi)$ for which

$$\sum_{\chi} A_{\chi\rho} n_\chi$$

is an integer different from zero.

APPENDIX

CONCERNING FUNCTIONS OF REAL VARIABLES

We have need of concepts and theorems about functions of real variables which cannot as yet be found in general textbooks on infinitesimal calculus. For this reason we shall assemble sundry information about this matter, and use as a basis, the text of C. Carathéodory, Vorlesungen über reelle Funktionen, 2nd Edition, 1927 (quoted in what follows by "Carath").

MEASURABILITY

1. Let A be a given point set in the k dimensional Cartesian space $\mathfrak{R}_k$, $(k = 1, 2, 3, \ldots)$. (In the sense of Lebesgue), the set is either measurable or non-measurable. In the first case, a well determined measure mA, is assigned to it. For this measure, a non-negative number with the inclusion of 0 and $+\infty$ is specified. If $mA = 0$, we say that the set is a null set. In place of the expression, "on the point set B with the exception of a null set", we may also say, "almost everywhere in B". In order to operate with null sets, it is essential that the union of finitely or countably infinitely many null sets be again a null set as they are.

2. If an affine transformation

$$\overline{x_\chi} = a_{\chi 0} + \sum_{\lambda=1}^{k} a_{\chi\lambda} x_\lambda \quad , \qquad \chi = 1, 2, \ldots, k \quad ,$$

is carried out, each measurable point set goes over into a measurable point set, and the measure number is multiplied by the absolute value of the transformation determinant. It is evident therefore that null sets go over into null sets (Carath. §335, Theorem 2).

3. On a measurable point set A in the $(x_1, \ldots, x_k)$ space, let us erect in the $(x_1, \ldots, x_k; y_1)$ space, the cylinder

$$- 1 < y_1 < 1 \quad .$$

This cylinder is again measurable (with the measure equal to $2 \cdot m(A)$); cf. Carath. §375, Theorem 1. By the transformation $\bar{x}_\chi = x_\chi$; $\bar{y}_1 = Ny_1$ and by letting $N \longrightarrow \infty$ (the limit set of measurable point sets is also measurable), we obtain the result that the cylinder A_1:

$$- \infty < y_1 < \infty$$

is also measurable. By induction from ℓ to $\ell + 1$, we obtain finally that the cylinder A_ℓ

$$- \infty < y_\lambda < \infty \qquad \lambda = 1, 2, \ldots, \ell$$

erected on the base A in the space $(x_1, \ldots, x_k; y_1, \ldots, y_\ell)$, is likewise measurable. If A was a (k-dimensional) null set, then A_ℓ is also a $(k+\ell)$-dimensional null set.

4. Let $f(x_1, \ldots, x_k)$ be an arbitrary function defined on a measurable point set E of $\mathfrak{R}_k$. For the present, we envisage real numbers as values of the function in the interval $- \infty \leq f \leq + \infty$. To distinguish between measurable and non-measurable functions, it is sufficient for example for the measurability of f, that for each real a, each subset M_a of E for which

$$f(x_1, \ldots, x_k) > a$$

be measurable. In conjunction with 3 and 2, the following then easily results.

If $f(x_1, \ldots, x_k)$ is measurable in $\mathfrak{R}_k$, then it is also measurable as a function of the $(x_1, \ldots, x_k; y_1, \ldots, y_\ell)$-space. If $\ell = k$, then by 2, the function

$$f(y_1 - x_1, y_2 - x_2, \ldots, y_k - x_k)$$

is also measurable in $\mathfrak{R}_{k+k}$.

5. Sums, differences, and products of measurable functions, and (by the characterization given in 4) the square root of a non-negative measurable function, are also measurable. We use these results to arrive at the following.

If f and g are measurable functions, then

$$\sqrt{f^2 + g^2}$$

is also a measurable function (hence in the special case $g = 0$, the function $|f|$) is measurable.

If f_1 and f_2 are measurable functions in $\mathfrak{R}_k$, then

$$f_1(x_1, \ldots, x_k) \cdot f_2(y_1 - x_1, \ldots, y_k - x_k) \tag{1}$$

is measurable in $\mathfrak{R}_{k+k}$.

Let $f(x)$ be measurable on E. For real c, we define $f_c(x)$ as follows. Let $f_c(x) = f(x)$ for those points for which $f(x) \leqq c$ and let $f_c(x) = c$ for those points for which $f(x) > c$. Then $f_c(x)$ is also measurable, as is evident from the relation

$$f_c = \frac{f + c}{2} - \left|\frac{f - c}{2}\right| .$$

A similar conclusion is valid if, in the definition of f_c the ">" sign is replaced by the "<" sign. More generally if for $a < b$, f_{ab} is defined so that $f_{ab} = f$ for those points for which $a \leqq f \leqq b$, and $f_{ab} = a$ or b for those points for which $f < a$ or $f > b$, then f_{ab} is again measurable.

6. A definition due to Carath. is the following. Two functions f and φ defined on the same point set are called "equivalent" if they agree almost everywhere. Sums, products and limit functions of equivalent functions are again equivalent. Equivalent functions are simultaneously measurable. By 3-5 we arrive at the following. If the functions f_1 and f_2 defined in $\mathfrak{R}_k$ are equivalent to the functions φ_1 and φ_2, then (1) is equivalent to

$$\varphi_1(x_1, \ldots, x_k) \cdot \varphi_2(y_1 - x_1, \ldots, y_k - x_k) .$$

SUMMABILITY

7. Among the functions measurable on A, the ones which are summable play a special role. In the text we have avoided the term "summable", and used instead the more familiar name "integrable". We now, however, make the following observation. If the point set is bounded, then the two terms have exactly the same meaning. But if A is unbounded, then "summable" and "absolutely integrable" mean the same. Hence the term not absolutely integrable which occurs frequently in the text does not fall under the concept "summability" as used by Carath. To preserve the restricted usage of Carath., we shall hereafter in this appendix use the term summability. The following theorems are valid (Carath. §414-445).

1) To each function f which is summable on E, there corresponds a finite number

$$\int_E f \, dx$$

which is called the integral of f over E. (If $k \geq 2$, then the dx under the integral sign is to be replaced by $dx_1 \ldots dx_k$.) If f is measurable and bounded (for example if $f = c$), and E has a finite measure, then f is summable and

$$\int_E c \, dx = c \int_E dx = c \cdot mE \quad .$$

2) Equivalent functions are simultaneously summable, and yield the same values for the integral. Among the functions equivalent to a summable one, there are always functions which are finite (at each point of E).

3) In order that f be summable, it is necessary and sufficient that f itself be measurable and $|f|$ be summable. Hence

$$\left| \int_E f \, dx \right| \leq \int_E |f| \, dx \quad . \tag{2}$$

4) If for the summable function f, an equivalent one exists whose values lie in the interval (g, G), then

$$g \cdot mE \leq \int_E f \, dx \leq G \cdot mE \quad , \tag{3}$$

(Carath. §406, Theorem 3). In particular if $f \geq 0$ and measurable, then

$$\int_E f \, dx \geq 0 \quad . \tag{4}$$

5) If f_1 and f_2 are finite and summable functions, and λ_1 and λ_2 are finite constants, then $\lambda_1 f_1 + \lambda_2 f_2$ is summable. Further

$$\int_E (\lambda_1 f_1 + \lambda_2 f_2) \, dx = \lambda_1 \int_E f_1 \, dx + \lambda_2 \int_E f_2 \, dx \quad .$$

If in particular, $f_2 = f - f_1$, then in conjunction with (4), the following results. If $f_1 \leq f$ and f_1 and f are summable, then

$$\int_E f_1 \, dx \leq \int_E f \, dx \quad .$$

A similar conclusion is valid if "<" is replaced by ">". Hence it follows, if f and φ are summable and $|f| \leq \varphi$, that

$$\left| \int_E f \, dx \right| \leq \int_E \varphi \, dx \quad .$$

6) If f is finite and summable, and φ is bounded $(|\varphi| \leq G)$ and measurable, then $f\varphi$ is measurable. It also follows from the above that

$$(5) \qquad \left| \int_E f\varphi \, dx \right| \leq G \int_E |f| \, dx \quad .$$

This last relation is a generalization of (2) and (3).

7) If $f_1, f_2, f_3, \ldots$ are measurable functions and

$$|f_n| \leq \varphi \quad , \qquad n = 1, 2, 3, \ldots,$$

where φ is summable, and if the sequence of functions f_n is convergent, then

$$\lim_{n \to \infty} \int_E f_n \, dx = \int_E \lim_{n \to \infty} f_n \, dx \quad .$$

In place of the assumption that the sequence of functions f_n converges everywhere, it is sufficient to assume that it converges almost everywhere. For if at non-convergent points, the functions are replaced by zero values, then the new sequence converges everywhere, and one can now use 6 and 7.2).

8) The assumption 7) will be satisfied if

$$f_n = gh_n$$

where g is summable and the h_n are measurable and uniformly bounded in n and convergent almost everywhere. In this case, therefore

$$\lim_{n \to \infty} \int_E gh_n \, dx = \int_E g \lim_{n \to \infty} h_n \, dx \quad .$$

In particular, if E has a finite measure (say is bounded), then one can set $g = 1$ and obtain

$$\lim_{n \to \infty} \int_E h_n \, dx = \int_E \lim_{n \to \infty} h_n \, dx \quad .$$

9) If the functions f_n are summable and increase monotonically to f, and if the integral of f_n is bounded, then f is also summable. In this case

$$\lim_{n \to \infty} \int_E f_n \, dx = \int_E \lim f_n \, dx \quad .$$

10) Let the functions $f(\tau_1, \ldots, \tau_k)$ be summable in $\mathfrak{R}_k$. Divide the variables τ_χ arbitrarily into two groups, and call one $x_1, \ldots, x_m$; the other $y_1, \ldots, y_n$, $(m + n = k)$. Then there is a null set e of $\mathfrak{R}_n$ such that the repeated integral

$$\int_{\mathfrak{R}_n - e} dy \int_{\mathfrak{R}_m} f \, dx$$

exists. Its value is

$$\int_{\mathfrak{R}_{m+n}} f d\tau \quad .$$

If the variables τ_χ are divided arbitrarily into groups

$$z_\pi = (\tau_{\pi 1}, \tau_{\pi 2}, \ldots, \tau_{\pi k_\pi}) \qquad \pi = 1, 2, \ldots, p$$

$$k_\pi \geq 1, \quad k_1 + k_2 + \ldots + k_p = k \quad ,$$

then there is a function $\varphi(\tau_1, \ldots, \tau_k)$ which is equivalent to $f(\tau_1, \ldots, \tau_k)$ and for which the integral

$$(7) \qquad \int_{\mathfrak{R}_{k_1}} dz_1 \int_{\mathfrak{R}_{k_2}} dz_2 \cdots \int_{\mathfrak{R}_{k_p}} \varphi dz_p$$

exists; (Carath. §551 and §554 only formulates the case for

$$p = k; \quad k_1 = k_2 = \ldots = k_p = 1 \quad ,$$

but the proof follows literally for arbitrary k_π). The value of (7) is

$$\int_{\Re_k} \varphi d\tau ,$$

which in turn agrees again with

$$\int_{\Re_k} f d\tau .$$

Let the function $f(\tau_1, \ldots, \tau_k)$ be measurable in $\Re_k$. In order that it be also summable in $\Re_k$, the following requirements are sufficient: 1. that a function ψ exist which is equivalent to f, 2. that there be a majorant of $\psi (\varphi \geqq |\psi|)$, possibly the function $\varphi = |\psi|$ itself, and 3. that there be a variable grouping z_π of such a kind that the integral (7) exists. This important collection of theorems, which goes back to Fubini, is the basis for the operations with integrals. We have frequently referred to it in the text, especially for the proof of the Faltung rule. In the proof of the Faltung rule, parts 5 and 6 (of this appendix) are employed in addition.

DIFFERENTIABILITY

8. Let $f(x)$ be representable, on a finite or infinite linear interval $[a, b]$, in the form

$$f(x) = c + \int_a^x f_1(\xi)\, d\xi ,$$

where $f_1(x)$ is measurable in $[a, b]$. The function $f(x)$ is continuous, and at almost all points of $[a, b]$, possesses a derivative in the usual sense which is equivalent to $f_1(x)$. In the text, we called such a function $f(x)$ simply <u>differentiable</u> and the function $f_1(x)$ its <u>derivative</u>. If among the functions equivalent to $f_1(x)$, there is a continuous one (there can be at most one such), then $f(x)$ is called continuously differentiable, and by the derivative of $f(x)$ we mean only the continuous function $f_1(x)$. If this last function is differentiable (in the above sense), then $f(x)$ is called 2-times differentiable. It is easy to extend our meaning to functions $f(x)$ which we call r-times differentiable or r-times continuously differentiable.

If $f(x)$ and $g(x)$ are both differentiable in (a, b), then

$f(x)g(x)$ is also differentiable in (a, b), and the derivative of the last function is equivalent to $fg' + f'g$. The formula for partial integration

$$\int_a^b fg'\, dx = f(b)g(b) - f(a)g(a) - \int_a^b f'g\, dx$$

is also valid.

9. Let $\varphi(x_1, \ldots, x_k)$ be summable in $\mathfrak{R}_k$. Consider the quantity

$$F(\Delta) = \int_\alpha^\beta \varphi(x_1, \ldots, x_k)\, dx$$

for an arbitrary interval

$$\Delta : \alpha_\chi < x_\chi < \beta_\chi \qquad \chi = 1, 2, \ldots, k \quad .$$

The function $F(\Delta)$ is a totally continuous interval function (Carath. Chapt. IX). Hence the expression

$$\frac{1}{h_1 \ldots h_k} \int_0^h \varphi(x_1 + h_1, \ldots, x_k + h_k)\, dh \quad ,$$

say for $h_1 = h_2 = \ldots = h_k = h$, is convergent to $\varphi(x_1, \ldots, x_k)$ as $h \longrightarrow 0$ for almost all $(x_1, \ldots, x_k)$ (Carath. §446, Theorem 3).

APPROXIMATION IN THE MEAN

10. Let $f(x_1, \ldots, x_k)$ be summable in $\mathfrak{R}_k$. To each $\epsilon > 0$, one can determine a continuous function $\varphi(x_1, \ldots, x_k)$ which vanishes outside a finite interval, such that

$$\int |f - \varphi|\, dx < \frac{\epsilon}{2} \quad .$$

Since a continuous function of this kind can be approximated arbitrarily closely by a function which always has a constant value in finitely many intervals parallel to the axes, and otherwise vanishes (step function), there is to each $\epsilon > 0$, a step function ψ for which

$$\int |f - \psi|\, dx < \epsilon \quad . \tag{8}$$

11. Let f_n, $n = 1, 2, \ldots$ be summable in $\Re_k$. From

$$\lim_{\substack{m \to \infty \\ n \to \infty}} \int |f_m - f_n| \, dx = 0 \tag{9}$$

the existence of a summable function f follows for which

$$\lim_{n \to \infty} \int |f - f_n| \, dx = 0 \quad . \tag{10}$$

Any two functions f which satisfy (10) are equivalent to each other. If (10) holds, and the functions f_n converge to φ almost everywhere on a point set E, then f and φ are equivalent in E.

If the functions $f_n(x)$ are summable on E and as $n \longrightarrow \infty$ converge almost everywhere to a function $f(x)$ which is likewise summable, and if the f_n have a summable majorant say, then

$$\lim_{n \to \infty} \int_E |f - f_n| \, dx = 0 \quad . \tag{11}$$

11a. The proofs which we shall produce for the theorems in 10. and 11. utilize, to a considerable extent, the theories developed in Carath.

Let a sequence of functions $f_n(x)$ be given which satisfy (9). For any measurable point set E of $\Re_k$, we consider the quantities

$$J_n(E) = \int_E f_n \, dx \quad . \tag{12}$$

Hence

$$|J_m(E) - J_n(E)| \leqq \int |f_m - f_n| \, dx \quad ,$$

and the sequence (12) is therefore convergent as $n \longrightarrow \infty$. If we denote this limit by

$$J(E) \quad ,$$

we have the more precise result that to each ϵ, there exists an n such that

$$|J(E) - J_n(E)| \leqq \frac{\epsilon}{2} \quad , \tag{13}$$

for all point sets E.

We now make the following assertion. There is a summable function $f(x)$ such that, analogous to (12),

$$J(E) = \int_E f\, dx \quad . \tag{14}$$

In order that such a function $f(x)$ exist, it is necessary and sufficient that the quantity $J(E)$ have the following two properties (cf. Carath., Chapter IX):

α) $J(E_1 + E_2) = J(E_1) + J(E_2)$

β) to each $\epsilon > 0$, there is a $\delta > 0$ such that

$$|J(E)| \leqq \epsilon \quad \text{for} \quad mE \leqq \delta \quad .$$

Property α) is verifiable immediately. Because of the summability of $f_n(x)$, the property β) applies to the quantity $J_n(E)$, and because of (13), the property carries over from $J_n(E)$ to $J(E)$. - We shall show that the function $f(x)$ thus obtained satisfies (10). By (12) and (14), we have

$$\lim_{n \to \infty} \int_E f_n\, dt = \int_E f\, dt \quad . \tag{15}$$

In particular, if any interval

$$x_\chi \leqq t_\chi \leqq x_\chi + h_\chi \qquad \chi = 1, 2, \ldots, k$$

is taken for E, and the functions

$$F_{nh} = \frac{1}{h_1 \ldots h_k} \int_0^h f_n(x_1 + \xi, \ldots, x_k + \xi_k)\, d\xi$$

$$F_h = \frac{1}{h_1 \ldots h_k} \int_0^h f(x_1 + \xi_1, \ldots, x_k + \xi_k)\, d\xi \tag{16}$$

introduced, then

$$\lim_{n \to \infty} F_{nh}(x_1, \ldots, x_k) = F_h(x_1, \ldots, x_k) \quad . \tag{17}$$

With the aid of the theorem of Fubini, we obtain

$$\int |F_{mh} - F_{nh}| \, dx \leqq \frac{1}{h_1 \cdots h_k} \int_0^h d\xi \int |f_m(x_1 + \xi_1, \ldots, x_k + \xi_k) - f_n| \, dx$$

$$\leqq \int |f_m - f_n| \, dx \quad .$$

If we now let $m \longrightarrow \infty$, we have that to each ϵ, there is an n such that the function

$$G_h(x_1, \ldots, x_k) = |F_h - F_{nh}| \tag{18}$$

satisfies

$$\int G_h \, dx \leqq \epsilon \quad , \tag{19}$$

for all combinations of the quantities $h > 0$. If by chance $h_1 = \ldots = h_k = h$, then the non-negative function G_h converges to the function

$$G = |f - f_n|$$

for almost all x as $h \longrightarrow 0$, and (10), which must be still proved, will have been attained when we show that the limit of (19) is

$$\int G \, dx \leqq \epsilon \quad . \tag{20}$$

By 7, 8), this is certainly true if all the functions G_h vanish outside of an interval

$$-p \leqq x_\chi \leqq p \quad , \tag{21}$$

and inside the interval satisfy the common relation

$$G_h \leqq k \quad .$$

In the general case, we change the function G_h by this means to a function G_{hp} which is replaced by zero values outside of (21), and by the value p everywhere inside (21) where it exceeds the value p. One sees immediately that the limit function

$$G_p = \lim_{h \to 0} G_{ph} \quad ,$$

which now satisfies (20), increases monotonically to G as $p \longrightarrow \infty$. By 7, 9) therefore

$$\lim_{p \to \infty} \int G_p \, dx = \int G \, dx \leq \epsilon$$

Q.E.D.

From

$$\int |f - g| \, dx \leq \int |f - f_n| \, dx + \int |g - f_n| \, dx \quad ,$$

it is evident that there can be only one function f which satisfies (10).

If (10) holds and the functions f_n converge to φ almost everywhere on E, then the functions $f - f_n$ are convergent to $f - \varphi$ almost everywhere. If one sets

$$G_h = |f - f_{1/h}| \quad ,$$

the above mentioned corollary concerning G_h results, namely that if

$$\int_E G_h \, dx \leq \epsilon \tag{22}$$

then

$$\int_E |f - \varphi| \, dx \leq \epsilon \quad . \tag{23}$$

Because of (10), the quantity ϵ in (22) can be taken arbitrarily small. Hence (23) holds for $\epsilon = 0$, and f and φ are therefore equal to each other almost everywhere on E. In a similar manner, one proves the last assertion in 11.

For the proof of 10. one first introduces the function $f_n(x)$ which vanishes outside the interval $-n \leq x_\chi \leq n$, has the value n where $f(x) \geq n$, has the value $-n$ where $f(x) \leq -n$, and otherwise agrees with $f(x)$. It is evident that

$$|f(x) - f_n(x)| \leq |f| \quad ,$$

and on the other hand that

$$\lim_{n \to \infty} |f(x) - f_n(x)| = 0 \quad ,$$

for almost all x.

By 7,7), therefore, for each ϵ there is an n such that

$$\int |f - f_n| \, dx \leqq \frac{\epsilon}{4} \quad . \tag{24}$$

We now form the functions F_{nh} in accordance with (16) and with them, the functions

$$G_h = |f_n - F_{nh}| \qquad h \leqq h_0 \quad .$$

The functions G_h vanish outside of an interval independent of h. Inside these intervals they are uniformly bounded, and moreover, for almost all x,

$$\lim_{h \to 0} G_h = 0 \quad .$$

By 7, 8), there is therefore a suitable h such that

$$\int G_h \, dx \leqq \frac{\epsilon}{4} \quad . \tag{25}$$

From (24) and (25), it follows for the function

$$\varphi = F_{nh} \quad ,$$

continuous and vanishing outside of a finite interval, that

$$\int |f - \varphi| \, dx \leqq \frac{\epsilon}{2} \quad .$$

COMPLEX VALUED FUNCTIONS

12. Let

$$f(x_1, \ldots, x_k) = f_1(x_1, \ldots, x_k) + if_2(x_1, \ldots, x_k)$$

be a given function on E where f_1 and f_2 are real valued. We call f summable or measurable, if f_1 and f_2 are summable or measurable. From

$$|f| \leqq \sqrt{f_1^2 + f_2^2} \leqq |f_1| + |f_2|$$

it follows that $|f|$ is also summable or measurable.

The observations of 1 to 7, 3) carry over fairly easily. Only the relation

$$\left|\int_E f\, dx\right| \leqq \int |f|\, dx \tag{26}$$

is not trivial. The following however is trivial

$$\left|\int_E f\, dx\right| = \left|\int_E f_1\, dx + i\int_E f_2\, dx\right| \leqq \int_E |f_1|\, dx + \int_E |f_2|\, dx \leqq 2\int_E |f|\, dx \quad .$$

For the proof of the more exact relation (26), it may be assumed that $E = \Re_k$. If it were not, f can be extended outside of E by zero values. If f is constant on an interval A, and otherwise vanishes, then the correctness of (26) is evident. Further, if the validity of (26) is known for two functions f and g, each of which vanishes wherever the other is different from zero, then (26) is readily seen to hold for the sum $f + g$. Hence (26) holds for each step function. For each ϵ, a step function

$$\psi = \psi_1 + i\psi_2$$

can now be determined such that

$$\int |f_1 - \psi_1|\, dx \leqq \epsilon, \qquad \int |f_2 - \psi_2|\, dx \leqq \epsilon \quad .$$

From

$$\left|\int (f - \psi)\, dx\right| \leqq \int |f_1 - \psi_1|\, dx + \int |f_2 - \psi_2|\, dx \quad ,$$

and

$$\left|\int (|f| - |\psi|)\, dx\right| \leqq \int |f - \psi|\, dx \quad ,$$

we obtain, by letting $\epsilon \longrightarrow 0$, the validity of (26) for each complex valued function f.

Everything which follows thereafter carries over without difficulty to complex valued functions.

EXTENSION OF FUNCTIONS

13. Let two points β, γ be given on the α axis, and numbers b_ρ, c_ρ; $\rho = 0, 1, \ldots, r - 1$. Then there is an analytic function in (β, γ), actually a polynomial $P(\alpha)$ for which

$$P^{(\rho)}(\beta) = b_\rho, \quad P^{(\rho)}(\gamma) = c_\rho; \qquad \rho = 0, 1, \ldots, r-1 .$$

See for instance A. A. Markoff, Differenzenrechnung, Leipzig, 1896.

14. Let $\Gamma(\alpha)$ be a given $(r-1)$-times continuously differentiable function in an interval (β, γ), and β', γ' given numbers for which $\beta' < \beta$, $\gamma < \gamma'$. By 13. there is in (β', β) an analytic function $\Gamma_\beta(\alpha)$ which together with its $r-1$ derivatives, vanishes for $\alpha = \beta'$, and for $\alpha = \beta$ joins continuously with $\Gamma(\alpha)$. Similarly there is in (γ, γ') an analytic function $\Gamma_\gamma(\alpha)$ which has an analogous behavior for the points γ' and γ. If to the function pieces Γ_β, Γ, Γ_γ, one adds the function values zero on the intervals $[-\infty, \beta']$ and $[\gamma', +\infty]$, there results a function $G(\alpha)$ which is $(r-1)$-times continuously differentiable in the entire interval $[-\infty, \infty]$. If the function $\Gamma(\alpha)$ is actually r-times differentiable (in the sense of 8.) in (β, γ), then the function $G(\alpha)$ is also in $[-\infty, \infty]$.

15. In particular, if the function $\Gamma(\alpha)$ has the constant value c, then it is even possible to make the functions Γ_β and Γ_γ monotonic, say by selecting

$$\Gamma_\beta = ce^{-\left(\frac{x-\beta'}{\beta-x}\right)^r}, \qquad \Gamma_\gamma = ce^{-\left(\frac{x-\gamma'}{x-\gamma}\right)^r} .$$

In this case, one can even construct the functions Γ_β and Γ_α as monotonic function pieces in such a way that the resulting function $G(\alpha)$ has derivatives of arbitrarily high order. However we shall not discuss this question further.

16. If two functions $P(\alpha)$ and $Q(\alpha)$ are given, each r-times differentiable on two half lines $[-\infty, \beta')$ and $(\gamma', +\infty]$, then by 13. they can be extended to a function $H(\alpha)$ which is r-times differentiable in $[-\infty, \infty]$. One can even demand of the function $H(\alpha)$ that it differ from zero in (β', γ'), provided only $P(\beta') \neq 0$ and $Q(\gamma') \neq 0$. One then conceives this extension brought about by means of a polynomial $P(\alpha)$, possibly in accordance with 13. From the outset, $P(\alpha)$ can have finitely many zeros in $[\beta', \gamma']$. We shall now modify $P(\alpha)$ in such a way that the altered function is again r-times differentiable in the neighborhood of each of its zeros, but is different from zero there. For the sake of simplicity, we assume that a selected zero has the value $\alpha_0 = 0$. In a definite neighborhood which contains no further zeros of $P(\alpha)$, one can write

$$P(\alpha) = c\alpha^p(1 + f(\alpha) + ig(\alpha)) \qquad p > 0 ,$$

where $f(\alpha)$ and $g(\alpha)$ are polynomials with real coefficients and $f(\alpha)$ is taken absolutely < 1 on a subinterval

$$|\alpha| \leq \alpha_1 (< 1/2) \tag{27}$$

of our neighborhood. We add to $P(\alpha)$ that function which in (27) has the value

$$ci(\alpha_1^2 - \alpha^2)^{r+1} ,$$

and vanishes otherwise. The extended function is r-times continuously differentiable, and apart from a factor c, can be written in (27) as

$$\alpha^p(1 + f(\alpha)) + i[\alpha^p g(\alpha) + (\alpha_1^2 - \alpha^2)^{r+1}] .$$

Since the real part vanishes only for $\alpha = 0$, while the imaginary part is $\neq 0$, the function has no zeros in (27).

SUMMATION OF REPEATED INTEGRALS

17. If an integrable function $\varphi(\alpha)$, defined in $[0, \infty]$ approaches a limit φ as $\alpha \longrightarrow \infty$, then the function

$$\Phi(\alpha) = \frac{1}{\alpha}\int_0^\alpha \varphi(\beta)\, d\beta \tag{28}$$

also converges to φ as $\alpha \longrightarrow \infty$. This fact is recognized without difficulty, if the substitution

$$\varphi(\alpha) = \varphi + \epsilon(\alpha) \qquad \epsilon(\alpha) \longrightarrow 0$$

is considered in (28).

18. Let an integrable function

$$\varphi(\alpha_1, \ldots, \alpha_k) \tag{29}$$

be given in the octant

$$0 < \alpha_\chi < \infty \qquad \chi = 1, \ldots, k ,$$

which converges to a limit φ as $\alpha_\chi \longrightarrow \infty$. The function

$$\Phi(\alpha_1, \ldots, \alpha_k) = \frac{1}{\alpha_1 \cdots \alpha_k} \int_0^{\alpha} \varphi(\beta_1, \ldots, \beta_k) \, d\beta \tag{30}$$

does not then need to converge as $\alpha_\chi \longrightarrow \infty$ (provided $\varphi(\alpha_1, \ldots, \alpha_k)$ is not bounded in the entire octant). But if it does converge, then its limit has the same value φ. I have produced a proof of this statement in a paper in the Mathematischen Zeitschrift 35, 122/6, 1932. It is true that the summation mean value

$$\Phi_{n_1 \ldots n_k} = \frac{1}{n_1 \cdots n_k} \sum_{\nu_\chi = 0}^{n_\chi - 1} \varphi(\nu_1, \ldots, \nu_k)$$

is treated instead of the integral mean value (30). However the argument therein carries over very easily to (30).

Now let $\psi(\gamma_1, \ldots, \gamma_k)$ be a function defined and integrable in the whole γ_χ-space, and let the integral

$$\varphi(\beta_1, \ldots, \beta_k) = \int_{-\beta}^{\beta} \psi(\gamma_1, \ldots, \gamma_k) \, d\gamma \tag{31}$$

have a limit as $\beta_\chi \longrightarrow \infty$. If we insert (31) in (30), we obtain

$$\int_{-\alpha}^{\alpha} \prod_{\chi} \left(1 - \left| \frac{\beta_\chi}{\alpha_\chi} \right| \right) \psi(\beta_1, \ldots, \beta_k) \, d\beta \quad . \tag{32}$$

If therefore (32) also approaches a limit, then the two limits agree.

REMARKS - QUOTATIONS

1. A basic historical-genetic introduction to the theory of trigonometric integrals appears in the Enzyklopädie der mathematischen Wissenschaften, Article II A12, by H. Burkhardt entitled, Trigonometrische Reihen und Integrale (to about 1850). In this connection, Chapt. V especially, "Das Fouriersche Integral" should be considered. We shall quote this fundamental original work in what follows, briefly by

"Burkhardt"

along with pages or formula number. In particular, we shall refer to "Burkhardt" in cases of remarkable special integrals because of its source references. Prior to the encyclopedia article, the same author initially made a very comprehensive collection of material in the Jahresbericht der Deutschen Mathematiker-Vereinigung X, 1 and 2, 1909, under the title: Entwicklungen nach oszilierenden Funktionen und Integration der Differentialgleichungen der mathematischen Physik, which also considered exhaustively, amongst others, the occurrence of Fourier integrals in physics. Yet this report is still not arranged systematically from a mathematical point of view, and moreover because of its considerable size not easy to use.

The oldest textbook concerning Fourier integrals (and in certain respects the only one up to year 1931) is O. Schlömlich, Analytische Studien, second addition, Leipzig, 1848.

A remarkable contribution to the story of Fourier integrals is given by A. Pringsheim, especially to the question regarding the extent to which the name of Fourier integrals can be legitimately associated with J. J. Fourier in connection with the Fourier integral theorem, Jahresbericht der Deutschen Mathematiker-Vereinigung 16, 2-16, 1907, and in connection with the validity of the limit of the Fourier integral formula, Math. Ann. 68, 307-408, 1910. We shall quote the last work by

"Pringsheim I"

It contains the first exact criteria for the validity of the Fourier integral formula and the Fourier integral theorem. It was amplified later in a supplement which appeared in Math. Ann. 71, 289-298, 1912. We quote it in what follows by

"Pringsheim II"

The results of Pringsheim, and later additional generalizations by other writers are completely reproduced in the textbook of L. Tonelli, Serie trigonometriche, Bologna 1928. We quote it hereafter, especially pp. 402-433, by

"Tonelli"

We still name in addition the book by E. W. Hobson, Theory of Functions of a Real Variable, 2nd Edition, Vol. 2, 1926, quoted hereafter by

"Hobson"

-, especially Chapter X, where the theory of Plancherel transformations and summation theories are treated at great length.

Long before Pringsheim, the first precise result concerning the validity of the Fourier integral formula,- and indeed concerning "summability" instead of direct convergence was advanced by A. Sommerfeld in his dissertation: Über die willkürlichen Funktionen in der mathematischen Physik, Königsberg, 1891.

The book, Fourier Integrals for Practical Application, by George A. Campbell and Ronald M. Foster, Bell Telephone Lab., New York, 1931, contains a very varied selection of definite Fourier integrals.

At the proper place, we shall quote additional citations.

2. Cf. Burkhardt, p. 1085, 6. The complex manner of writing originated with A. Cauchy, Burkhardt, p. 1086.

3. This theorem is known under the name of the Riemann-Lebesgue lemma. The theorem has been proved for Riemann integrable functions by B. Riemann, Über die Darstellbarkeit einer willkürlichen Funktion durch eine trigonometrische Reihe, Works, 2nd Edition, p. 253-255, and for general functions by H. Lebesgue, Lecons sur les séries trigonométriques, Paris, 1906, p. 61.

4. The integrals (4) are, "the only case which can be settled by the direct calculation of the indefinite integral and the insertion of the limit," cf. Burkhardt, p. 1098.

5. Pringsheim I.

6. The name, "transform" goes back in the last analysis to Cauchy, see Burkhardt, p. 1098, although Cauchy speaks of the "reciprocal" function.

7. H. Lebesgue, Bulletin de la Société mathématique de France, 38, 184 to 200, 1910.

8. If a function is of bounded variation on a finite or infinite interval, then it can be represented as the difference of two bounded monotonic functions. It is customary to consider this problem in the theory of real functions only in the case of a finite interval; nevertheless the case where the interval is infinite is very easily obtained for it, cf. for example Tonelli, p. 411, footnote.

9. The theorem stems from G. Darboux, Mémoire sur l'approximation des Fonctions de très grands nombres, Journ. de Math. 3rd Series, 4, 5-56, 1878, - cf. say the representation by Tonelli, pp. 226-228. - For more general theorems, cf. say A. Haar, Über asymptotische Entwicklung von Funktionen, Math. Ann. 96, 69-107, 1927; in addition the works of Du Bois - Reymond and Hardy, quoted under [20].

10. Cf. say Ch.-J. de la Vallée Poussin, Cours d'analyse, sixth edition, Vol. II, 1928, p. 30.

11. Formula (3) stems from J. Fr. Pfaff and L. Mascheroni, cf. Burkhardt, p. 1123. - Concerning the different methods for the evaluation of this formula, cf. A. Berry, Messenger of Mathematics, Series 2, 37, 61-2, 1907, and E. L. Nanson, the same, 113-4.

12. Formula (5) first appears in G. Bidone and J. J. Fourier, formula (8) in Fourier, cf. Burkhardt, p. 963.

In regard to the discontinuity factor, cf. [64].

13. Formulas (13) and (14) stem originally from Ph. Kelland and J. Dienger, cf. Burkhardt, p. 1121.

14. G. H. Hardy, Proceedings of the London Mathematical Society, Series (1), 23, 16-40, and 55-91, 1901/2; 35, 81-107, 1902/3; Series (2), 7, 181-208, 1908/9.

15. Formula (3) and more general ones are due to Cauchy.

16. Formulas (4) and (5) are from L. P. Gilbert, Mémoires couronnées de l'Académie R. des Sciences de Bruxelles, XXX 1, 1-52, 1861.

17. For literature, cf. Article II C 11 by E. Hilb and O. Szasz, pp. 1239/43 in the Enzyklopädie der mathematischen Wissenschaften, and Hobson, Chapter VII.

18. For the naming after L. Fejér, cf. [42].

19. H. Lebesgue, cited in [3], p. 12/14. - The same, p. 95-96, for the proof also of Theorem 4 and indeed for a function which vanishes outside of a finite interval.

20. This basic formula is due to P. G. Lejeune-Dirichlet and is named after him, cf. Burkhardt, p. 1036/8, - P. du Bois-Reymond and G. H. Hardy have investigated the Dirichlet integral for the case where the function $f(x)$ itself showed singularities at $x = 0$.

21. Pringsheim I

22. Announced in Pringsheim II.

23. The "Wiener formula" in the general wording of Theorem 9 is apparently new. Only formula (13) occurs in Wiener - cf. N. Wiener, Math. Zeitschrift 24, 575-616, 1926 - and the corresponding formula with the kernel

$$\frac{\sin^4 nx}{n^3 x^4} \quad \text{instead of} \quad \frac{\sin^2 nx}{nx^2}$$

cf. N. Wiener, Journal of Mathematics and Physics, 5, 99-121, 1926, in particular §1. Our proofs of Theorem 9 are a generalization of a proof by Wiener for formula (13) in N. Wiener, Journal of the London Mathematical Society 2, 118-123, 1927. A note of S. Bochner and G. H. Hardy, the same 1, 240-244, led the way to the last note, in which the Wiener formula was expressed as such for the first time. In Wiener it occurred only implicitly. M. Jacob, the same 3, 182-187, 1928, gave a generalization of formula (13) in another direction.

24. To name the formula exclusively after S. D. Poisson is hardly justifiable, cf. Burkhardt, p. 1339/42.

25. For Theorems 10 and 10a cf. say J. R. Wilton, Journ. of the London Mathematical Society, 5, 276-280, 1930.

26. By Cauchy; Burkhardt, p. 1341.

27. Cf. O. Schlömilch, Beitrage zur Theorie der bestimmten Integrale, Jena 1843, p. 20. For a previous similar formula by Pfaff, cf. Burkhardt, p. 939.

28. By Poisson, cf. Burkhardt, p. 1341, formula (1713).

29. By Poisson, cf. Burkhardt, p. 1130/31. For an evaluation,

cf. say Schlömilch, cited in [27], §6 ff.

30. Formula (23) is by C. J. Malmsten, cf. Burkhardt, p. 942, formula (451).

31. Pringsheim I. - Our assumptions 1) and 2) are indispensable. There are very beautiful criteria by Pringsheim, H. W. Yuong, H. Hahn among others, for which we refer to the representation by Tonelli, where additional literature is also quoted.

32. Cf. citation in [6].

33. P. S. de Laplace gave Formula (12), cf. Burkhardt, p. 1100/01.

34. Cf. Burkhardt, p. 115.

35. For a proof of (5) and for allowing $k \longrightarrow 0$, cf. say G. F. Meyer, Vorlesungen über die Theorie der bestimmten Integrale, Leipzig, 1871, p. 164-188.

36. Formula (6) stems from Cauchy; Burkhardt, p. 1110/11.

37. Cf. Burkhardt, p. 1119/20.

38. This proof stems principally from Fourier, cf. Burkhardt, p. 1117.

39. Cf. the summarizing paper of G. H. Hardy and E. C. Titchmarsh, Quarterly Journal of Mathematics, Oxford Series 1, 196-231, 1930.

40. Instead of "Faltung", one can also say "composition". For the story of the concept cf. G. Doetsch, Überblick über Gegenstand und Methode der Funktionalanalysis, Jahresbericht der Deutschen Mathematiker-Vereinigung, 36, 1-30, 1927, especially from p. 19.

41. A generalization of the theorem by S. Pollard, Proceedings of the London Mathematical Society, Series (2), 25, 451-468, 1926; M. Jacob, Mathematische Annalen 97, 663-674, 1927, and A. Zygmund, the same 99, 562-589, 1928.

42. Cf. Sommerfeld, cited in [1]. The further special case $\varphi(\alpha) = 1 - |\alpha|$ for $|\alpha| < 1$ and $= 0$ for $|\alpha| > 1$, stems essentially from L. Fejér, Mathematische Annalen 58, 51-69, 1904, who applied this summation method to Fourier series. The summation for a general function $\varphi(\alpha)$ stems from G. H. Hardy, Transactions of the Cambridge Philosophical Society XXI, 431, 1912. Hardy's functions $\varphi(\alpha)$ are more special than ours, but in return, the result is applicable not only to functions $f(x)$ which are absolutely integrable as $x \longrightarrow \pm\infty$. In addition to the representation by Hobson in regard to Cesàro-summability for arbitrary exponents, cf. the work of M. Jacob, Bulletin International de l'Académie Polonaise de Cracovie, Series A: Sciences mathématiques, p. 40-74, 1926, and Mathematische Zeitschrift 29, 20-33, 1928; S. Pollard, Proceedings of

the Cambridge Philosophical Society, XXIII, 373-382, 1927.

43. Formula (24) stems from Cauchy, cf. Meyer cited in [35], p. 205-208.

44. By E. Catalan, Burkhardt, p. 1105.

45. Formula (29) and a generalization of it has been found by Dirichlet, cf. Burkhardt, p. 1114.

46. Formulas (30), (31) and (32) stem from Cauchy, Burkhardt, p. 850 to 851.

47. Cf. Meyer, cited in [35], p. 310-313.

48. Formula (14) stems from O. Schlömilch and A. de Morgan, cf. Burkhardt, p. 1145-47. - Our proof is in principle the one by G. H. Hardy, Proceedings of the London Mathematical Society, Series (1), 35, 81-103, 1902/03, especially p. 96. - For a function theory proof by L. Kronecker, see his Vorlesungen, Vol. 1, Bestimmte Integrale, 1894, p. 199-214.

49. Cf., for example, G. H. Hardy, Transactions of the Cambridge Philosophical Society, 21, 39-86, 1912.

50. By Laplace; Burkhardt, p. 1126-1127.

51. By Cauchy; Burkhardt, p. 1128, formula (951).

52. After A. M. Legendre; Burkhardt, p. 1142-43.

53. After O. Schlömilch; Burkhardt, p. 1154.

54. For formulas (13) and (14) cf. Meyer, cited in [35], p. 286-289 and p. 319-324.

55. By E. C. J. von Lommel; Watson, p. 48. The inversion integral is contained in a general integral formula of H. Weber and P. Schafheitlin; Watson, p. 389-415.

55a. By H. Weber; Watson, p. 405.

56. By L. Gegenbauer; Watson, p. 50, formula (3).

57. By N. J. Sonine and Gegenbauer; Watson, p. 415, formula (1).

58. By E. G. Gallop; Watson, p. 422.

59. Cf. Watson, p. 150, formula (1).

60. Formula (13) is contained in a general formula of S. Ramanujan; Watson, p. 449. Numerous other (not necessarily related to Bessel functions) Fourier integrals and inversions of such integrals stem from Ramanujan. These are distributed in various papers of his collected works (Collected Papers of S. Ramanujan, Cambridge, 1927). We shall meet several later in §35.

61. By F. G. Mehler and Sonine; Watson, p. 169-170.

62. By A. B. Basset; Watson, p. 172.

63. Formulas (19) and (20) go back to O. Heavisidie; Watson, p. 388.

64. Dirichlet's application of the discontinuity factors forms the source of this theorem, cf. §4, 4; for the evaluation of certain definite integrals, cf. Burkhardt, p. 1321-1322. Formula (7) of our theorem stems from Dirichlet, the general formula stems from Schlömilch, cited in [1], p. 160-181.

65. Concerning this, cf. say O. Perron, Die Lehre von den Kettenbrüchen, 2nd Ed., 1929, p. 362-367.

66. Cf. P. Lévy, Calcul des Probabilites, 1925, p. 163-172.

67. The same, p. 192-195.

68. This is an important theorem in probability theory by P. Lévy, the same, p. 197 ff., and previously in Comptes rendus de l'académie des sciences de Paris, 175, 854-856, 1922. Cf. also M. Jacob, the same, 188, 541-543, 754-757, 1929. The proofs of the theorem are by G. Pólya, Mathematische Zeitschrift 18, 96-108, 1923, and by F. P. Cantinelli, Rendiconti del Circolo matematico di Palermo, 52, 416, 1928.

69. The definition of the concept of the positive-definite functions and the further consideration of such functions, but not our Theorem 23 in the full extent, stems from M. Mathias, Mathematische Zeitschrift, 16, 101-125, 1923. - Preceding it was a note by F. Bernstein, Mathematische Annalen, 85, 155-159, 1922.

70. Pólya, cited in [68].

71. Mathias, cited in [69]. The special case $\rho = 4$ was treated previously in another manner by F. Bernstein, Mathematische Annalen, 79, 265-268, 1918.

72. Lévy, cited in [66], p. 252-277.

73. Cf. H. Bohr, Acta Mathematica, 45, 29-127, 1925.

74. This proof is essentially that of N. Wiener, Proceedings of the London Mathematical Society, Series (2), 27, 483-496, 1928.

75. For the considerations of this chapter and the corresponding part of the next chapter, I have been stimulated by the paper of N. Wiener, The Operational Calculus, Mathematische Annalen, 95, 557-585, 1926.

76. In regard to the general case, cf. my note: Beitrag zur absoluten Konvergenz fastperiodischer Fourierreihen, Jahresbericht der Deutschen Mathematiker-Vereinigung, 39, 52-54, 1930.

77. G. H. Hardy and E. C. Titchmarsh, Proceedings of the London Mathematical Society, Series (2), 23, 1-26, 1925, and a correction, the same 24, XXXI to XXXIII, 1925 (Records for February 1925). In addition by the same authors, the same 30, 95-106, 1930, and E. Hopf, Journal of the London Mathematical Society, 4, 23-27, 1929.

78. Generalized Fourier integrals have been examined quite systematically for the first time by H. Hahn; cf. the work in Acta Mathematica, 49, 301-353, 1926 and Sitzungsberichte der Akademie der Wissenschaften in Wien, Mathematisch-Naturwissenschaftliche Klasse, Section II, 134, 449 to 470, 1925. Somewhat later, generalized Fourier integrals were examined by N. Wiener, cited in [23], but methodically different from H. Hahn. - Both these authors have carried through the generalization nearly to the extent of our class $\mathfrak{F}_2$; I gave a generalization of higher **function** classes in Mathematische Annalen, 97, 635-662, 1927. - An extension of Fourier integrals for the purpose of the extension of the theorem of Plancherel, cf. §40 and §41, was discussed by I. C. Burkhill, Proceedings of the London Mathematical Society, Series (2), 25, 513-529, 1926.

79. For generalization of the uniqueness theorem in addition to the generalized Fourier-integrals by Hahn, Wiener and Burkhill, cf. M. Jacob, Mathematische Annalen, 100, 278-294, 1928.

80. Lebesgue, cited in [3], p. 15-16.

81. For theorems proper concerning convergence at isolated points, cf. [31], especially H. Hahn cited in [78].

82. H. Bohr, cited in [73].

83. This result stems from Erh. Schmidt, Mathematische Annalen, 70, 499-524, 1911. G. Hoheisel examined a generalization of the Schmidt method for the case where the $\alpha_{\rho\sigma}$ are polynomials in x, Mathematische Zeitschrift, 35-99, 1922.

84. E. Picard, Annales de l'Ecole Normale Supérieure, 3rd Séries, 28, 313 to 324, 1911.

85. For literature for this and the following paragraphs, cf. Doetsch, cited in [40].

86. For more general criteria for the validity of the inverse formula and its summation, and for the examples which follow, cf. especially G. H. Hardy, Messenger of Mathematics, 47, 178-189, 1918. The remarkable example 2) stems from G. Ramanujan, the same, 46, 10-18, 1915, especially formula (2). I. C. Burkhill treated the case where both inverse formulas are Stieltjes integrals, Proceedings of the Cambridge Philosophical Society, 23, 356-360, 1926.

87. T. V. Stachó, Mathematische und Naturwissenschaftliche Berichte aus Ungarn, 33, 20-32, 1926.

88. G. Doetsch, Sitzungsberichte der Preussichen Akademie der Wissenschaften, Phys-mathem. Klasse, 1930, X.

89. Cf. G. Pólya, Journal Für die reine und angewandte Mathematik, 158, 6-18, 1927, and E. C. Titchmarsh, Proceedings of the London Mathematical Society, Series (2), 25, 283-302, 1925.

90. Doetsch, cited in [40] and Haar, cited in [9].

91. For this paragraph in all its particulars, cf. Doetsch, cited in [40], and in addition G. H. Hardy, Messenger of Mathematics, 49, 85-91, 1920; 50, 165-171, 1921.

92. For formula (3), see G. N. Watson, The Theory of Bessel Functions, Cambridge, 1922, p. 386.

93. Cf. E. T. Whittaker and G. N. Watson, A Course of Modern Analysis, 3rd Edition, 1920, p. 289.

94. By F. Bernstein, cf. citation in [91].

95. Cf. once more G. Doetsch, Mathematische Zeitschrift, 32, 587-599, 1930.

96. The theorem stems from Hj. Mellin (but he proved it under very narrow assumptions) and from H. Hamburger and M. Fujiwara. Cf. Doetsch, cited in [40] and Haar, cited in [9].

97. The oldest systematic competent study of conjugate trigonometric integrals was O. Beau, Untersuchungen auf dem Gebiete der trigonometrischen Reihen und trigonometrischen Integrale, Leipzig, 1883. For a more recent work, cf. E. C. Titchmarsh, Proceedings of the London Mathematical Society, Series (2), 24, 109-130, 1924, and Journal of the London Mathematical Society, 5, 89-91, 1930.

98. A. Weinstein, Rendiconti delle R. Accademia Nazionale dei Lincei, Classe di Scienze fisiche, matematiche e naturale, Bd. V, Series 6^a, 1. Sem. 1927 and Comptes Rendus des Séances de l'Académie des Sciences de Paris, 184, 479-499, 1927, and papers from the Mathematischen Seminar der Hamburgischen Universität, 6, 263-264, 1928. - For the solution itself, cf. again G. Hoheisel, Jahresbericht der Deutschen Mathematiker-Vereinigung, 39, 54-58, 1930.

99. This theorem stems from E. Fischer, F. Riesz, I. Radon, and its extension to infinite intervals by Pia Nalli. For a comprehensive representation cf. Hobson, p. 246-249.

100. The content of Theorems 51 to 55 stems from M. Plancherel, Rendiconti del Circolo Matematico di Palermo, 30, 289-339, 1910 and

Proceedings of the London Mathematical Society, Series (2), 24, 62-70, 1925. The original Plancherel Theory is very general. The observations for the special case of the Fourier and Hankel integrals were specialized by E. C. Titchmarsh, Proceedings of the Cambridge Philosophical Society, 11, 463-473, 1923 (and a correction in 12, 1924), and Journal of the London Mathematical Society, 1, 195-196, 1927. The case of the Fourier integral was treated anew by F. Riesz in Acta litterarum ac scientiarum (Scientiarum matematicarum), Szeged, 3, 235-241, 1927.

101. M. Plancherel, E. C. Titchmarsh among others; cf. [31], and Hobson, p. 748-751, gave criteria for the convergence of integrals in the usual sense.

102. Cf. G. H. Hardy and E. C. Titchmarsh, cited in [77].

103. I. C. Burkhill studied the case where the differential $f(x)\,dx$ was replaced in the integral (15') by the more general $d\varphi(x)$, cf. citation in [86]; also similarly in the case of the Hankel integral. For this, cf. S. Izumi, The Tohoku Mathematical Journal, 29, 266-277, 1928.

104. E. C. Titchmarsh, Proceedings of the London Mathematical Society, Series (2), 23, 279-281, 1925 and Journal of the London Mathematical Society, 2, 148-150, 1927; A. C. Berry, Annals of Mathematics, Series 2, 32, 227-238 and 830-838, 1931. - Cf. also Hobson, p. 742-747.

105. For generalizations, cf. G. H. Hardy and E. C. Titchmarsh, Journal of the London Mathematical Society, 6, 44-48, 1930.

106. Cf. [100] and I. C. Burkhill, cited in [103]. For literature concerning the topic prior to Plancherel cf. Watson, cited in [92]. For generalizations of the Fourier integral theorems based on other than Hankel kernels, cf. M. H. Stone, Mathematische Zeitschrift, 28, 654-676, 1928.

107. Cf. Watson, cited in [92], p. 406, formula (8).

108. This proof by E. Czuber, Monatshefte für Mathematik und Physik, 2, 119-124, 1891.

109. For $k = 2, 3$, by Poisson and Cauchy, Burkhardt, p. 1165-1173. The theorem is also not new for k arbitrary.

110. By Poisson and Cauchy, Burkhardt, p. 1164.

111. Cf. Watson, cited in [92], Chapt. 13.47.

112. Cf. [17].

113. Cf. Ch. H. Müntz, Mathematische Annalen, 90, 279- 291, 1923 and Sitzungsberichte der Berliner Mathematischen Gesellschaft 1925, p. 81-93.

114. W. Thomson Lord Kelvin, Papers on Electrostatics and

Magnetism, Second Edition, 1884, p. 112-125.

115. Cf. A. C. Berry, cited in [104].

116. Our definition of monotone functions of several variables should amount essentially to that by G. H. Hardy, Quarterly Journal of Mathematics, 37, 55-60, 1905/06. For literature concerning the subject and for analogues to our Theorem 65 in the case of repeated Fourier series, cf. H. Hahn, Theorie der reellen Funktionen, Bd. I, Berlin 1921, p. 539-547, Hobson, p. 702-711, and Tonelli, Chapter IX.

117. The Poisson summation formula in several variables was first studied by Ch. H. Müntz, cited in [113], then by L. I. Mordell, cf. [120]. For the present criteria and generalizations, cf. my note in Mathematische Annalen, 106, 56-63, 1932.

118. For this formula, including formula (13), cf. say Tonelli, p. 487.

119. This is a theorem concerning Fourier series in several variables. Cf. Tonelli, p. 486-500.

120. Considerable further generalizations than (19) play of late an important role in analytic number theory. For literature, cf. L. I. Mordell, Proceedings of the London Mathematical Society, Series 2, 32, 501-556, 1931.

121. C. L. Siegel, Mathematische Annalen, 87, 36-38, 1922.

AUTHOR'S SUPPLEMENT

MONOTONIC FUNCTIONS, STIELTJES INTEGRALS AND HARMONIC ANALYSIS

INTRODUCTION

Consider the totality of all monotonically increasing[1] functions of one variable. It is known that for this function class the following notions of equality and convergence are very appropriate. Two functions are called "essentially equal" if they agree at their points of continuity. A sequence of functions $f_n(x)$ is called "essentially convergent" if there exists a function $f_o(x)$ such that at each point of continuity of $f_o(x)$ the sequence of numbers $f_n(x)$ converges to the number $f_o(x)$. The importance of these concepts of equality and convergence consists in the fact that the following compactness theorem is then valid. Each infinite set of uniformly bounded functions contains an essentially convergent subsequence[2], and the limit function of an essentially convergent sequence remains unchanged if each function of the sequence is replaced by an essentially equal one.

The compactness theorem can be formulated more profitably if the point functions $f(x)$ are replaced by interval functions $f(x; y) \equiv f(y) - f(x)$. An interval is called a continuity interval of the function $f(x; y)$ if small changes of the boundaries of the interval produce little change in the function value itself. Correspondingly one defines the concepts "essentially equal" and "essentially equivalent." The compactness theorem will then read as follows. Each infinite set of uniformly bounded (non-negative, additive) interval functions contains an essentially convergent subsequence, and the limit function of an essentially convergent sequence remains unchanged if each function of the sequence is replaced by an essentially equal one.

The basic advantage of this approach is that the compactness

---

[1] or monotonically decreasing.

[2] This theorem is usually named after E. Helly, Sitzungsberichte der Wiener Akademie [Proceedings of the Vienna Academy] 121 (1921), p. 265-297. The terms "essentially equal" and "essentially convergent" are due to A. Wintner, Spektraltheorie der unendlichen Matrizen, 1929, p. 77.

theorem and other facts about interval functions and the Rieman-Stieltjes integrals[3] connected with them can be carried over verbatim for functions of several variables, although the discontinuity character of the interval functions in the case of more variables is considerably more complicated than in the case of one variable. It will be evident from the results of this work that this formulation is an appropriate one.

A theorem of Fourier-Stieltjes integrals useful in probability theory is the following. If for a sequence of monotonic functions $V_n(\alpha)$, the corresponding "characteristic functions"

$$f_n(x) = \int_{-\infty}^{\infty} e^{ix\alpha} dV_n(\alpha)$$

converge uniformly in each finite x-interval, then the functions $V_n(\alpha)$ are essentially convergent. We shall prove this theorem not only for several variables, but shall also free it from the requirement of uniform convergence. It will be sufficient that the functions $f_n(x)$ converge for almost all x and at the origin. In addition we shall give a necessary <u>and sufficient</u> condition that a function can be written[4] in the form

$$f(x) = \int_{-\infty}^{\infty} e^{ix\alpha} dV(\alpha) \quad .$$

By reason of it, we shall then be able to formulate and prove, in an especially lucid manner, a theorem of Norbert Wiener on spectral analysis of a rather general kind of function. Let $g(x)$ be a square integrable function (of one variable). If this function is also square integrable in the infinite region, then Plancherel has shown that it can be represented by a Fourier integral

$$g(x) \sim \int e^{ix\alpha} \Gamma(\alpha)\, d\alpha \quad .$$

If, however, it is not square integrable in the infinite region, but instead is periodic or more generally almost periodic in its entire domain, then it possesses a Fourier <u>series</u>

---

[3] The theory of Stieltjes integrals in several variables was first systematically investigated by J. Radon, Wiener Berichte 122 (1913), Cf. also A. Kolmogoroff, Untersuchungen über den Integralbegriff, Math. Annalen 103 (1930), p. 654-696.

[4] For functions of <u>one</u> variable it has already been given in the author's Vorlesungen über Fouriersche Integrale, Leipzig, Akademische Verlagsgesellschaft, 1932, in particular Chapter IV. In what follows we shall quote this book by "Fourier Integrals".

$$g(x) \sim \sum_{\nu} e^{ix\alpha_\nu} \Gamma(\alpha_\nu) \quad .$$

The "spectral intensity" of $g(x)$ is that interval function which in the one case belongs to the (monotonic) function

$$V(\alpha) = \int_0^\alpha |\Gamma(\alpha)|^2 \, d\alpha \quad ,$$

and in the other to the function

$$V(\alpha) = \sum_{\alpha_\nu = 0}^{\alpha} |\Gamma(\alpha_\nu)|^2 \quad .$$

In both cases, it is a bounded, monotonic function. If in addition one introduces the function

$$(0,1) \qquad f(x) = \int_{-\infty}^{\infty} e^{ix\alpha} dV(\alpha) \quad ,$$

then in the one case, we have

$$f(x) = \int_{-\infty}^{\infty} g(x + \xi)\overline{g(\xi)} \, d\xi \quad ,$$

and in the other

$$(0,2) \qquad f(x) = \lim_{T \to \infty} \frac{1}{2T} \int_{-T}^{T} g(x + \xi)\overline{g(\xi)} \, d\xi \quad .$$

Now N. Wiener[5] has proved the following result (and also generalized it to several variables). If $g(x)$ is an arbitrary square integrable function in the finite region, regarding whose behavior in the infinite region it is known only that the (finite) limit (0, 2) exists for all x, then one can assign to it in a meaningful manner, a bounded monotonic function $V(\alpha)$ as its spectral function. More precisely stated, this assignment means that the relation (0, 1) holds for almost all x (and correspondingly for more variables) between the faltungs-function $f(x)$ and the spectral function $V(\alpha)$. Due to a very original (unpublished) idea of M. Riesz, we shall replace the convergence denominator $2T$ in the integral (0, 2) by an arbitrary

[5] N. Wiener, Acta Mathematica 55 (1930), pp.117-258.

positive monotonically increasing function $\rho(t)$ for which

$$\lim_{T \to \infty} \frac{\rho(T+1)}{\rho(T)} = 1 \quad .$$

That is, we shall prove the existence of the spectral function $V(\alpha)$ under the more general[6] assumption that the limit

$$f(x) = \lim_{T \to \infty} \frac{1}{\rho(T)} \int_{-T}^{T} g(x + \xi)\overline{g(\xi)}\, d\xi$$

exists for all x. Even the existence of this limit will be needed only for a discrete sequence of values $T_1, T_2, T_3, \ldots$. This very comprehensive formulation of the Wiener theorem has the merit that it also includes the Plancherel case when $\rho(T) \equiv 1$. It therefore can be looked at, in its way, as an actual generalization of the Fourier integral formula.

I. MONOTONIC FUNCTIONS

§1. DEFINITION OF THE MONOTONIC FUNCTIONS

1.1. We take as a basis a Euclidean space of several dimensions. We shall denote its dimension number by k, and an index which takes on the values 1 to k, by χ. A point of this space will be denoted by $\alpha, \beta, \ldots, \alpha_1, \ldots, x, y, \ldots$ etc., and its individual components by upper accents. Hence

$$\varphi(\alpha), \ \varphi(\alpha', \alpha'', \ldots, \alpha^{(k)}) \quad \text{and} \quad \varphi(\alpha^{(\chi)})$$

denote one and the same function.

The coordinate system of the space is fixed. By an interval we mean the point set

$$(1,\ 11) \qquad \alpha^{(\chi)} \leq x^{(\chi)} < \beta^{(\chi)} \qquad \chi = 1, 2, \ldots, k \quad .$$

The values $-\infty$ and $+\infty$ will also be admitted[7] for the numbers $\alpha^{(\chi)}$ and $\beta^{(\chi)}$ if they are interval boundaries. Consequently the whole space, which we denote by $\mathfrak{R}$, also belongs to the intervals, as well as all "octants", "half spaces", "parallel slices", etc. As a rule we shall denote an interval by the letter $\mathfrak{J}$, and the interval (1, 11) sometimes more precisely

[6] A beginning of this generalization which goes back to M. Jacob, can already be found in N. Wiener, op. cit.

[7] If a left interval boundary is $-\infty$, then the sign $\leq$ in (1, 11) is to be replaced by $<$.

by the symbol $\mathfrak{I}(\alpha; \beta)$.

Besides the intervals, we shall consider, when the contrary is not expressly stressed, only such point sets $\mathfrak{P}$ which can be expressed as the sum of finitely many (disjoint) intervals. By a point set, we shall simply mean a set constituted in such a manner. It is not difficult to see that sums and differences, union and intersection of two point sets (of our kind) are again point sets (of our kind).

1.2. To each bounded interval $\mathfrak{I}$, it is possible to assign a well determined number $\varphi(\mathfrak{I})$ which has the following two properties.

I. Additivity, i.e., if

$$\mathfrak{I} = \mathfrak{I}_1 + \mathfrak{I}_2 + \dots + \mathfrak{I}_n ,$$

then

$$\varphi(\mathfrak{I}) = \varphi(\mathfrak{I}_1) + \varphi(\mathfrak{I}_2) + \dots + \varphi(\mathfrak{I}_n) .$$

II. (Bounded) monotonicity, i.e., there is a constant M such that

$$0 \leq \varphi(\mathfrak{I}) \leq M . \tag{1.21}$$

We shall denote such functions $\varphi(\mathfrak{I})$, and only such[8] as interval functions.

If one represents a bounded point set $\mathfrak{P}$ as a sum of intervals $\mathfrak{I}_\nu$, in two different ways, then by the additivity property the number

$$\sum_\nu \varphi(\mathfrak{I}_\nu)$$

is independent of the special decomposition used. For this reason, the range of definition of the function $\varphi(\mathfrak{I})$ can be extended to bounded point sets $\mathfrak{P}$, in such a way that the above two properties are preserved, i.e.,

$$\varphi\left(\sum_{\nu=1}^{n} \mathfrak{P}_\nu\right) = \sum_{\nu=1}^{n} \varphi(\mathfrak{P}_\nu) , \tag{1, 22}$$

and

$$0 \leq \varphi(\mathfrak{P}) \leq M , \tag{1,23}$$

8 We shall turn in a later statement to interval functions of different algebraic signs.

where M is the same number as in (1, 21). The term "monotonic" which appears in Property II can be justified as follows. Let $\mathfrak{P}_1 \subset \mathfrak{P}_2$. By I,

$$\varphi(\mathfrak{P}_1) + \varphi(\mathfrak{P}_2 - \mathfrak{P}_1) = \varphi(\mathfrak{P}_2) \quad .$$

Because of II we hence have

$$\varphi(\mathfrak{P}_1) \leq \varphi(\mathfrak{P}_2) \quad \text{if} \quad \mathfrak{P}_1 \subset \mathfrak{P}_2 \quad ,$$

and this is a relation of monotinicity literally. We note that if we admit the empty set and for the empty set $\mathfrak{P}$ define $\varphi(\mathfrak{P})$ as zero, then everything thus far asserted remains valid.

1.3. It is now also possible to expand the range of definition of $\varphi(\mathfrak{J})$ to unbounded point sets. Let $\mathfrak{P}$ be an arbitrary point set. We take a monotonic sequence of cubes $\mathfrak{W}_n$ which converge to $\mathfrak{R}$, and consider the intersections

$$\mathfrak{P}_n = \mathfrak{P} \cdot \mathfrak{W}_n \quad .$$

Since

$$\varphi(\mathfrak{P}_n) \leq \varphi(\mathfrak{P}_{n+1}) \leq M$$

the limit

$$\lim \varphi(\mathfrak{P}_n) \tag{1.31}$$

exists. Indeed it has the important property that for bounded $\mathfrak{P}$ it agrees with the original value $\varphi(\mathfrak{P})$, and as is easily seen, it is independent of the special choice of the sequence of cubes. For this reason, as is easily verified, if the number $\varphi(\mathfrak{P})$ is defined by the limit (1, 31) the laws (1, 22), (1, 23) and (1, 24) remain in force.

1.4. For each cube we have

$$\varphi(\mathfrak{W}) + \varphi(\mathfrak{R} - \mathfrak{W}) = \varphi(\mathfrak{R}) \quad .$$

By the definition of $\varphi(\mathfrak{R})$, in conjunction with (1, 24), we obtain the following. To each $\epsilon > 0$, there is a cube $\mathfrak{W}$ such that

$$0 \leq \varphi(\mathfrak{P}) \leq \epsilon \quad \text{if} \quad \mathfrak{P} \subset (\mathfrak{R} - \mathfrak{W})$$

1.5. Let $\varphi(\alpha)$ be a point function (which has an unique value at each point). For each <u>bounded</u> interval (1, 11), we form the sum consisting of the 2^k terms

$$(1,51) \qquad \sum_{\epsilon_\chi=0}^{1} (-1)^{\epsilon_1+\dots+\epsilon_k} \varphi(\beta^{(\chi)} + \epsilon_\chi(\alpha^{(\chi)} - \beta^{(\chi)})) \quad .$$

For example if $k = 2$, it becomes

$$\varphi(\beta', \beta'') - \varphi(\alpha', \beta'') - \varphi(\beta', \alpha'') + \varphi(\alpha', \alpha'') \quad ;$$

and for general k it becomes

$$\int_{\alpha'}^{\beta'} \dots \int_{\alpha^{(k)}}^{\beta^{(k)}} \frac{d^k\varphi}{d\gamma' \dots d\gamma^{(k)}} d\gamma' \dots d\gamma^{(k)} \quad ,$$

whenever the partial derivatives under the integral sign exist and happen to be continuous. We denote the expression (1, 51) by

$$(1, 52) \qquad \varphi(\alpha; \beta)$$

or also by

$$(1, 53) \qquad \varphi(\mathfrak{J}) \quad .$$

It is easy to recognize that this function $\varphi(\mathfrak{J})$ has the property I above. If in addition it possesses property II which we interpret as an assumption on the generating function $\varphi(\alpha)$, then we call $\varphi(\alpha)$ a monotonic (point) function and (1, 52) or (1, 53) the corresponding interval function.

1.6. For each interval function[9] $\varphi(\mathfrak{J})$, there exists a monotonic function to which it belongs. If, for example, one assigns to each point α the "octant"

$$x^{(\chi)} < \alpha^{(\chi)} \qquad \chi = 1, 2, \dots, k \quad ,$$

and denote it by

$$(1, 61) \qquad \mathfrak{A}(\alpha) \equiv \mathfrak{A}(\alpha', \alpha'', \dots, \alpha^{(k)}) \quad ,$$

then

$$(1, 62) \qquad \varphi(\mathfrak{A}(\alpha)) \quad ,$$

as can easily be verified, is such a function.

Besides this function, there are still other point functions to which $\varphi(\mathfrak{J})$ belongs. But we shall show that the function (1, 62) because

---

9 We recall that by interval function we mean only the kind described in 1, 2.

of a certain uniqueness property, is exceptional. By 1.4, it has, for example, the following property. If for a fixed χ, the component $\alpha^{(\chi)}$ is allowed to decrease to $-\infty$, then the function is uniformly convergent to zero regardless of the values of the other components. We call a function of this kind normalized. Conversely let $\varphi(\alpha)$ be a normalized function. If in the sum (1, 51), the point β is held fixed, and all the components $\alpha^{(\chi)}$ are allowed to decrease to $-\infty$, then all elements of the sum, apart from $\varphi(\beta', \beta'', \ldots, \beta^{(k)})$, go to zero. That is

$$\varphi(\beta) = \lim_{\alpha^{(\chi)} \to -\infty} \varphi(\alpha; \beta) .$$

But since

$$\lim_{\alpha^{(\chi)} \to -\infty} \varphi(\alpha; \beta) = \varphi(\mathfrak{A}(\beta)) ,$$

it follows that $\varphi(\beta)$ agrees with function $\varphi(\mathfrak{A}(\beta))$ defined in (1, 62).

We shall, as a rule, assume that the monotonic functions are normalized.

§2. CONTINUITY INTERVALS

2.1. For each bounded interval (1, 11), we consider for each (sufficiently small) $\epsilon > 0$, the interior interval

$$\mathfrak{E}_\epsilon: \qquad a^{(\chi)} + \epsilon \leqq x^{(\chi)} < b^{(\chi)} - \epsilon \qquad \chi = 1, \ldots, k ,$$

the surrounding interval

$$\mathfrak{A}_\epsilon: \qquad a^{(\chi)} - \epsilon \leqq x^{(\chi)} < b^{(\chi)} + \epsilon \qquad \chi = 1, \ldots, k ,$$

and the boundary layer

$$\mathfrak{G}_\epsilon = \mathfrak{A}_\epsilon - \mathfrak{E}_\epsilon .$$

For each interval function we have

$$\varphi(\mathfrak{E}_\epsilon) \leqq \varphi(\mathfrak{J}) \leqq \varphi(\mathfrak{A}_\epsilon) . \tag{2, 11}$$

We call the function $\varphi(\mathfrak{J})$ continuous in $\mathfrak{J}$, — and $\mathfrak{J}$ a continuity interval of $\varphi(\mathfrak{J})$, — if the limit relations

$$\lim_{\epsilon \to 0} \varphi(\mathfrak{E}_\epsilon) = \varphi(\mathfrak{J}) = \lim_{\epsilon \to 0} \varphi(\mathfrak{A}_\epsilon) , \tag{2, 12}$$

which are equivalent to the single relation

$$\lim_{\epsilon \to 0} \varphi(\mathfrak{G}_\epsilon) = 0 \quad ,$$

hold.

2.2. We call a system of intervals everywhere dense, if to each bounded interval $\mathfrak{J}$ of the space, there are intervals of the system, whose corner points differ arbitrarily little from the corner points of $\mathfrak{J}$, and we call it finely meshed if in addition, we can subdivide each interval of the system in other intervals of the system of arbitrarily small diameter.

THEOREM 1. Among the common continuity intervals of finitely or countably infinitely many interval functions, there is a finely meshed system.

PROOF. Let $\varphi(\mathfrak{J})$ be an interval function and $\varkappa$ a fixed index. For an arbitrary number $\gamma^{(\varkappa)}$, we denote by $\mathfrak{H}(\gamma^{(\varkappa)})$ the half space

$$x^{(\varkappa)} < \gamma^{(\varkappa)} \quad ,$$

and we form the function of one variable

$$\varphi(\mathfrak{H}(\gamma^{(\varkappa)})) \quad . \tag{2, 21}$$

By (1, 24), it is monotonically increasing, and hence possesses at most countably many discontinuity points. We denote these by

$$\gamma_1^{(\varkappa)}, \ \gamma_2^{(\varkappa)}, \ \ldots \quad . \tag{2, 22}$$

If therefore $\gamma^{(\varkappa)}$ differs from the countably many numbers (2, 22), then to each $\delta > 0$, there is an $\epsilon > 0$ such that the value of $\varphi(\mathfrak{P})$ is less than δ over the parallel slices

$$\gamma^{(\varkappa)} - \epsilon < x^{(\varkappa)} < \gamma^{(\varkappa)} + \epsilon \quad .$$

We form the values (2, 22) for each $\varkappa$, and call the (k-1)-dimensional hyperplanes

$$x^{(\varkappa)} = \gamma_\nu^{(\varkappa)} \quad ,$$

the exceptional planes of $\varphi(\mathfrak{J})$ or $\varphi(\alpha)$.

If for a bounded interval, none of the 2^k corners falls in an exceptional plane, or what amounts to the same thing, if no end point of the interval falls in an exceptional plane, then to each $\eta > 0$, one can determine a boundary layer $\mathfrak{G}_\epsilon$ of the interval, for which $\varphi(\mathfrak{G}_\epsilon) \leqq \eta$. Hence each such interval is a continuity interval. The assertion of our theorem therefore results immediately from the fact that the exceptional planes of countably many interval functions are likewise countable, Q.E.D.

By means of the representation (1, 62), one finds that each (normalized) monotonic function $\varphi(\alpha)$ is continuous at all points which do not lie in an exceptional plane. Hence

THEOREM 2. Except for the points of countably many planes parallel to the axes each monotonic function $\varphi(\alpha)$ is continuous.

2.3. We call two interval functions (or the corresponding monotonic functions) "essentially equal" or only "unessentially different" if each continuity interval of one function is also a continuity interval of the other, and their values agree in the continuity intervals.

It is easy to prove

THEOREM 3. If two interval functions agree in an everywhere dense interval system, they are essentially equal.

By Theorems 1 and 3, one concludes that our equality concept, just as one would desire of any equality concept, does possess the transitive property: If φ_1 and φ_2, and φ_2 and φ_3 are essentially equal, then so also are φ_1 and φ_3.

2.4. For the broadening of our equality concept, we shall utilize a result of H. Hahn, and introduce for the remainder of this section, wholly arbitrary point sets $\mathfrak{M}$ of our spaces instead of only such point sets $\mathfrak{P}$ which are the sums of finitely many intervals.

Let $\psi(\mathfrak{J})$ be a non-negative bounded function which is defined for all bounded intervals $\mathfrak{J}$, and has the following property. If an interval $\mathfrak{J}$ is divided into (disjoint) intervals $\mathfrak{J}_1, \ldots, \mathfrak{J}_n$, then

$$\psi\left(\sum_{\nu=1}^{n} \mathfrak{J}_\nu\right) \leqq \sum_{\nu=1}^{n} \psi(\mathfrak{J}_\nu) \tag{2, 41}$$

where the sum on the right is assumed to be bounded.

Now let $\mathfrak{O}$ be an arbitrary open point set. We consider the disjoint intervals $\mathfrak{J}_1, \ldots, \mathfrak{J}_n$ which together with their boundaries are contained in $\mathfrak{O}$. We then form the sum

$$\sum_{\nu=1}^{n} \psi(\mathfrak{J}_\nu)$$

and take the upper limit of this sum. We denote this upper limit by $\varphi(\mathfrak{O})$. If $\mathfrak{M}$ is now an arbitrary point set we understand by $\varphi(\mathfrak{M})$ the lower bound of $\varphi(\mathfrak{O})$ for all open point sets $\mathfrak{O}$ containing $\mathfrak{M}$. This set function $\varphi(\mathfrak{M})$ is what H. Hahn calls a "content"[10]. It has among others the property that for each sequence of mutually disjoint Borel sets $\mathfrak{B}_\nu$ we have

$$\varphi\left(\sum_\nu \mathfrak{B}_\nu\right) = \sum_\nu \varphi(\mathfrak{B}_\nu) \quad .$$

In particular our intervals are Borel sets. For example, since the interval (1, 11) is the intersection of the open point sets $(n = 1, 2, 3, \ldots)$

$$\alpha^{(\chi)} - \frac{1}{n} < x^{(\chi)} < \beta^{(\chi)} \qquad \chi = 1, \ldots, k \quad ,$$

we have among others

$$\varphi\left(\sum_{\nu=1}^{n} \mathfrak{J}_\nu\right) = \sum_{\nu=1}^{n} \varphi(\mathfrak{J}_\nu) \quad ,$$

that is, the function $\varphi(\mathfrak{J})$ is an interval function (in our sense).

Let $\psi(\mathfrak{J})$ be an arbitrary interval function (in our sense). By the procedure just described, there is attached to it a well determined content which we call its <u>dominant function</u>. It is important to note that the two interval functions $\psi(\mathfrak{J})$ and $\varphi(\mathfrak{J})$ are essentially equal. In fact, because of the additivity of $\psi(\mathfrak{J})$ it is easy to verify, in the notation of 2.1, that

$$\text{(2, 42)} \qquad \varphi(\mathfrak{C}_\epsilon) \leq \psi(\mathfrak{J}) \leq \varphi(\mathfrak{A}_\epsilon)$$

is valid for each bounded interval $\mathfrak{J}$. If now $\mathfrak{J}$ is a continuity interval of $\varphi(\mathfrak{J})$, then (2, 12) is valid and (2, 42) implies

$$\varphi(\mathfrak{J}) = \psi(\mathfrak{J}) \quad ,$$

as claimed.

---

[10] H. Hahn, Reelle Funktionen, Springer 1921, p. 453 and 448. In defining the generating function $\psi(\mathfrak{J})$, H. Hahn views the intervals as closed, and not just left sided closed as are ours, cf. 1.1., but this difference is of no consequence. We remark that we will continue to assume that an interval is left sided closed as heretofore.

If two interval functions $\psi_1(\mathfrak{J})$ and $\psi_2(\mathfrak{J})$ are essentially equal then their dominant functions $\varphi_1(\mathfrak{M})$ and $\varphi_2(\mathfrak{M})$ are identically equal. For the proof it is sufficient to show that for any open set $\mathfrak{L}$

$$\varphi_1(\mathfrak{L}) = \varphi_2(\mathfrak{L}) \quad . \tag{2, 43}$$

By definition, $\varphi_1(\mathfrak{L})$ is the upper limit of

$$\sum_{\nu=1}^{n} \psi_1(\mathfrak{J}_\nu)$$

for disjoint intervals $\mathfrak{J}_\nu$ which with their closure are contained in $\mathfrak{L}$. Now, this upper limit remains unchanged if in addition we require of the intervals $\mathfrak{J}_\nu$ that they be continuity intervals of $\psi_1(\mathfrak{J})$. For on the basis of observations made in 2.2, we can associate with each interval $\mathfrak{J}_\nu$ a continuity interval $\bar{\mathfrak{J}}_\nu$ of $\psi_1(\mathfrak{J})$ in such a manner that it encloses $\mathfrak{J}_\nu$ but differs from it arbitrarily little and that the union of these $\bar{\mathfrak{J}}_\nu$ is decomposable into disjoint continuity intervals of $\psi_1(\mathfrak{J})$. A similar reasoning applies to $\varphi_2(\mathfrak{L})$, and since $\psi_1(\mathfrak{J})$ and $\psi_2(\mathfrak{J})$ are equal in continuity intervals, (2, 43) now results. Conversely if $\varphi_1(\mathfrak{M})$ and $\varphi_2(\mathfrak{M})$ are identical, then $\psi_1(\mathfrak{J})$ and $\psi_2(\mathfrak{J})$ are essentially equal because they are both essentially equal to $\varphi_1(\mathfrak{M})$. The following theorem is therefore valid.

THEOREM 4. In order that two interval functions be essentially equal, it is necessary and sufficient that their dominant functions be identical.

2.5. We make in our k-dimensional space $\mathfrak{R}$ an orthogonal transformation and denote the new coordinates by $\tilde{x}$, the whole space of the new coordinates by $\tilde{\mathfrak{R}}$, the intervals in $\tilde{\mathfrak{R}}$ by $\tilde{\mathfrak{J}}$, and point sets in it by $\tilde{\mathfrak{P}}$.

If an interval function $\psi(\mathfrak{J})$ in $\mathfrak{R}$ and an interval function $\chi(\tilde{\mathfrak{J}})$ in $\tilde{\mathfrak{R}}$ are given, there is from the outset no measure by which to equate the functions to each other. But by introduction of dominant functions, we can by Theorem 4 generalize our definition of equality as follows. We call the functions $\psi(\mathfrak{J})$ and $\chi(\tilde{\mathfrak{J}})$ <u>essentially equal</u>, if their dominant functions are identical. This "equality" also possesses the property of transitivity.

§3. SEQUENCES OF MONOTONIC FUNCTIONS

3.1. Let a sequence of interval functions

$$\varphi_1(\mathfrak{J}),\ \varphi_2(\mathfrak{J}),\ \varphi_3(\mathfrak{J}),\ \ldots \tag{3, 11}$$

be given which is constituted as follows. There is an additional function $\varphi(\mathfrak{J})$ such that the sequence (3, 11) converges to $\varphi(\mathfrak{J})$ in each continuity interval of $\varphi(\mathfrak{J})$. Every other function $\psi(\mathfrak{J})$ which stands in the same relation to the sequence (3, 11) as the function $\varphi(\mathfrak{J})$, is evidently, only unessentially different from this latter function. The sequence (3, 11) is then called <u>essentially convergent</u> and the function $\varphi(\mathfrak{J})$ its limit function.

THEOREM 5. If each function $\varphi_n(\mathfrak{J})$, in an essentially convergent sequence (3, 11), is replaced by a function $\Phi_n(\mathfrak{J})$ unessentially different, then the new sequence is likewise essentially convergent and the limit function is the same.

PROOF. Let $\mathfrak{J}_0$ be a continuity interval of $\varphi(\mathfrak{J})$ and $\bar{\mathfrak{J}}$ a common continuity interval of all functions $\varphi_n(\mathfrak{J})$ and $\Phi_n(\mathfrak{J})$ and $\varphi(\mathfrak{J})$. If $\bar{\mathfrak{J}} \subset \mathfrak{J}_0$, then

$$\varphi_n(\bar{\mathfrak{J}}) = \Phi_n(\bar{\mathfrak{J}}) \leqq \Phi_n(\mathfrak{J}_0) \quad ,$$

and hence

$$\varphi(\bar{\mathfrak{J}}) \leqq \underline{\lim}_{n \to \infty} \Phi_n(\mathfrak{J}_0) \quad .$$

But since the intervals $\bar{\mathfrak{J}}$ are everywhere dense, it further follows from this that

$$\varphi(\mathfrak{J}_0) \leqq \underline{\lim}_{n \to \infty} \Phi_n(\mathfrak{J}_0) \quad .$$

Similarly one finds

$$\overline{\lim}_{n \to \infty} \Phi_n(\mathfrak{J}_0) \leqq \varphi(\mathfrak{J}_0) \quad ,$$

and therefore that

$$\varphi(\mathfrak{J}_0) = \lim_{n \to \infty} \Phi_n(\mathfrak{J}_0) \quad .$$

Q.E.D.

3.2. THEOREM 6. From each sequence of functions (3, 11) which is uniformly bounded,

$$\varphi_n(\mathfrak{J}) \leqq M \quad ,$$

it is possible to choose a subsequence which is essentially convergent.

And if the original sequence itself is not essentially convergent, then it contains two essentially convergent subsequences whose limit functions are not essentially equal.

PROOF. We consider all such bounded intervals $\bar{\mathfrak{J}}$, where all the coordinates of all corner points are rational numbers. The totality of the intervals $\bar{\mathfrak{J}}$ is a finely meshed system. Since the number of these intervals is countable, one can choose a subsequence

$$\varphi_{n_p}(\mathfrak{J}) \tag{3, 21}$$

which converges on the intervals $\bar{\mathfrak{J}}$. The function

$$\psi(\bar{\mathfrak{J}}) = \lim \varphi_{n_p}(\bar{\mathfrak{J}}) \tag{3, 22}$$

evidently has the property

$$\psi\left(\sum_{\nu=1}^{n} \bar{\mathfrak{J}}_\nu\right) = \sum_{\nu=1}^{n} \psi(\bar{\mathfrak{J}}_\nu) \quad .$$

Starting with $\psi(\bar{\mathfrak{J}})$ we form $\varphi(\mathfrak{M})$ just as we did in (2.4). That is, $\varphi(\mathfrak{O})$ is the upper limit of

$$\sum_{\nu=1}^{n} \psi(\bar{\mathfrak{J}}_\nu)$$

for disjoint intervals $\bar{\mathfrak{J}}_1, \ldots, \bar{\mathfrak{J}}_n$ which with their closure are contained in $\mathfrak{O}$ and $\varphi(\mathfrak{M})$ is the lower limit of $\varphi(\mathfrak{O})$ for all open sets $\mathfrak{O}$ containing $\mathfrak{M}$. This function $\varphi(\mathfrak{M})$ is again a content in the sense that the statements of H. Hahn remain suitably in force if the generating function $\psi(\bar{\mathfrak{J}})$ is defined on a finely meshed interval system $\bar{\mathfrak{J}}$ only.

Of the interval function $\varphi(\mathfrak{J})$ thus gained, we shall show that the sequence (3, 21) converges essentially to it. Let $\mathfrak{J}$ be a bounded interval and $\mathfrak{E}_\epsilon$ an inscribed interval. There is an interval $\bar{\mathfrak{J}}_1$ of our system such that $\mathfrak{E}_\epsilon$ is contained wholly in the interior of $\bar{\mathfrak{J}}_1$ and $\bar{\mathfrak{J}}_1$ is contained wholly in the interior of $\mathfrak{J}$. If, in addition, one inserts an open set between $\mathfrak{E}_\epsilon$ and $\bar{\mathfrak{J}}_1$, one finds that

$$\varphi(\mathfrak{E}_\epsilon) \leq \psi(\bar{\mathfrak{J}}_1) \leq \varphi(\mathfrak{J}) \quad .$$

Similarly there is an interval $\bar{\mathfrak{J}}_2$ between $\mathfrak{J}$ and $\mathfrak{A}_\epsilon$ such that

$$\varphi(\mathfrak{J}) \leqq \psi(\bar{\mathfrak{J}}_2) \leqq \varphi(\mathfrak{A}_\epsilon) \quad .$$

Since

$$\psi(\bar{\mathfrak{J}}_1) \leqq \underline{\lim}\ \varphi_{n_p}(\mathfrak{J}) \leqq \overline{\lim}\ \varphi_{n_p}(\mathfrak{J}) \leqq \psi(\bar{\mathfrak{J}}_2) \quad ,$$

we have

$$\varphi(\mathfrak{G}_\epsilon) \leqq \underline{\lim}\ \varphi_{n_p}(\mathfrak{J}) \leqq \overline{\lim}\ \varphi_{n_p}(\mathfrak{J}) \leqq \varphi(\mathfrak{A}_\epsilon) \quad .$$

If $\mathfrak{J}$ is now a continuity interval of $\varphi(\mathfrak{J})$, then (2, 11) is valid. Hence

$$\varphi(\mathfrak{J}) = \lim \varphi_{n_p}(\mathfrak{J}) \quad ,$$

thus proving the first part of our theorem.

If the entire sequence does not converge essentially then there is a continuity interval $\mathfrak{J}_0$ of the above function $\varphi(\mathfrak{J})$, in which the sequence $\varphi_n(\mathfrak{J}_0)$ does not converge to $\varphi(\mathfrak{J}_0)$, (otherwise $\varphi(\mathfrak{J})$ would be the limit function of the sequence). There is therefore in any case, another subsequence which converges in the interval $\mathfrak{J}_0$, but not to $\varphi(\mathfrak{J}_0)$. From this we now choose a subsequence which is essentially convergent but its limit function is, in $\mathfrak{J}_0$, different from $\varphi(\mathfrak{J}_0)$, Q.E.D.

3.3. If a sequence $\varphi_n(\mathfrak{J})$ converges essentially to a function $\varphi(\mathfrak{J})$, then it is not necessary for the total variation $\varphi_n(\mathfrak{R})$ to converge to the total variation $\varphi(\mathfrak{R})$. But if this also occurs, then we call the sequence <u>compactly convergent</u>.

3.4. For a given $\epsilon > 0$, one can determine a (sufficiently large) cube $\mathfrak{W}$, in which $\varphi(\mathfrak{J})$ is continuous, such that

$$\varphi(\mathfrak{R} - \mathfrak{W}) \leqq \epsilon \quad . \tag{3, 41}$$

If now $\varphi_n(\mathfrak{J})$ converges essentially to $\varphi(\mathfrak{J})$, then

$$\lim_{n \to \infty} \varphi_n(\mathfrak{W}) = \varphi(\mathfrak{W}) \quad .$$

Thus, for the relation

$$\lim_{n \to \infty} \varphi_n(\mathfrak{R}) = \varphi(\mathfrak{R})$$

to hold it is necessary and sufficient that

(3, 42) $$\overline{\lim_{n\to\infty}}\ \varphi_n(\mathfrak{R} - \mathfrak{W}) \leq \epsilon \quad .$$

We have therefore the following criterion. In order that an essentially convergent sequence converge compactly, it is necessary and sufficient that to each $\epsilon > 0$, there exist a cube $\mathfrak{W}$, for which the inequalities (3, 41) and (3, 42) hold.

II. STIELTJES INTEGRALS

§4. DEFINITION AND IMPORTANT PROPERTIES

4.1. Let a monotonic function $\varphi(\alpha)$ with its corresponding interval function $\varphi(\mathfrak{J})$ be given; also an everywhere defined continuous function $\psi(\alpha)$ which is bounded throughout, i.e.,

(4, 11) $$|\psi(\alpha)| \leqq H \quad ,$$

and a point set $\mathfrak{P}$ which for the present, we assume is bounded. Since the function $\psi(\alpha)$ is uniformly continuous on each bounded point set it is possible to define[11] in the obvious manner, following the Riemann procedure, the Stieltjes integral

(4, 12) $$J = \int_{\mathfrak{P}} \psi(\alpha) d\varphi(\alpha) \quad .$$

If one breaks up $\mathfrak{R}$ into finitely many disjoint intervals $\mathfrak{J}_\nu$ and takes from each interval $\mathfrak{J}_\nu$ a point α_ν, then the sum

(4, 13) $$\sum = \sum_{\nu=1}^{n} \psi(\alpha_\nu)\varphi(\mathfrak{J}_\nu)$$

differs from a certain limit value by an arbitrary small amount, provided only that the maximum diameter of the interval $\mathfrak{J}_\nu$ is sufficiently small. This limit value is actually the value of the integral (4, 12).

The following rules of computation are valid:

(4, 14) $$\int_{\mathfrak{P}} 1 d\varphi(\alpha) = \varphi(\mathfrak{P}) \quad ,$$

(4, 15) $$\int_{\mathfrak{P}} c\psi(\alpha) d\varphi(\alpha) = c \int_{\mathfrak{P}} \psi(\alpha) d\varphi(\alpha) \quad ,$$

[11] J. Radon, op. cit[3]. The function $\psi(\alpha)$ can also be complex valued.

$$(4,\ 16)\qquad \int_{\mathfrak{P}} \psi(\alpha)d\varphi(\alpha) = \sum_{\mu=1}^{m} \int_{\mathfrak{P}_\mu} \psi(\alpha)d\varphi(\alpha) \qquad \text{if} \quad \mathfrak{P} = \sum_{\mu=1}^{m} \mathfrak{P}_\mu ,$$

$$(4,\ 17)\qquad \int_{\mathfrak{P}} \psi_1(\alpha)d\varphi(\alpha) \leqq \int_{\mathfrak{P}} \psi_2(\alpha)d\varphi(\alpha) \qquad \text{if} \quad \psi_1(\alpha) \leqq \psi_2(\alpha)$$

$$(4,\ 18)\qquad \left|\int_{\mathfrak{P}} \psi(\alpha)d\varphi(\alpha)\right| \leqq G\varphi(\mathfrak{P}) \qquad \text{if} \quad |\psi(\alpha)| \leqq G ,$$

$$(4,\ 19)\qquad \left|\int_{\mathfrak{P}} \psi(\alpha)d\varphi(\alpha)\right| \leqq \int_{\mathfrak{P}} |\psi(\alpha)|d\varphi(\alpha) ,$$

and

$$(4,\ 10)\qquad \int_{\mathfrak{P}} \left(\sum_{\nu} \psi_\nu\right) d\varphi(\alpha) = \sum_{\nu} \int_{\mathfrak{P}} \psi_\nu d\varphi(\alpha) ,$$

where the summation, in the last equality, can also extend over infinitely many terms, provided the series

$$\sum_{\nu} \psi_\nu$$

converges uniformly in $\mathfrak{P}$.

4.2. By (4, 15) and (4, 16), one can subsequently determine an estimate of the difference of the integral (4, 12) and the approximating sums (4, 13). If δ denotes the maximum diameter of the interval $\mathfrak{J}_\nu$, and if $\omega(\delta)$ denotes a bound for the oscillation of $\varphi(\alpha)$ for any two points of $\mathfrak{P}$ of maximum distance δ, then

$$|J - \Sigma| \leqq \omega(\delta)\varphi(\mathfrak{P}) .$$

4.3. If in addition the function $\varphi(\alpha)$ is dependent on a parameter λ, $\psi = \psi(\alpha, \lambda)$ and is uniformly convergent in (α, λ) then

$$(4,\ 31)\qquad J(\lambda) = \int_{\mathfrak{P}} \psi(\alpha,\ \lambda)d\varphi(\alpha)$$

is uniformly convergent in λ. In fact,

$$|J(\lambda_1) - J(\lambda_2)| \leq \int_{\mathfrak{P}} |\psi(\alpha, \lambda_1) - \psi(\alpha, \lambda_2)| d\varphi(\alpha)$$

$$\leq G(\lambda_1, \lambda_2)\varphi(\mathfrak{P})$$

where

$$G(\lambda_1, \lambda_2) = \text{supremum } |\psi(\alpha, \lambda_1) - \psi(\alpha, \lambda_2)| \quad , \qquad \alpha \in \mathfrak{P} \ .$$

If λ stands for several ordinary parameters and takes on the values of an interval $\mathfrak{L}$ of the λ-space then, since $J(\lambda)$ is continuous, the Riemann integral

$$\int_{\mathfrak{L}} J(\lambda)\, d\lambda \quad , \tag{4, 32}$$

exists. We are going to show that the integrations with respect to λ and α can be interchanged, so that (4, 32) has the value

$$\int_{\mathfrak{P}} \left[\int_{\mathfrak{L}} \psi(\alpha, \lambda)\, d\lambda \right] d\varphi(\alpha) \quad . \tag{4, 33}$$

If $\mathfrak{L}$ is divided into disjoint intervals $\mathfrak{L}_\nu$ of sufficiently small diameter, then (4, 32) can be approximated by a sum

$$\sum_{\nu=1}^{n} J(\lambda_\nu) m(\mathfrak{L}_\nu) \tag{4, 34}$$

in which λ_ν is a point of $\mathfrak{L}_\nu$ and $m(\mathfrak{L}_\nu)$ the Euclidean volume of $\mathfrak{L}_\nu$. By (4, 31), (4, 34) has the value

$$\int_{\mathfrak{P}} \left[\sum_{\nu=1}^{n} \psi(\alpha, \lambda_\nu) m(\mathfrak{L}_\nu) \right] d\varphi(\alpha) \quad .$$

But the integrand

$$\sum_{\nu=1}^{n} \psi(\alpha, \lambda_\nu) m(\mathfrak{L}_\nu) \tag{4, 35}$$

is now an approximation sum of the quantity

$$(4,\ 36) \qquad \int_{\mathfrak{L}} \psi(\alpha,\ \lambda)\ d\lambda = J(\lambda) \quad ,$$

and since $\psi(\alpha,\ \lambda)$ is uniformly continuous, the difference $\rho(\alpha)$ of the quantities (4, 35) and (4, 36) is, by 4.2, smaller than an ϵ independently of α. Our assertion then follows from

$$\left| \int_{\mathfrak{P}} \rho(\alpha) d\varphi(\alpha) \right| \leq \epsilon\varphi(\mathfrak{P}) \quad .$$

4.4. We shall now define the integral (4, 12) for general point sets $\mathfrak{P}$. We introduce the point set $\mathfrak{P}_n$ as in 1.3 and define (4, 12) by the limit

$$\lim_{n \to \infty} \int_{\mathfrak{P}_n} \psi(\alpha) d\varphi(\alpha) \quad .$$

That the limit exists and is independent of the special sequence $\mathfrak{P}_n$ follows from

$$\left| \int_{\mathfrak{P}_{n+p}} \psi(\alpha) d\varphi(\alpha) - \int_{\mathfrak{P}_n} \right| \leq H\varphi(\mathfrak{P}_{n+p} - \mathfrak{P}_n) \leq H\varphi(\mathfrak{R} - \mathfrak{P}_n) \quad .$$

In particular, the integral formed over the whole space

$$\int_{\mathfrak{R}} \psi(\alpha) d\varphi(\alpha)$$

exists. We will write for this

$$\int \psi(\alpha) d\varphi(\alpha)$$

without specifying the pointset over which to integrate.

It is easy to see that the rules in 4.1 remain valid for general $\mathfrak{P}$.

We now introduce two "theorems of identification".

4.5. Let $v(\alpha)$ be a non-negative function, Lebesgue integrable over the whole space. Then as is well known, the Lebesgue integral

$$(4,\ 51) \qquad \varphi(\mathfrak{P}) = \int_{\mathfrak{P}} v(\alpha)\ d\alpha$$

is an interval function (according to our definition), and what is more $\varphi(\mathfrak{P})$ is absolutely continuous, by which we mean that to each ϵ there is a δ such that

$$\varphi(\mathfrak{P}) \leqq \epsilon \quad \text{if} \quad m(\mathfrak{P}) \leqq \delta \ .$$

Conversely, if an interval function $\varphi(\mathfrak{P})$ is absolutely continuous, then it is representable in the form (4, 51). The first "theorem of identification" states that in the last case, the Stieltjes integral

$$\int_{\mathfrak{P}} \psi(\alpha) d\varphi(\alpha) \tag{4, 52}$$

is equal to the Lebesgue integral

$$\int_{\mathfrak{P}} \psi(\alpha) v(\alpha)\, d\alpha \ .$$

It is sufficient to prove this assertion for bounded $\mathfrak{P}$. We approximate (4, 52) by a sum (4, 13) and note that

$$\left| \psi(\alpha_\nu)\varphi(\mathfrak{J}_\nu) - \int_{\mathfrak{J}_\nu} \psi(\alpha) v(\alpha)\, d\alpha \right| \leqq \int_{\mathfrak{J}_\nu} |\psi(\alpha_\nu) - \psi(\alpha)| \cdot v(\alpha)\, d\alpha$$

$$\leqq \omega(\delta) \int_{\mathfrak{J}_\nu} v(\alpha)\, d\alpha \leqq \omega(\delta)\varphi(\mathfrak{J}_\nu) \ ,$$

where $\omega(\delta)$ is defined as in 4.2. Our assertion then follows.

4.6. Let $\varphi(\alpha)$ now be a general monotonic function and $\psi(\alpha)$ a non-negative continuous bounded function. The function

$$\chi(\mathfrak{P}) = \int_{\mathfrak{P}} \psi(\alpha) d\varphi(\alpha) \ , \tag{4, 61}$$

[which agrees with $\varphi(\mathfrak{P})$ if $\psi(\alpha) \equiv 1$], is an interval function, and our second "theorem of identification" reads as follows. If $\Psi(\alpha)$ is another continuous bounded function, then the integrals

$$\int_{\mathfrak{P}} \Psi(\alpha) d\chi(\alpha) \tag{4, 62}$$

and

$$\int_{\mathfrak{P}} [\Psi(\alpha)\psi(\alpha)]d\varphi(\alpha) \tag{4, 63}$$

are equal to each other. For its proof we make use of the following conclusion from the rules of computation in 4.1. Between the upper and lower limits of $\psi(\alpha)$ in $\mathfrak{P}$, there is an in between value ψ_0 such that

$$\int_{\mathfrak{P}} \psi(\alpha)d\varphi(\alpha) = \psi_0\varphi(\mathfrak{P}) \quad .$$

Hence for bounded $\mathfrak{P}$, each approximation sum

$$\sum_{\nu=1}^{n} \Psi(\alpha_\nu)\chi(\mathfrak{J}_\nu)$$

of (4, 62) can be written in the form

$$\sum_{\nu=1}^{n} \Psi(\alpha_\nu)\psi_\nu\varphi(\mathfrak{J}_\nu) \quad , \tag{4, 64}$$

where ψ_ν is an in between value of $\psi(\alpha)$ in $\mathfrak{J}_\nu$. Because of the uniform continuity of $\psi(\alpha)$, (4, 64) itself differs arbitrarily little from the sum

$$\sum_{\nu=1}^{n} \Psi(\alpha_\nu)\psi(\alpha_\nu)\chi(\mathfrak{J}_\nu) \quad ,$$

which is itself an approximation sum of (4, 63), Q.E.D.

§5. UNIQUENESS AND LIMIT THEOREMS

5.1. THEOREM 7. If the interval functions $\varphi_1(\mathfrak{J})$ and $\varphi_2(\mathfrak{J})$ are essentially equal, then

$$\int \psi(\alpha)d\varphi_1(\alpha) = \int \psi(\alpha)d\varphi_2(\alpha) \quad .$$

PROOF. Because of Theorem 1 and by reason of the definitions, the integrals

$$(5,\ 11) \qquad \int \psi(\alpha)d\varphi_1(\alpha) \ , \qquad \int \psi(\alpha)d\varphi_2(\alpha)$$

can be approximated arbitrarily closely by sums

$$(5,\ 12) \qquad \sum_{\nu=1}^{n} \psi(\alpha_\nu)\varphi_1(\mathfrak{J}_\nu) \ , \qquad \sum_{\nu=1}^{n} \psi(\alpha_\nu)\varphi_2(\mathfrak{J}_\nu) \ ,$$

in which the intervals $\mathfrak{J}_\nu$ are common continuity intervals of $\varphi_1(\mathfrak{J})$ and $\varphi_2(\mathfrak{J})$. But these sums are equal term for term; therefore the integrals (5, 11) are also equal.

5.2. It can now be shown that our integrals are independent of the special coordinate system.

GENERALIZATION OF THEOREM 7. If the given interval functions $\varphi_1(\mathfrak{J})$ and $\varphi_2(\widetilde{\mathfrak{J}})$ are essentially equal in different coordinate systems, then

$$\int \psi(\alpha)d\varphi_1(\alpha) = \int \psi(\widetilde{\alpha})d\varphi_2(\widetilde{\alpha}) \ .$$

PROOF. Let $\varphi(\mathfrak{M})$ be a common dominant function of $\varphi_1(\mathfrak{J})$ and $\varphi_2(\widetilde{\mathfrak{J}})$. By Theorem 7, we have left to show only that

$$\int \psi(\alpha)d\varphi(\alpha) = \int \psi(\widetilde{\alpha})d\varphi(\widetilde{\alpha}) \ .$$

But the proof of this fact proceeds, entirely analogously, as it does for the invariance of the Riemann integral under orthogonal transformations. We shall only sketch its proof. We approximate the left integral by a sum

$$(5,\ 21) \qquad \sum_{\nu=1}^{n} \psi(\alpha_\nu)\varphi(\mathfrak{J}_\nu) \ ,$$

which is held fixed and in which the $\mathfrak{J}_\nu$ are continuity intervals. It is now possible to introduce an approximating sum

$$(5,\ 22) \qquad \sum_{\mu=1}^{m} \psi(\widetilde{\alpha}_\mu)\varphi(\widetilde{\mathfrak{J}}_\mu)$$

in which the maximal diameter of $\widetilde{\mathfrak{J}}_\mu$ is sufficiently small, and which is such that the expressions (5, 21) and (5, 22) differ by a quantity whose absolute value is $\leq \omega(\delta)\varphi(\mathfrak{R})$ where $\omega(\delta)$ is defined as in 4.2.

5.3. THEOREM 8. If for each continuous bounded function $\psi(\alpha)$

$$\int \psi(\alpha)d\varphi_1(\alpha) = \int \psi(\alpha)d\varphi_2(\alpha) ,$$

then the interval functions $\varphi_1(\mathfrak{J})$ and $\varphi_2(\mathfrak{J})$ are essentially equal.

PROOF. Let $\mathfrak{J}$ be a bounded interval and E_ϵ an inscribed interval. Then there evidently exists a continuous function $v(\alpha)$ for which

$$\begin{aligned} v(\alpha) &= 1 && \text{if} \quad \alpha \subset E_\epsilon \\ v(\alpha) &= 0 && \text{if} \quad \alpha \subset \mathfrak{R} - \mathfrak{J} \\ 0 \leqq v(\alpha) &\leqq 1 && \text{if} \quad \alpha \subset \mathfrak{J} - E_\epsilon . \end{aligned}$$

By the rules of computation in 4.1

$$\int v(\alpha)d\varphi_1(\alpha) = \varphi_1(\mathfrak{J}) - \vartheta_1\varphi_1(\mathfrak{J} - E_\epsilon) \qquad 0 \leqq \vartheta_1 \leqq 1 ,$$

$$\int v(\alpha)d\varphi_2(\alpha) = \varphi_2(\mathfrak{J}) - \vartheta_2\varphi_2(\mathfrak{J} - E_\epsilon) \qquad 0 \leqq \vartheta_2 \leqq 1 .$$

By subtraction, there results

$$|\varphi_1(\mathfrak{J}) - \varphi_2(\mathfrak{J})| \leqq \varphi_1(\mathfrak{J} - E_\epsilon) + \varphi_2(\mathfrak{J} - E_\epsilon) .$$

If therefore $\mathfrak{J}$ is a common continuity interval of $\varphi_1(\mathfrak{J})$ and $\varphi_2(\mathfrak{J})$, then as $\epsilon \longrightarrow 0$, we have

$$\varphi_1(\mathfrak{J}) - \varphi_2(\mathfrak{J}) = 0 ,$$

Q.E.D.

5.4. THEOREM 9.[12] If the interval functions

$$(5, 41) \qquad \varphi_1(\mathfrak{J}), \quad \varphi_2(\mathfrak{J}), \quad \varphi_3(\mathfrak{J}), \quad \ldots$$

converge compactly to a function $\varphi_0(\mathfrak{J})$, then

$$(5, 42) \qquad \lim_{m \to \infty} \int \psi(\alpha)d\varphi_m(\alpha) = \int \psi(\alpha)d\varphi_0(\alpha) .$$

[12] For $k = 1$ due to Helly, op. cit.

PROOF. Because of the compactness of the convergence, if $\mathfrak{J}$ is sufficiently large, the integral

$$\int_{\mathfrak{J}} \psi(\alpha) d\varphi_m(\alpha) \tag{5, 43}$$

differs sufficiently little from

$$\int \psi(\alpha) d\varphi_m(\alpha)$$

uniformly for $m = 0, 1, 2, 3, \ldots$. Let $\mathfrak{J}$ now be a common continuity interval of the functions $\varphi_0, \varphi_1, \varphi_2, \varphi_3, \ldots$ which can be divided into arbitrarily fine intervals of the same kind. By the evaluation in 4.2, the integral (5, 43) can be approximated uniformly for $m = 0, 1, 2, 3, \ldots$ by a sum

$$\sum_m = \sum_{\nu=1}^{n} \varphi(\alpha_\nu) \varphi_m(\mathfrak{J}_\nu) \quad .$$

But for fixed $\mathfrak{J}_\nu$

$$\lim_{m \to \infty} \sum_m = \sum_0 \quad .$$

Hence our assertion follows.

5.5. THEOREM 10.[12] Let a sequence (5, 41) be given. If for each bounded continuous function $\psi(\alpha)$, the number

$$\psi_m = \int \psi(\alpha) d\varphi_m(\alpha)$$

converges to a (finite) number ψ_0, then the sequence (5, 41) is essentially convergent.

PROOF. If $\psi(\alpha) = 1$, then $\psi_m = \varphi_m(\mathfrak{R})$, and therefore the sequence (5, 41) is uniformly bounded. Let $\varphi_{m_p}(\alpha)$, by Theorem 6, be any essentially convergent subsequence, and denote the limit function by $\chi_p(\alpha)$. From the proof of Theorem 9, we infer, for each continuous function $v(\alpha)$ which vanishes outside a bounded interval, that

$$v_0 = \lim \int v(\alpha) d\varphi_{m_p}(\alpha) = \int v(\alpha) d\chi_p(\alpha) \quad .$$

If now a second subsequence $\varphi_{m_q}(\alpha)$ converges essentially to a function $\chi_q(\alpha)$, then also

$$v_0 = \int v(\alpha) dx_q(\alpha) \quad .$$

By comparing the two, we have

$$\int v(\alpha) dx_p(\alpha) = \int v(\alpha) dx_q(\alpha) \quad .$$

Hence by the proof of Theorem 8, we infer that x_p and x_q are essentially equal. By the second part of Theorem 6, the sequence (5, 41) is therefore essentially convergent, Q.E.D.

III. HARMONIC ANALYSIS

§6. FOURIER-STIELTJES INTEGRALS

6.1. Let α and x be two points of our (k-dimensional) space. For the number

$$\sum_{\kappa=1}^{\kappa} \alpha^{(\kappa)} x^{(\kappa)}$$

which occurs frequently, we introduce a shortened notation,

$$\alpha x \ .$$

We shall denote monotonic functions in what follows by $V(\alpha)$, $W(\alpha)$, $U(\alpha)$, etc., and call them <u>distribution functions</u>. To each distribution function $V(\alpha)$, we attach the function

$$f(x) = \int e^{ix\alpha} dV(\alpha) \quad , \tag{6, 11}$$

and call it the <u>characteristic function</u> belonging to $V(\alpha)$. It is defined at all points of the x-space. By a characteristic function we shall (simply) mean a function $f(x)$ which can be determined by means of a distribution function in the manner specified.

In particular, for non-negative functions $v(\alpha)$, integrable over $\mathfrak{R}$, the functions

$$\int e^{ix\alpha} v(\alpha)\, d\alpha \quad ,$$

belong to the characteristic functions.

6.2. The characteristic functions are

1. positive-additive, i.e., if f_1 and f_2 are characteristic functions, then $c_1 f_1 + c_2 f_2$ is also a characteristic function, where $c_1 \geqq 0$ and $c_2 \geqq 0$;

2. bounded, and what is more

$$|f(x)| \leq f(0) = V(\mathfrak{R}) \quad , \tag{6, 21}$$

3. "hermitian", i.e.,

$$\overline{f(-x)} = f(x) \quad , \tag{6, 22}$$

and

4. uniformly continuous.

This last property comes about in the following manner. If with cubes W_n, which converge to $\mathfrak{R}$, one forms the functions

$$f_n(x) = \int_{W_n} e^{ix\alpha} dV(\alpha) \quad , \tag{6, 23}$$

then by 4.3, each of them is uniformly continuous. And because

$$|f - f_n| \leq \int_{\mathfrak{R} - W_n} |e^{ix\alpha}| dV(\alpha) = V(\mathfrak{R} - W_n) \quad ,$$

it follows that they converge uniformly to f.

6.3. We will base our analysis on the fact that the relation

$$g(\lambda) = \int e^{i\lambda\alpha} v(\alpha)\, d\alpha \tag{6, 31}$$

has the Fourier inverse relation

$$v(\alpha) = \frac{1}{(2\pi)^k} \int e^{-i\alpha\lambda} g(\lambda)\, d\lambda \tag{6, 32}$$

and this holds in the following specific setting. Assume that the given function $v(\alpha)$ is absolutely integrable in $\mathfrak{R}$. Therefore the integrand on the right of (6, 31) exists and the transform $g(\lambda)$ is a continuous bounded function. Now, assume further that this function $g(\lambda)$ is, for its part, also absolutely integrable in $\mathfrak{R}$. Then the integral on the right of (6, 32) represents a bounded continuous function, which we denote, for the moment, by $v_1(\alpha)$. A theorem from the theory of Fourier integrals[13] now states that for almost all α, the functions $v_1(\alpha)$ and $v(\alpha)$ agree. If the original function $v(\alpha)$ is therefore modified, if need be, on a set of measure zero, [a modification which does not change the integral (6, 31),

[13] "Fourier Integrals" Theorem 60.

i.e., the value of the function $g(\alpha)$] then for <u>all</u> points α, (6, 32) results.

6.4. THEOREM 11. Let

$$f(x) = \int e^{ix\alpha} dV(\alpha) \tag{6, 41}$$

be a given general characteristic function, and

$$g(\xi) = \int e^{i\xi\alpha} v(\alpha)\, d\alpha \quad , \tag{6, 42}$$

a "special" characteristic function which is absolutely integrable over the whole space.

Then

$$\frac{1}{(2\pi)^k} \int f(x - \xi) g(\xi)\, d\xi = \int e^{ix\alpha} v(\alpha) dV(\alpha) \quad . \tag{6, 43}$$

REMARK. By 6.3, $v(\alpha)$ itself is continuous and bounded. Therefore the integral on the right of (6, 43) is, according to our definition, a Stieltjes integral. By 4.6, the function

$$W(\mathfrak{J}) = \int_{\mathfrak{J}} v(\alpha) dV(\alpha)$$

is a distribution function, and our theorem is equivalent with that statement that the function on the left of (6, 43) is its characteristic function.

PROOF. Fy 4.3, for fixed x and any bounded interval $\mathfrak{L}$ of the λ-space we have

$$\frac{1}{(2\pi)^k} \int_{\mathfrak{L}} f_n(x - \lambda) g(\lambda)\, d\lambda = \int_{W_n} e^{ix\alpha} \left[\frac{1}{(2\pi)^k} \int_{\mathfrak{L}} e^{-i\lambda\alpha} g(\lambda) d\lambda \right] dV(\alpha) \tag{6,44}$$

for each approximating function (6, 23) of $f(x)$. But since for large $\mathfrak{L}$ the expression in brackets differs but little from

$$\frac{1}{(2\pi)^k} \int e^{-i\alpha\lambda} g(\lambda)\, d\lambda \equiv v(\alpha)$$

uniformly in α it follows, by (6, 44) that

$$\frac{1}{(2\pi)^k} \int f_n(x - \lambda) g(\lambda)\, d\lambda = \int_{W_n} e^{ix\alpha} v(\alpha) dV(\alpha) \quad .$$

We now let $n \longrightarrow \infty$.

THEOREM 12. Let $V(\alpha)$ be a given distribution function, and $v(\alpha)$ a non-negative, continuous, bounded function integrable over the whole space.

The function

$$w(\beta) = \int v(\beta - \alpha) dV(\alpha)$$

is again non-negative, continuous, bounded and integrable over $\Re$, and the product of the characteristic functions

$$f(x) = \int e^{ix\alpha} dV(\alpha), \qquad g(x) = \int e^{ix\alpha} v(\alpha)\, d\alpha$$

is again a characteristic function, namely

$$(6,51) \qquad f(x)g(x) = \int e^{ix\alpha} w(\alpha)\, d\alpha \quad .$$

PROOF. Take a cube A_n of the α-space and a cube B_m of the β-space each converging monotonically to the whole space. The functions

$$w_n(\beta) = \int_{A_n} v(\beta - \alpha) dV(\alpha)$$

are non-negative, continuous, bounded and monotonically increasing to $w(\beta)$. Therefore $w(\beta)$ is non-negative. From

$$|w(\beta) - w_n(\beta)| \leqq G \cdot V(\Re - A_n)$$

(where G denotes a bound of $v(\alpha)$), it follows that $w_n(\beta)$ converges uniformly to $w(\beta)$. Hence $w(\beta)$ is continuous and bounded. Further

$$\int_{B_m} w_n(\beta)\, d\beta = \int_{A_n} \left[\int_{B_m} v(\beta - \alpha)\, d\beta \right] dV(\alpha)$$

$$\leqq \int \left[\int v(\beta - \alpha)\, d\beta \right] dV(\alpha) = \int v(\beta)\, d\beta \cdot V(\Re) \; .$$

From this it follows, as $n \longrightarrow \infty$, that

$$\int_{B_m} w(\beta)\, d\beta \leqq \int v(\beta)\, d\beta \cdot V(\Re) \quad ,$$

from which we infer that $w(\beta)$ is integrable over the whole space.

We have in addition

$$(6, 52) \quad \int_{B_m} e^{ix\beta} w_n(\beta)\, d\beta = \int_{A_n} \left[\int_{B_m} e^{i(\beta-\alpha)x} v(\beta - \alpha)\, d\beta \right] e^{ix\alpha} dV(\alpha) \quad ,$$

and since for all α out of A_n

$$\lim_{m \to \infty} \int_{B_m} e^{i(\beta-\alpha)x} v(\beta - \alpha)\, d\beta = \int e^{i(\beta-\alpha)x} v(\beta - \alpha)\, d\beta = g(x)$$

uniformly, there results by (6, 52)

$$(6, 53) \qquad \int e^{ix\beta} w_n(\beta)\, d\beta = g(x) \cdot \int_{A_m} e^{ix\alpha} dV(\alpha) \quad .$$

Now since

$$|e^{ix\beta} w_n(\beta)| \leq w(\beta) \quad ,$$

and $w(\beta)$ is integrable over $\Re$, one can allow $n \longrightarrow \infty$ under the integral sign in the left of (6, 53). The result is (6, 51).

6.6. We shall make use of the fact that if

$$v(\beta) = e^{- \sum_{\kappa=1}^{k} (\beta^{(\kappa)})^2} \quad ,$$

then the function

$$g(x) = \int e^{ix\alpha} v(\alpha)\, d\alpha$$

has the value

$$\pi^{k/2} e^{- 1/4 \sum_{\kappa=1}^{k} (x^{(\kappa)})^2}$$

§7. UNIQUENESS AND LIMIT THEOREMS

7.1. THEOREM 13. If the distribution functions $V_1(\alpha)$ and $V_2(\alpha)$ are essentially equal, then their characteristic functions are identically equal,

$$\int e^{ix\alpha}dV_1(\alpha) = \int e^{ix\alpha}dV_2(\alpha) \quad .$$

PROOF. Follows from Theorem 7.

THEOREM 14. Conversely, if the functions

$$f_1(x) = \int e^{ix\alpha}dV_1(\alpha), \qquad f_2(x) = \int e^{ix\alpha}dV_2(\alpha)$$

are identically equal (because of their continuity, it is already sufficient for this requirement that they agree for an everywhere dense set of points), then their distribution functions are essentially equal.

Expressed differently, the attachment of the distribution functions and characteristic functions is an uniquely reversible one.

PROOF. We form the equation (6, 43) for $f(x) \equiv f_1$ and $f(x) \equiv f_2$, set $x = 0$, and equate the right sides. This gives

$$\int v(\alpha)dV_1(\alpha) = \int v(\alpha)dV_2(\alpha)$$

for all functions $v(\alpha)$ which satisfy the assumptions of Theorem 11. But to these functions belong all those functions which possess partial derivatives up to the 2 kth order and vanish outside a bounded interval.[14] Now the functions $v(\alpha)$ which occur in the proof of Theorem 8 may be presumed to have derivatives of arbitrarily high order and we can now reason as we did for Theorem 8.

7.2. THEOREM 15. If the functions

$$V_1(\alpha),\ V_2(\alpha),\ V_3(\alpha),\ \ldots \tag{7, 21}$$

converge compactly to a function $V_0(\alpha)$, then the functions

$$f_m(x) = \int e^{ix\alpha}dV_m(\alpha) \qquad m = 1, 2, 3, \ldots \tag{7, 22}$$

converge uniformly in each finite x-interval to

$$f_0(x) = \int e^{ix\alpha}dV_0(\alpha) \quad . \tag{7. 23}$$

[14] "Fourier Integrals", §44.4.

PROOF. In the proof of Theorem 9, we replace $\psi(\alpha)$ by the function $e^{ix\alpha}$, and note that then all valuations hold uniformly in each bounded x-interval.

THEOREM 16. Conversely if the sequence (7, 22), is uniformly bounded in x, i.e.,

$$|f_m(x)| \leq M \tag{7, 24}$$

and for almost all x converges to a limit $f_*(x)$, then the sequence (7, 21) is essentially convergent. If one denotes this limit function by $V_0(\alpha)$, then

$$f_*(x) = f_0(x) \tag{7, 25}$$

for almost all x.

REMARK. Because

$$|f_m(x)| \leq f_m(0) = V_m(\mathfrak{R}) \quad , \tag{7, 26}$$

assumption (7, 24) is equivalent to

$$V_m(\mathfrak{R}) \leq M \quad . \tag{7, 27}$$

Because $f_m(0) = V_m(\mathfrak{R})$, this condition is then always satisfied if the sequence (7, 22) also converges at the point $x = 0$ (to a finite number); in particular therefore if the functions (7, 22) converge for all x.

7.3. PROOF. At first we make the additional assumption that the sequence (7, 21) is essentially convergent to a function V_0. Because of this fact, we shall show that (7, 25) is valid. We take the functions $v(\alpha)$ and $g(x)$ from 6.6, and form with them the functions

$$F_m(x) = f_m(x)g(x) \qquad m = 0, 1, 2, 3, \ldots \tag{7, 31}$$

$$F_*(x) = f_*(x)g(x) \tag{7, 32}$$

$$w_m(\beta) = \int v(\beta - \alpha)dV_m(\alpha) \qquad m = 0, 1, 2, 3, \ldots \tag{7, 33}$$

By Theorem 12,

$$F_m(x) = \int e^{ix\alpha}w_m(\alpha)\, d\alpha \qquad m = 0, 1, 2, 3, \ldots \tag{7, 34}$$

and by hypothesis

$$(7,\ 35)\qquad \lim_{m \to \infty} F_m(x) = F_*(x) \qquad \underline{\text{for almost all}}\ \ x\ .$$

For each common continuity interval $\mathfrak{J}$ of all $V_m(\alpha)$. we have

$$\lim_{m \to \infty} \int_{\mathfrak{J}} v(\beta - \alpha) dV_m(\alpha) = \int_{\mathfrak{J}} v(\beta - \alpha) dV_o(\alpha) \ .$$

Further

$$\left| \int_{\mathfrak{R} - \mathfrak{J}} v(\beta - \alpha) dV_m(\alpha) \right| \leq \delta M \ ,$$

where δ means, for fixed β, the maximum of $v(\beta - \alpha)$ for all α in $\mathfrak{R} - \mathfrak{J}$. Since δ becomes arbitrarily small for large $\mathfrak{J}$, there results

$$(7,\ 36)\qquad \lim_{m \to \infty} w_m(\beta) = w_o(\beta) \ .$$

By (7, 24)

$$|F_m(x)| \leq Mg(x) \ ,$$

and since $g(x)$ is integrable over the whole space, we have from (7, 35), by a theorem of Lebesgue, that

$$(7,\ 37)\qquad \lim_{m \to \infty} \int |F_m(x) - F_*(x)|^{\cdot} dx = 0 \ .$$

Making use of the first half of Theorem 12, we have by 6.4, the converse to (7, 34) namely

$$w_m(\alpha) = \frac{1}{(2\pi)^k} \int e^{-i\alpha x} F_m(x)\, dx \ .$$

By (7, 36) and (7, 37) and by letting $m \longrightarrow \infty$, there results

$$w_o(\alpha) = \frac{1}{(2\pi)^k} \int e^{-i\alpha x} F_*(x)\, dx \ .$$

The converse of this reads

$$F_*(x) = \int e^{ix\alpha} w_o(\alpha)\, d\alpha \qquad \underline{\text{for almost all}}\ \ x\ ,$$

or differently written, cf. (7, 32) and (7, 34),

$$f_*(x)g(x) = f_0(x)g(x) \qquad \underline{\text{for almost all}} \quad x.$$

And since $g(x) \neq 0$, (7, 25) results.

We now assume as given only the hypothesis of the theorem. If the sequence $V_m(\alpha)$ were not essentially convergent, then there would be two subsequences which would converge to two essentially different functions $V_{01}(\alpha)$ and $V_{02}(\alpha)$. By the same proofs would

$$f_*(x) = \int e^{ix\alpha} dV_{01}(\alpha) \qquad \underline{\text{for almost all}} \quad x$$

and

$$f_*(x) = \int e^{ix\alpha} dV_{02}(\alpha) \qquad \underline{\text{for almost all}} \quad x .$$

Therefore would

$$\int e^{ix\alpha} dV_{01}(\alpha) = \int e^{ix\alpha} dV_{02}(\alpha) \qquad \underline{\text{for almost all}} \quad x$$

in contradiction to Theorem 14.

7.4. THEOREM 17. If the functions

$$f_m(x) = \int e^{ix\alpha} dV_m(\alpha)$$

converge, for almost all x, to a continuous function $f_*(x)$, then the sequence $V_m(\alpha)$ is compactly convergent to a function $V_0(\alpha)$, and hence

$$f_*(x) = f_0(x) \qquad \underline{\text{for all}} \quad x .$$

PROOF. Since the assumptions of Theorem 16 are satisfied, the sequence $V_m(\alpha)$ is essentially convergent to $V_0(\alpha)$, and hence $f_*(x) = f_0(x)$ for almost all x. Since now $f_*(x)$ is continuous by hypothesis and $f_0(x)$ is continuous from the outset, it follows that $f_*(x) = f_0(x)$ for all x. In particular if $x = 0$, we have

$$\lim_{m \to \infty} \int dV_m(\alpha) = \int dV_0(\alpha) ,$$

and thus the compactness of the convergence results.

We are now in the position to prove the following theorem.

THEOREM 18. The product of two arbitrary characteristic functions is a characteristic function.

PROOF. Let us call the given functions $f_1(x)$ and $f_2(x)$. With the functions $v(\alpha)$ and $g(x)$ of 6.6, we form the functions

$$v_n(\alpha) = n^k v(n\alpha', n\alpha'', \ldots, n\alpha^{(k)})$$

(7, 51)

$$g_n(x) = g\left(\frac{x'}{n}, \frac{x''}{n}, \ldots, \frac{x^{(k)}}{n}\right) .$$

Then again the relation

$$g_n(x) = \int e^{ix\alpha} v_n(\alpha)\, d\alpha$$

holds, and the pair of functions $v_n(\alpha)$, $g_n(x)$, satisfy Theorem 12. If we apply this theorem first to the functions f_1 and g_n and subsequently to the functions f_2 and $f_1 g_n$ we find that the function

$$g_n(x) f_1(x) f_2(x) \tag{7, 52}$$

is a characteristic function. As $n \longrightarrow \infty$, the sequence (7, 51) is convergent to the continuous function $f_1(x) f_2(x)$. Hence by Theorem 17, the last function is also a characteristic function.

§8. POSITIVE-DEFINITE FUNCTIONS

8.1. We call a function $f(x)$, which is defined for all points of our space, positive-definite if it

1. is bounded and continuous,
2. is "hermitian", i.e.,

$$\overline{f(-x)} = f(x) \quad , \tag{8, 11}$$

3. satisfies the following decisive condition. For any points $x_1, x_2, \ldots, x_m$ $(m = 2, 3, 4, \ldots)$ and any numbers $\rho_1, \rho_2, \ldots, \rho_m$

$$\sum_{\mu=1}^{m} \sum_{\nu=1}^{m} f(x_\mu - x_\nu)\rho_\mu \rho_\nu \geq 0 \quad . \tag{8, 12}$$

From 3, in conjunction with the continuity of $f(x)$, we obtain the following. For each function $\rho(x)$, continuous in a bounded interval $\mathfrak{J}$,

$$\int_{\mathfrak{J}}\int_{\mathfrak{J}} f(x - y)\rho(x)\overline{\rho(y)}\, dx\, dy \geq 0 \tag{8, 13}$$

(the integral is a 2k-dimensional one). This inequality results if the integral (8, 13) is approximated by a suitable Riemann sum to which (8, 12) is applicable. If $\rho(x)$ is continuous and absolutely integrable in the whole space, then as $\mathfrak{J} \longrightarrow \mathfrak{R}$ we have by (8, 13),

$$\int\int f(x - y)\rho(x)\overline{\rho(y)}\, dx\, dx \geq 0 \quad . \tag{8, 14}$$

8.2. Each characteristic function

$$f(x) = \int e^{ix\alpha} dV(\alpha)$$

is positive-definite. In fact properties 1. and 2. are known to us. In regard to 3., the left side of (8, 12) has the value

$$\int T(\alpha) dV(\alpha) \quad ,$$

where

$$T(\alpha) = \sum_{\mu=1}^{m} \sum_{\nu=1}^{m} e^{i(x_\mu - x_\nu)\alpha} \rho_\mu \overline{\rho_\nu} = \left| \sum_{\mu=1}^{m} e^{ix_\mu \alpha} \rho_\mu \right|^2 \geq 0 \quad ,$$

and is therefore ≥ 0.

But the converse is also valid, namely that each positive-definite function is a characteristic function.

THEOREM 19. In order that a function $f(x)$ can be written in the form

$$\int e^{ix\alpha} dV(\alpha) \quad ,$$

it is necessary and sufficient that it be positive-definite.

PROOF. Let $f(x)$ therefore be positive-definite. We have to show that it is a characteristic function. We shall assume at first that $f(x)$ is also absolutely integrable over $\mathfrak{R}$. With each continuous function $g(x)$ absolutely integrable over $\mathfrak{R}$, we can then form the function

$$F(\xi) = \frac{1}{(2\pi)^k} \int\int f(\xi - x - y) g(x) \overline{g(-y)} \, dx \, dy$$

$$= \frac{1}{(2\pi)^k} \int\int f(\xi + x - y) g(-x) \overline{g(-y)} \, dx \, dy \quad .$$

It is easy to see that it is continuous, and by (8, 14) that

(8, 21) $$F(0) \geq 0 \quad .$$

Further $F(\xi)$ is absolutely integrable over $\mathfrak{R}$.[15] If one forms

(8, 22) $$E(\alpha) = \frac{1}{(2\pi)^k} \int e^{-ix\alpha} f(x) \, dx$$

and

(8, 23) $$\Gamma(\alpha) = \frac{1}{(2\pi)^k} \int e^{-ix\alpha} g(x) \, dx \quad ,$$

then as is easily computed[14]

$$|\Gamma(\alpha)|^2 E(\alpha) = \frac{1}{(2\pi)^k} \int e^{-i\xi\alpha} F(\xi) d(\xi) \quad .$$

In addition, if $\Gamma(\alpha)$ is absolutely integrable, the converse, cf. 6.4,

$$F(\xi) = \int e^{i\xi\alpha} |\Gamma(\alpha)|^2 E(\alpha) \, d\alpha$$

is valid. Since $F(\xi)$ is continuous, it is valid for <u>all</u> points. Because of (8, 21), for $\xi = 0$, we obtain

(8, 24) $$\int |\Gamma(\alpha)|^2 E(\alpha) \, d\alpha \geq 0 \quad .$$

By (8, 11), $E(\alpha)$ is real, and since to the functions $\Gamma(\alpha)$, which can be formed with the aid of (8, 23) by means of an absolutely integrable continuous function $g(x)$, also belong those functions which differ from zero at a prescribed point α_0 and vanish outside of a prescribed neighborhood of α_0, it therefore follows from (8, 24) that

$$E(\alpha) \geq 0 \quad .$$

But if the function $f(x)$, as in our case, is absolutely integrable and

---

15 "Fourier Integrals" §43.3.

bounded and the function $E(\alpha)$ defined by (8, 22) is non-negative, then[13] $E(\alpha)$ is also integrable over the whole space. Hence the converse to (8, 22),

$$f(x) = \int e^{ix\alpha} E(\alpha)\, d\alpha$$

holds. Therefore, as we asserted, $f(x)$ is a characteristic function.

Let $f(x)$ now be an arbitrary positive-definite function. It is easily verifiable that for each function $\gamma(\alpha)$, positive and integrable over $\Re$, the function

$$f(x) \int e^{ix\alpha} \gamma(\alpha)\, d\alpha \tag{8, 25}$$

is also positive-definite. But the functions

$$f(x) g_n(x) \tag{8, 26}$$

also belong to the functions (8, 25), where $g_n(x)$ is the function (7, 51). Moreover the function (8, 26) is absolutely integrable. It is therefore, by the proofs already given, a characteristic function. Further as $n \longrightarrow \infty$, the sequence (8, 26) converges to the continuous function $f(x)$. By Theorem 17, this last function is therefore also a characteristic one, Q.E.D.

§9. SPECTRAL DECOMPOSITION OF SQUARE INTEGRABLE FUNCTIONS

9.1. Let $g(\xi)$ be a square integrable function in $\Re$. Then one can form the function

$$f(x) = \int g(x + \xi) \overline{g(\xi)}\, d\xi \quad .$$

Hence

$$|f(x)|^2 \leq \int |g(x + \xi)|^2\, d\xi \cdot \int |g(\xi)|^2\, d\xi = f(0)^2 \quad ,$$

and $f(x)$ is therefore bounded. Further

$$|f(x) - f(x_1)|^2 \leq \int |g(x + \xi) - g(x_1 + \xi)|^2\, d\xi \cdot \int |g(\xi)|^2\, d\xi \quad .$$

If x is held fixed, then by a theorem on square integrable functions, the first factor on the right is convergent to zero as $x_1 \longrightarrow x$. Hence $f(x)$ is continuous. Because

$$\overline{f(-x)} = \int \overline{g(-x+\xi)}g(\xi)\,d\xi = \int \overline{g(t)}g(t+x)\,dt = f(x) \quad ,$$

it follows that $f(x)$ is hermitian. Moreover

$$f(x-y) = \int g(x-y+\xi)\overline{g(\xi)}\,d\xi = \int g(x+t)\overline{g(y+t)}\,dt \quad ,$$

and therefore

$$\sum_{\mu=1}^{m}\sum_{\nu=1}^{m} f(x_\mu - x_\nu)\rho_\mu\overline{\rho_\nu} = \int \left|\sum_{\mu=1}^{m} g(x_\mu + t)\rho_\mu\right|^2 dt \geqq 0 \quad .$$

Hence, taking everything into consideration, $f(x)$ is positive-definite and by Theorem 19, is therefore representable in the form

$$f(x) = \int e^{ix\alpha}dV(\alpha) \quad ,$$

(cf. the Introduction). This statement can now be considerably generalized.

9.2. THEOREM 20. Let a function $g(\xi)$ which is square integrable over each bounded interval possess the following properties. There exists

1. A sequence of bounded domains

$$G_1, G_2, G_3, \dots$$

which converge monotonically to $\mathfrak{R}$ such that for each two the distance between their boundaries is ≥ 1.

2. a sequence of positive numbers

$$c_1, c_2, c_3, \dots$$

for which

$$\lim_{n\to\infty} \frac{c_{n+1}}{c_n} = 1 \quad , \tag{9, 21}$$

such that the functions

$$F_n(x) = \frac{1}{c_n}\int_{G_n} g(x+\xi)\overline{g(\xi)}\,d\xi \tag{9, 22}$$

converge, as $n \longrightarrow \infty$, to a finite limit function

(9, 23) $$F(x) = \lim_{n \to \infty} F_n(x) \quad ,$$

for all x (or for almost all x and for $x = 0$).

Then there exists a distribution function $V(\alpha)$ such that

(9, 24) $$F(x) = \int e^{ix\alpha} dV(\alpha) \qquad \underline{\text{for almost all}} \quad x \quad .$$

PROOF. We set

$$g_n(\xi) \begin{cases} g(\xi) & \text{if } \xi \subset G_n \\ 0 & \text{if } \xi \not\subset G_n \end{cases} ,$$

and consider the functions

(9, 25) $$f_n(x) = \frac{1}{c_n} \int g_n(x + \xi) \overline{g_n(\xi)} \, d\xi \quad .$$

For a fixed x we denote by H_n those points which are at a distance less than $|x|$ from a boundary point of G_n (where $|x|$ denotes the distance of the point x from the origin), and by K_n those points which from a point of H_n are at a distance less than $|x|$.

If we take into account the actual domain of integration for the expression (9, 25) we obtain

$$c_n^2 \, |f_n(x) - F_n(x)|^2 \leq \int_{H_n} |g(x + \xi)|^2 \, d\xi \cdot \int_{H_n} |g(\xi)|^2 \, d\xi \quad ,$$

and from this that

$$|f_n(x) - F_n(x)| \leq \frac{1}{c_n} \int_{K_n} |g(\xi)|^2 \, d\xi \quad .$$

We have assumed that the two domains G_n have a border distance $\geqq 1$. If now p denotes a fixed whole number $\geqq 2|x|$, then

$$K_n \subset G_{n+p} - G_{n-p} \quad .$$

Hence

$$\int_{K_n} |g(\xi)|^2 \, d\xi \subset \int_{G_{n+p}} - \int_{G_{n-p}} .$$

Therefore

$$|f_n(x) - F_n(x)| \leqq \frac{c_{n+p}}{c_n} \frac{1}{c_{n+p}} \int_{G_{n+p}} |g(\xi)|^2 \, d\xi$$

$$- \frac{c_{n-p}}{c_n} \frac{1}{c_{n-p}} \int_{G_{n-p}} |g(\xi)|^2 \, d\xi .$$

But by our assumption

$$\lim_{n \to \infty} \frac{1}{c_{n \pm p}} \int_{G_{n \pm p}} |g(\xi)|^2 \, d\xi = F(0)$$

$$\lim_{n \to \infty} \frac{c_{n \pm p}}{c_n} = 1 .$$

Hence

$$\lim_{n \to \infty} |f_n(x) - F_n(x)| = 0 .$$

For the same points therefore for which (9, 23) is valid, it is also true that

$$F(x) = \lim_{n \to \infty} f_n(x) .$$

But the functions $f_n(x)$ are by 9.1, characteristic functions. Hence our assertion (9, 24) is now an immediate consequence of Theorem 16.

SYMBOLS

	Page
$\mathfrak{F}_0$	54
$\mathfrak{F}_k$	138, 194
$\mathfrak{F}$	178, 209
$\mathfrak{F}^1_k$, $\mathfrak{F}^2_k$	216
$\mathfrak{M}\{\ \}$	35, 38
$\mathfrak{P}$	81
$\mathfrak{T}_0$	56
$\mathfrak{T}_k$	140
$r[\ ]$	108, 166
$O(\ )$, $o(\ )$	10
$\|^-\|^k$	143
$\asymp$ and $\overset{k}{\asymp}$	80, 139

INDEX

associative law 157

bounded variation 10

characteristic functions
 of operators 105, 134
 of distributions 317

————, special 318

compactly convergent 306

conjugate functions 208

continuity, interval of 299

difference — differential equation,
 definition of 104

Dirichlet discontinuous factor 16

————, generalization of .... 74 - 77

distribution functions 80

————, convergence of 87

distribution functions in several variables ... 316

"Division", "Divisor" 108

dominant function 302

equivalent 80, 139

"equivalent" multipliers 114

essentially equal 301

Euler integral 51

Faltung 56, 189, 237

finely meshed 300

Fresnel integrals 51

hermitian function 81, 92

Interval function 295

Interval of continuity 299

Page

Inversion; Inversion formulas.................. 46 - 49
"Isolation"..................................... 113
k-continuous.................................... 203
k-continuous boundary values.................... 203
k-convergent.................................... 145
k-equivalent.................................... 139, 199
k-transform..................................... 141, 199
linear class of functions....................... 54
Liouville, theorem of........................... 201
"Mean value".................................... 35 - 38
Mellin formulas................................. 186
"Multiplication", multiplier.................... 108, 166
normalized functions............................ 299
Operator.. 106
operator equation............................... 175
positive-definite functions..................... 92
—————, in several variables.......... 325
self-reciprocal functions....................... 54
"Solution" of a functional equation............ 106-7, 179
Sommerfeld Integral............................. 59
Theta functions, a functional equation of...... 193
"trivial" function.............................. 144, 200, 209
unessentially different......................... 301
Weber discontinuous integrals................... 71

www.ingramcontent.com/pod-product-compliance
Ingram Content Group UK Ltd.
Pitfield, Milton Keynes, MK11 3LW, UK
UKHW031027161224
452563UK00001BA/76